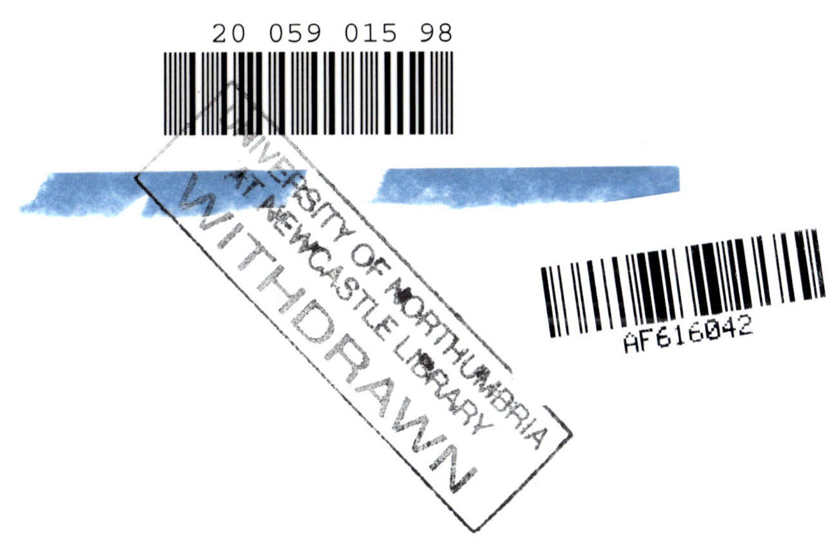

Clutches and Brakes

MECHANICAL ENGINEERING
A Series of Textbooks and Reference Books

Founding Editor

L. L. Faulkner

Columbus Division, Battelle Memorial Institute
and Department of Mechanical Engineering
The Ohio State University
Columbus, Ohio

1. *Spring Designer's Handbook*, Harold Carlson
2. *Computer-Aided Graphics and Design*, Daniel L. Ryan
3. *Lubrication Fundamentals*, J. George Wills
4. *Solar Engineering for Domestic Buildings*, William A. Himmelman
5. *Applied Engineering Mechanics: Statics and Dynamics*, G. Boothroyd and C. Poli
6. *Centrifugal Pump Clinic*, Igor J. Karassik
7. *Computer-Aided Kinetics for Machine Design*, Daniel L. Ryan
8. *Plastics Products Design Handbook, Part A: Materials and Components; Part B: Processes and Design for Processes*, edited by Edward Miller
9. *Turbomachinery: Basic Theory and Applications*, Earl Logan, Jr.
10. *Vibrations of Shells and Plates*, Werner Soedel
11. *Flat and Corrugated Diaphragm Design Handbook*, Mario Di Giovanni
12. *Practical Stress Analysis in Engineering Design*, Alexander Blake
13. *An Introduction to the Design and Behavior of Bolted Joints*, John H. Bickford
14. *Optimal Engineering Design: Principles and Applications*, James N. Siddall
15. *Spring Manufacturing Handbook*, Harold Carlson
16. *Industrial Noise Control: Fundamentals and Applications*, edited by Lewis H. Bell
17. *Gears and Their Vibration: A Basic Approach to Understanding Gear Noise*, J. Derek Smith
18. *Chains for Power Transmission and Material Handling: Design and Applications Handbook*, American Chain Association
19. *Corrosion and Corrosion Protection Handbook*, edited by Philip A. Schweitzer
20. *Gear Drive Systems: Design and Application*, Peter Lynwander
21. *Controlling In-Plant Airborne Contaminants: Systems Design and Calculations*, John D. Constance
22. *CAD/CAM Systems Planning and Implementation*, Charles S. Knox
23. *Probabilistic Engineering Design: Principles and Applications*, James N. Siddall
24. *Traction Drives: Selection and Application*, Frederick W. Heilich III and Eugene E. Shube
25. *Finite Element Methods: An Introduction*, Ronald L. Huston and Chris E. Passerello

26. *Mechanical Fastening of Plastics: An Engineering Handbook*, Brayton Lincoln, Kenneth J. Gomes, and James F. Braden
27. *Lubrication in Practice: Second Edition*, edited by W. S. Robertson
28. *Principles of Automated Drafting*, Daniel L. Ryan
29. *Practical Seal Design*, edited by Leonard J. Martini
30. *Engineering Documentation for CAD/CAM Applications*, Charles S. Knox
31. *Design Dimensioning with Computer Graphics Applications*, Jerome C. Lange
32. *Mechanism Analysis: Simplified Graphical and Analytical Techniques*, Lyndon O. Barton
33. *CAD/CAM Systems: Justification, Implementation, Productivity Measurement*, Edward J. Preston, George W. Crawford, and Mark E. Coticchia
34. *Steam Plant Calculations Manual*, V. Ganapathy
35. *Design Assurance for Engineers and Managers*, John A. Burgess
36. *Heat Transfer Fluids and Systems for Process and Energy Applications*, Jasbir Singh
37. *Potential Flows: Computer Graphic Solutions*, Robert H. Kirchhoff
38. *Computer-Aided Graphics and Design: Second Edition*, Daniel L. Ryan
39. *Electronically Controlled Proportional Valves: Selection and Application*, Michael J. Tonyan, edited by Tobi Goldoftas
40. *Pressure Gauge Handbook*, AMETEK, U.S. Gauge Division, edited by Philip W. Harland
41. *Fabric Filtration for Combustion Sources: Fundamentals and Basic Technology*, R. P. Donovan
42. *Design of Mechanical Joints*, Alexander Blake
43. *CAD/CAM Dictionary*, Edward J. Preston, George W. Crawford, and Mark E. Coticchia
44. *Machinery Adhesives for Locking, Retaining, and Sealing*, Girard S. Haviland
45. *Couplings and Joints: Design, Selection, and Application*, Jon R. Mancuso
46. *Shaft Alignment Handbook*, John Piotrowski
47. *BASIC Programs for Steam Plant Engineers: Boilers, Combustion, Fluid Flow, and Heat Transfer*, V. Ganapathy
48. *Solving Mechanical Design Problems with Computer Graphics*, Jerome C. Lange
49. *Plastics Gearing: Selection and Application*, Clifford E. Adams
50. *Clutches and Brakes: Design and Selection*, William C. Orthwein
51. *Transducers in Mechanical and Electronic Design*, Harry L. Trietley
52. *Metallurgical Applications of Shock-Wave and High-Strain-Rate Phenomena*, edited by Lawrence E. Murr, Karl P. Staudhammer, and Marc A. Meyers
53. *Magnesium Products Design*, Robert S. Busk
54. *How to Integrate CAD/CAM Systems: Management and Technology*, William D. Engelke
55. *Cam Design and Manufacture: Second Edition*; with cam design software for the IBM PC and compatibles, disk included, Preben W. Jensen
56. *Solid-State AC Motor Controls: Selection and Application*, Sylvester Campbell
57. *Fundamentals of Robotics*, David D. Ardayfio
58. *Belt Selection and Application for Engineers*, edited by Wallace D. Erickson
59. *Developing Three-Dimensional CAD Software with the IBM PC*, C. Stan Wei
60. *Organizing Data for CIM Applications*, Charles S. Knox, with contributions by Thomas C. Boos, Ross S. Culverhouse, and Paul F. Muchnicki

61. *Computer-Aided Simulation in Railway Dynamics*, by Rao V. Dukkipati and Joseph R. Amyot
62. *Fiber-Reinforced Composites: Materials, Manufacturing, and Design*, P. K. Mallick
63. *Photoelectric Sensors and Controls: Selection and Application*, Scott M. Juds
64. *Finite Element Analysis with Personal Computers*, Edward R. Champion, Jr., and J. Michael Ensminger
65. *Ultrasonics: Fundamentals, Technology, Applications: Second Edition, Revised and Expanded*, Dale Ensminger
66. *Applied Finite Element Modeling: Practical Problem Solving for Engineers*, Jeffrey M. Steele
67. *Measurement and Instrumentation in Engineering: Principles and Basic Laboratory Experiments*, Francis S. Tse and Ivan E. Morse
68. *Centrifugal Pump Clinic: Second Edition, Revised and Expanded*, Igor J. Karassik
69. *Practical Stress Analysis in Engineering Design: Second Edition, Revised and Expanded*, Alexander Blake
70. *An Introduction to the Design and Behavior of Bolted Joints: Second Edition, Revised and Expanded*, John H. Bickford
71. *High Vacuum Technology: A Practical Guide*, Marsbed H. Hablanian
72. *Pressure Sensors: Selection and Application*, Duane Tandeske
73. *Zinc Handbook: Properties, Processing, and Use in Design*, Frank Porter
74. *Thermal Fatigue of Metals*, Andrzej Weronski and Tadeusz Hejwowski
75. *Classical and Modern Mechanisms for Engineers and Inventors*, Preben W. Jensen
76. *Handbook of Electronic Package Design*, edited by Michael Pecht
77. *Shock-Wave and High-Strain-Rate Phenomena in Materials*, edited by Marc A. Meyers, Lawrence E. Murr, and Karl P. Staudhammer
78. *Industrial Refrigeration: Principles, Design and Applications*, P. C. Koelet
79. *Applied Combustion*, Eugene L. Keating
80. *Engine Oils and Automotive Lubrication*, edited by Wilfried J. Bartz
81. *Mechanism Analysis: Simplified and Graphical Techniques, Second Edition, Revised and Expanded*, Lyndon O. Barton
82. *Fundamental Fluid Mechanics for the Practicing Engineer*, James W. Murdock
83. *Fiber-Reinforced Composites: Materials, Manufacturing, and Design, Second Edition, Revised and Expanded*, P. K. Mallick
84. *Numerical Methods for Engineering Applications*, Edward R. Champion, Jr.
85. *Turbomachinery: Basic Theory and Applications, Second Edition, Revised and Expanded*, Earl Logan, Jr.
86. *Vibrations of Shells and Plates: Second Edition, Revised and Expanded*, Werner Soedel
87. *Steam Plant Calculations Manual: Second Edition, Revised and Expanded*, V. Ganapathy
88. *Industrial Noise Control: Fundamentals and Applications, Second Edition, Revised and Expanded*, Lewis H. Bell and Douglas H. Bell
89. *Finite Elements: Their Design and Performance*, Richard H. MacNeal
90. *Mechanical Properties of Polymers and Composites: Second Edition, Revised and Expanded*, Lawrence E. Nielsen and Robert F. Landel
91. *Mechanical Wear Prediction and Prevention*, Raymond G. Bayer

92. *Mechanical Power Transmission Components*, edited by David W. South and Jon R. Mancuso
93. *Handbook of Turbomachinery*, edited by Earl Logan, Jr.
94. *Engineering Documentation Control Practices and Procedures*, Ray E. Monahan
95. *Refractory Linings Thermomechanical Design and Applications*, Charles A. Schacht
96. *Geometric Dimensioning and Tolerancing: Applications and Techniques for Use in Design, Manufacturing, and Inspection*, James D. Meadows
97. *An Introduction to the Design and Behavior of Bolted Joints: Third Edition, Revised and Expanded*, John H. Bickford
98. *Shaft Alignment Handbook: Second Edition, Revised and Expanded*, John Piotrowski
99. *Computer-Aided Design of Polymer-Matrix Composite Structures*, edited by Suong Van Hoa
100. *Friction Science and Technology*, Peter J. Blau
101. *Introduction to Plastics and Composites: Mechanical Properties and Engineering Applications*, Edward Miller
102. *Practical Fracture Mechanics in Design*, Alexander Blake
103. *Pump Characteristics and Applications*, Michael W. Volk
104. *Optical Principles and Technology for Engineers*, James E. Stewart
105. *Optimizing the Shape of Mechanical Elements and Structures*, A. A. Seireg and Jorge Rodriguez
106. *Kinematics and Dynamics of Machinery*, Vladimír Stejskal and Michael Valášek
107. *Shaft Seals for Dynamic Applications*, Les Horve
108. *Reliability-Based Mechanical Design*, edited by Thomas A. Cruse
109. *Mechanical Fastening, Joining, and Assembly*, James A. Speck
110. *Turbomachinery Fluid Dynamics and Heat Transfer*, edited by Chunill Hah
111. *High-Vacuum Technology: A Practical Guide, Second Edition, Revised and Expanded*, Marsbed H. Hablanian
112. *Geometric Dimensioning and Tolerancing: Workbook and Answerbook*, James D. Meadows
113. *Handbook of Materials Selection for Engineering Applications*, edited by G. T. Murray
114. *Handbook of Thermoplastic Piping System Design*, Thomas Sixsmith and Reinhard Hanselka
115. *Practical Guide to Finite Elements: A Solid Mechanics Approach*, Steven M. Lepi
116. *Applied Computational Fluid Dynamics*, edited by Vijay K. Garg
117. *Fluid Sealing Technology*, Heinz K. Muller and Bernard S. Nau
118. *Friction and Lubrication in Mechanical Design*, A. A. Seireg
119. *Influence Functions and Matrices*, Yuri A. Melnikov
120. *Mechanical Analysis of Electronic Packaging Systems*, Stephen A. McKeown
121. *Couplings and Joints: Design, Selection, and Application, Second Edition, Revised and Expanded,* Jon R. Mancuso
122. *Thermodynamics: Processes and Applications*, Earl Logan, Jr.
123. *Gear Noise and Vibration*, J. Derek Smith
124. *Practical Fluid Mechanics for Engineering Applications,* John J. Bloomer
125. *Handbook of Hydraulic Fluid Technology*, edited by George E. Totten
126. *Heat Exchanger Design Handbook*, T. Kuppan

127. *Designing for Product Sound Quality*, Richard H. Lyon
128. *Probability Applications in Mechanical Design*, Franklin E. Fisher and Joy R. Fisher
129. *Nickel Alloys,* edited by Ulrich Heubner
130. *Rotating Machinery Vibration: Problem Analysis and Troubleshooting,* Maurice L. Adams, Jr.
131. *Formulas for Dynamic Analysis,* Ronald L. Huston and C. Q. Liu
132. *Handbook of Machinery Dynamics*, Lynn L. Faulkner and Earl Logan, Jr.
133. *Rapid Prototyping Technology: Selection and Application,* Kenneth G. Cooper
134. *Reciprocating Machinery Dynamics: Design and Analysis*, Abdulla S. Rangwala
135. *Maintenance Excellence: Optimizing Equipment Life-Cycle Decisions,* edited by John D. Campbell and Andrew K. S. Jardine
136. *Practical Guide to Industrial Boiler Systems*, Ralph L. Vandagriff
137. *Lubrication Fundamentals: Second Edition, Revised and Expanded*, D. M. Pirro and A. A. Wessol
138. *Mechanical Life Cycle Handbook: Good Environmental Design and Manufacturing,* edited by Mahendra S. Hundal
139. *Micromachining of Engineering Materials,* edited by Joseph McGeough
140. *Control Strategies for Dynamic Systems: Design and Implementation,* John H. Lumkes, Jr.
141. *Practical Guide to Pressure Vessel Manufacturing,* Sunil Pullarcot
142. *Nondestructive Evaluation: Theory, Techniques, and Applications,* edited by Peter J. Shull
143. *Diesel Engine Engineering: Thermodynamics, Dynamics, Design, and Control,* Andrei Makartchouk
144. *Handbook of Machine Tool Analysis,* Ioan D. Marinescu, Constantin Ispas, and Dan Boboc
145. *Implementing Concurrent Engineering in Small Companies,* Susan Carlson Skalak
146. *Practical Guide to the Packaging of Electronics: Thermal and Mechanical Design and Analysis*, Ali Jamnia
147. *Bearing Design in Machinery: Engineering Tribology and Lubrication,* Avraham Harnoy
148. *Mechanical Reliability Improvement: Probability and Statistics for Experimental Testing,* R. E. Little
149. *Industrial Boilers and Heat Recovery Steam Generators: Design, Applications, and Calculations*, V. Ganapathy
150. *The CAD Guidebook: A Basic Manual for Understanding and Improving Computer-Aided Design,* Stephen J. Schoonmaker
151. *Industrial Noise Control and Acoustics,* Randall F. Barron
152. *Mechanical Properties of Engineered Materials,* Wolé Soboyejo
153. *Reliability Verification, Testing, and Analysis in Engineering Design,* Gary S. Wasserman
154. *Fundamental Mechanics of Fluids: Third Edition*, I. G. Currie
155. *Intermediate Heat Transfer*, Kau-Fui Vincent Wong
156. *HVAC Water Chillers and Cooling Towers: Fundamentals, Application, and Operation*, Herbert W. Stanford III
157. *Gear Noise and Vibration: Second Edition, Revised and Expanded,* J. Derek Smith

158. *Handbook of Turbomachinery: Second Edition, Revised and Expanded,* edited by Earl Logan, Jr., and Ramendra Roy
159. *Piping and Pipeline Engineering: Design, Construction, Maintenance, Integrity, and Repair,* George A. Antaki
160. *Turbomachinery: Design and Theory,* Rama S. R. Gorla and Aijaz Ahmed Khan
161. *Target Costing: Market-Driven Product Design,* M. Bradford Clifton, Henry M. B. Bird, Robert E. Albano, and Wesley P. Townsend
162. *Fluidized Bed Combustion,* Simeon N. Oka
163. *Theory of Dimensioning: An Introduction to Parameterizing Geometric Models,* Vijay Srinivasan
164. *Handbook of Mechanical Alloy Design,* edited by George E. Totten, Lin Xie, and Kiyoshi Funatani
165. *Structural Analysis of Polymeric Composite Materials,* Mark E. Tuttle
166. *Modeling and Simulation for Material Selection and Mechanical Design,* edited by George E. Totten, Lin Xie, and Kiyoshi Funatani
167. *Handbook of Pneumatic Conveying Engineering,* David Mills, Mark G. Jones, and Vijay K. Agarwal
168. *Clutches and Brakes: Design and Selection, Second Edition,* William C. Orthwein
169. *Fundamentals of Fluid Film Lubrication: Second Edition,* Bernard J. Hamrock, Steven R. Schmid, and Bo O. Jacobson
170. *Handbook of Lead-Free Solder Technology for Microelectronic Assemblies,* edited by Karl J. Puttlitz and Kathleen A. Stalter
171. *Vehicle Stability,* Dean Karnopp

Additional Volumes in Preparation

Mechanical Wear Fundamentals and Testing: Second Edition, Revised and Expanded, Raymond G. Bayer

Engineering Design for Wear: Second Edition, Revised and Expanded, Raymond G. Bayer

Progressing Cavity Pumps, Downhole Pumps, and Mudmotors, Lev Nelik

Mechanical Engineering Software

Spring Design with an IBM PC, Al Dietrich

Mechanical Design Failure Analysis: With Failure Analysis System Software for the IBM PC, David G. Ullman

Clutches and Brakes

Design and Selection

Second Edition

William C. Orthwein
Southern Illinois University at Carbondale
Carbondale, Illinois, U.S.A.

MARCEL DEKKER, INC. NEW YORK • BASEL

Library of Congress Cataloging-in-Publication Data
A catalog record for this book is available from the Library of Congress.

ISBN: 0-8247-4876-X

This book is printed on acid-free paper.

Headquarters
Marcel Dekker, Inc., 270 Madison Avenue, New York, NY 10016, U.S.A.
tel: 212-696-9000; fax: 212-685-4540

Distribution and Customer Service
Marcel Dekker, Inc., Cimarron Road, Monticello, New York 12701, U.S.A.
tel: 800-228-1160; fax: 845-796-1772

Eastern Hemisphere Distribution
Marcel Dekker AG, Hutgasse 4, Postfach 812, CH-4001 Basel, Switzerland
tel: 41-61-260-6300; fax: 41-61-260-6333

World Wide Web
http://www.dekker.com

The publisher offers discounts on this book when ordered in bulk quantities. For more information, write to Special Sales/Professional Marketing at the headquarters address above.

Current printing (last digit):

10 9 8 7 6 5 4 3 2 1

PRINTED IN THE UNITED STATES OF AMERICA

To Helen, my adorable wife, who improved
my life by having been here

Preface to the Second Edition

Chapter 1, Friction Materials, has been rewritten for two reasons. The first is that graphical data of the sort found in the first edition can no longer be obtained from many of the lining manufacturers. It appears that this absence of graphical data is due to the Trial Lawyers Association curse that has made it risky to provide such data because it may be misinterpreted by technically illiterate judges and juries to place blame where there is no basis for it. The second reason is that asbestos is no longer used in brake and clutch lining materials manufactured in the United States. Thus, data for lining materials containing asbestos are obsolete.

Other changes in the second edition consist of correcting the misprints that have been discovered since the publication of the first edition, a corrected and expanded discussion of cone brakes and clutches, a simpler formulation of the torque from a centrifugal clutch, an update of antiskid control, the addition of a chapter dealing with fluid clutches and retarders, and a chapter dealing with friction drives.

The flowcharts in the first edition that were given as an aid to those readers who may have wished to write computer programs to simplify brake and clutch design have been eliminated in this edition. The availability of commercial numerical analysis programs that may be used in engineering design has eliminated most, if not all, of the need for engineers to write their own analytical programs. The two analytical programs used in the book are listed here with the addresses of their providers at this time. Suppliers for more extensive computer-aided engineering and design programs advertise in engineering magazines. Their addresses and capabilities may also be found in the *Thomas Register*, held by most engineering libraries, and they were available online in 2003 at www.thomasregister.com.

TK Solver from
Universal Technical
Systems, Inc.
1220 Rock St.
Rockford IL 61101
United States

Phone: 1 800 436 7887
e-mail: sales@uts.com

Mathcad 2001i from
MathSoft Engineering &
Education, Inc.
101 Main St.
Cambridge, MA 02142-1521
United States

Phone: 1 800 628 4223
e-mail: sales-info@mathsoft.com

Changes in ownership of many of the manufacturers of the products illustrated in this book have occurred since the publication of the first edition. Although the products available and their principles of operation generally have remained unchanged, the credit lines for some of these illustrations may refer to manufacturers' names that are no longer in use. Others may become obsolete in the future.

William C. Orthwein

Preface to the First Edition

This book has two objectives. The first is to bring together the formulas for the design and selection of a variety of brakes and clutches. The second is to provide flowcharts and programs for programmable calculators and personal computers to facilitate the application of often lengthy formulas and otherwise tedious iteration procedures indigenous to the clutch and brake design and selection process.

Formulas for the torque that may be expected from each of the brake or clutch configurations and the force, pressure, or current required to obtain this torque are derived and their application is demonstrated by example. Derivations are included to explicitly show the assumptions made and to delineate the role of each parameter in these governing relations so that the designer can more skillfully select these parameters to meet the demands of the problem at hand. Where appropriate, the resulting formulas are collected at the end of each chapter so that those not interested in their derivation may turn directly to the design and selection formulas.

Following the torque and force analysis for the sundry brake and clutch embodiments which dissipate heat, attention is directed to the calculation of the heat generated by these devices during the interval in which the speed is changing. Pertinent relations are derived and demonstrated for braking or accelerating of vehicles, conveyor belts, and hoists.

Calculation of the acceleration, temperature, and heat dissipation may be quite complicated and may be strongly dependent upon the location of the brake on the machine itself and upon the environment in which the machine is to operate. Discussion of acceleration, acceleration time, temperature, and heat dissipation are, therefore, limited to a common—and simple—brake configuration and to a standard environment of 20°C or 70°F, no wind, and no vibration.

Flowcharts follow the formula collection as appropriate to demonstrate their step-by-step application in arriving at the final design. They are written in the interactive mode (computer or calculator prompting for each variable and its increments) to permit use of programmable calculators and small personal computers for the comparison of several possible designs. With little modification they may be used as subprograms in larger computers having control programs to automate clutch and brake selection to whatever extent desired.

Even though the calculations may be lengthy, no flowcharts are given for those cases where branching is minimal (as in the case of acceleration or deceleration and heat dissipation calculations) where the reasoning is straightforward. It is intended that computer programs will be used for all but the simplest calculations.

William C. Orthwein

Contents

Introduction

It is the purpose of this book to briefly derive, where possible, the design formulas for the major types of clutches and brakes listed in the contents and to display an example of their use in a typical design. Some pertinent computer programs for longer formulas are listed in the references.

Each chapter is independent of the others, with the possible exception of Chapters 1 and 8, which are concerned with friction materials and with acceleration or deceleration time and heat dissipation during clutching and braking. The friction and pressure characteristic of friction materials used for brake and clutch linings and pads are discussed in Chapter 1 so that they may be available for applications in the following chapters. Chapter 8 deals with acceleration and heat dissipation considerations which apply to all chapters, and consequently draws upon the other chapters for brake types to be discussed in its examples. The logic to be delineated in that chapter is, however, contained entirely within that chapter, so that it may be read and understood without prior reading of any of the other chapters.

To Convert	To	Multiply by
pounds/in^2 (psi)	megapascals (MPa)	0.00689476
megapascals (MPa)	pounds/in^2 (psi)	145.03774
horsepower (hp)	kilowatts (kW)	0.7457
kilowatts (kW)	horsepower (hp)	1.34102
pounds (lb, force)	Newtons (N)	4.4482
Newtons (N)	pounds (lb, force)	0.2248

To Convert	To	Multiply by
Btu	calorie	251.995
calorie	Btu	0.003968

Since force and mass are misused in both systems it is necessary to use the acceleration of gravity to convert to proper units when confronted with incorrect usage, e.g., kg/cm^2. The acceleration of gravity in the two system of units is commonly taken to be

$$g = 32.1736 \text{ ft/s} \quad \text{Old Eglish}$$

$$= 9.80665 \text{ m/s} \quad \text{SI}$$

As implied by these previous numbers, we shall retain three or four places of significant digits in most calculations to minimize computational error. After all calculations are complete we shall round to the number of places that are practical for manufacture.

For those not familiar with SI stress and bearing pressure calculations, it may be well to point out that the Pascal is a rather awkward unit of stress, since

$$1 \text{ Pascal} = 1 \text{ N/m}^2$$

is an extremely small number in many applications. Two alternatives may be selected: to present pressure and stress in terms of atmospheres (atmospheric pressure at sea level) or in terms of megapascals, denoted by MPa. In the remainder of the book stress and bearing pressure in the SI system will be presented in terms of MPa because of the convenient relations

$$\text{N/mm}^2 = \text{MPa} \qquad \text{and} \qquad \text{MPa(mm}^2\text{)} = \text{N}$$

Since atmospheric pressure at sea level is often taken to be about 14.7 psi, it follows from the listing above that 1 MPa is approximately 10 atmospheric pressures. Conversion from MPa to atmosphere is, therefore, quite simple.

1

Friction Materials

Curves of the coefficient of friction as a function of load and of the speed differential between the lining and facings and their mating surface are no longer available from many manufacturers. Perhaps this is a consequence of the ease with which trial lawyers in the United States can collect large financial rewards for weak liability claims based upon often trivial, or unavoidable (due to physical limits on manufacturing tolerances), differences between published data and a particular specimen of the manufactured product. Furthermore, differences between published and operational coefficients of friction are beyond the control of the manufacturer because comparison of laboratory and operational data have shown that temperature, humidity, contamination, and utilization cycles of the machinery using these linings and facings can cause significant changes in the effective coefficient of friction at any given moment. Consequently, the coefficients of friction mentioned are nominal, the following discussion is in generic terms, and all curves shown should be understood to represent only the general character of the material under laboratory conditions.

The value of laboratory data is twofold, even though the data should not be used for design purposes. First, the data provides a comparison of the performance of different lining materials under similar conditions, such as given by the SAE 661 standard. Second, comparison of the laboratory data with field data for a particular type of machine for several different linings may suggest an empirical relationship that yields an approximate means of predicting the field performance of other lining materials based upon their laboratory data. A history of the comparison of field and laboratory data may, therefore serve as a starting point in the design of the prototype of a new machine of the same or similar type.

Field testing of a new machine by customers under the most adverse conditions is still necessary. Users often seem to devise abuses not envisioned by the design engineers.

I. FRICTION CODE

The usual range of the dynamic friction coefficients for those friction materials normally used in dry brake linings and pads is given in the Society of Automotive Engineers (SAE) coding standard SAE 866, which lists the code letters and friction coefficient ranges shown in Table 1 [1]. According to this code the first letter in the lining edge friction code indicates the normal friction coefficient and the second letter indicates the hot friction coefficient. Thus a lining material whose normal friction coefficient is 0.29 and whose hot friction coefficient is 0.40 would be coded as follows:

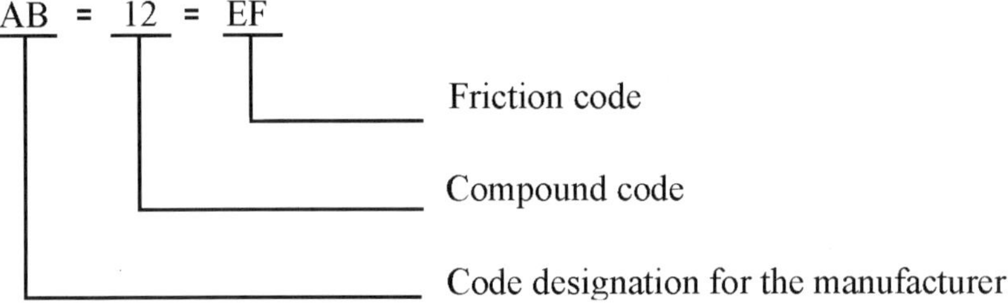

Temperatures for the normal and hot friction coefficients are defined in SAE J661, which also describes the measurement method to be used.

TABLE 1 Friction Identification System for Brake Linings and Brake Block for Motor Vehicles

Code letter	Friction coefficient
C	Not over 0.15
D	Over 0.15 but not over 0.25
E	Over 0.25 but not over 0.35
F	Over 0.35 but not over 0.45
G	Over 0.45 but not over 0.55
H	Over 0.55
Z	Unclassified

Static and dynamic coefficients of friction are usually different for most brake materials. If a brake is used to prevent shaft rotation during a particular operational phase, its stopping torque and heat dissipation are of secondary importance (i.e. a holding brake on a press); the static friction coefficient is the design parameter to be used. On the other hand, the pertinent design parameters are the dynamic friction coefficient and its change with temperature when a brake is designed for its stopping torque and heat dissipation when a rotating load is to be stopped or slowed.

Most manufacturers will provide custom compounds for the linings and facings within the general types that they manufacture if quantity requirements are met. In almost all applications it is suggested for all of these materials that the linings and facings run against either cast iron or steel with a surface finish of from 30 to 60 micro inches. Nonferrous metals are recommended only in special situations.

Effects of heating on the linings and facing discussed are expressed in terms of limiting temperatures or limiting power dissipated per unit area at the surface of the brake lining or clutch facing. Time is usually omitted, even though the surface temperature is determined by the power per unit area per unit time. This is because it is assumed that the power dissipation occurs over just a few seconds. More precise estimates, and only that, of the heat generated by the power dissipated in particular cases maybe had by using one of several heat transfer programs from suppliers of engineering software. It is for these reasons that prototype evaluation is always recommended.

II. WEAR

Hundreds of equations for wear may be found in the literature. These equations may depend a variety of factors, including the materials involved, the temperature, and the environment under consideration, i.e., the liquid or gas present, the formation of surface films, and so on [2]. Two of the relations that pertain to the following discussion are the specific wear rate and the wear rate.

The first of these, the specific wear rate, or wear coefficient, is a dimensional constant K that appears in the relation

$$\delta A = \upsilon = K\rho A d = KFd$$

From which υ may be written as

$$\upsilon = KFd \tag{2-1}$$

In these relations, δ represents the thickness of the lining material removed, υ is the volume material removed, K is a dimensional constant that is termed the specific wear rate or the wear coefficient, and p is the pressure acting over the surface area A that is in contact with the lining material. Force F is given by integral of the pressure acting on the specimen integrated over the area A over which it acts. Upon rewriting equation (2-1) to evaluate K we have that

$$K = \upsilon/(Fd) \tag{2-2}$$

Hence the units of K are lt^2/m where l,t, and m denote length, time, and mass, respectively. As a practical matter, if υ is millimeters cubed (mm^3), if force F is in newtons (N), and if the distance d is in meters (m), then the units of K become mm^3 N^{-1} m^{-1}, which explicitly shows the physical quantities involved, as in Figure 3.

The second relation that may be used by brake and clutch lining manufacturers to describe wear is

$$\Gamma = \upsilon Pt\Theta \tag{2-3}$$

in which Γ represents the wear rate, P is the power dissipated in the lining, and t is the time during which volume V was removed at temperature Θ. The units of Γ in equation (2-3) are those of the work (ml^2/t^2) required to remove a unit volume of material multiplied by the volume (l^3) removed.

Whenever the temperature is held constant during a test, the temperature variable Θ is suppressed. Since brake testing according to the SAE 661b standard is done at 200°F, the wear rate is often given by $\Gamma = \upsilon Pt$ and presented in the form $\upsilon = \Gamma/(\mathrm{Pt})$. Again, to be practical the wear rate divided by the product horsepower hours (hp hr) may be given in cubic inches (in.^{3}), as in Table 2 near the end of this chapter.

III. BRAKE FADE

Brake fade is a term that refers to the reduced effectiveness of many dry brakes as they become heated. A standard test described in SAE J661 outlines a procedure that uses controlled temperature drums and controlled brake lining pressure to stimulate brake fading as a basis of comparison of the brake fading characteristics of various lining materials. The equipment and temperatures are essentially identical to those used in estimating the coefficient of friction as a function of temperature. Only the presentation of the data is different, as shown in Figure 1. The fade test mode of presentation of data provides another indication of the recovery capability of the various lining materials. As with the previous test data, the fade test results are limited to a comparison of different lining materials for the test conditions only.

Limitation of the application of these data to preliminary design is emphasized because the friction coefficient is dependent upon the pressure,

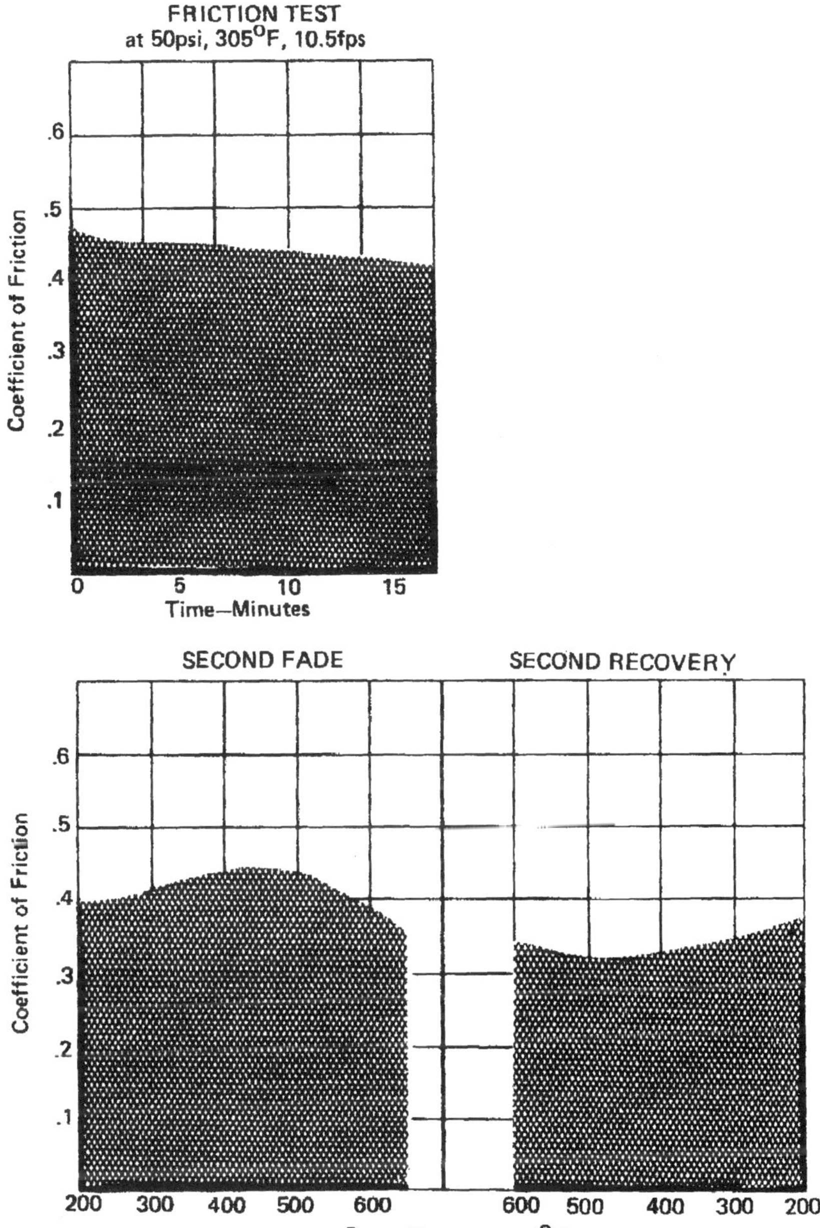

FIGURE 1 Display of brake lining fade test results. (Courtesy of Scan-Pac, Mequon, WI).

the temperature, and the relative velocities of the contracting surfaces, as noted earlier. Field tests are recommended before the production of any brake design because of the uncertainty usually associated with the variables involved in lining heating and in the cooling capability of the brake housing and any associated structure.

IV. FRICTION MATERIALS

Friction materials may be classified as either dry or wet. Wet friction lining materials are those that may operate in a fluid that is used for cooling because of the large amount of energy that must be dissipated during either braking or clutching. The fluids used are often motor oil or transmission fluids. Lining materials that cannot operate when immersed in a fluid are known as dry lining materials.

A. PTFE and TFE

At this time it appears that PTFE (polytetrafluoroethylene) and TFE (tetrafluoroethylene), both included under the trade name Teflon, are commonly used for brake linings [3]. PTFE exhibits a low coefficient of friction and is mechanically serviceable at about $\pm$ 260°C, is almost chemically inert, does not absorb water, and has good dimensional stability. Its weakness in shear stress is greatly improved by the addition of fillers, such as glass fibers. These fibers also increase its wear resistance and strength and increase its coefficient of friction by increasing its abrasiveness. The degree to which each of these properties is increased depends upon the amount, the physical dimensions, the orientation, and the nature of the material used as a filler [4].

Together these characteristics make PTFE brake pads useful for drag brakes in manufacturing processes, such as tape production, where the moving product must be held in tension during part of the manufacturing process. Likewise, PTFE clutch plates and linings that may be used whenever the transmitted torque should remain below a certain limit.

Laboratory measurements of the coefficients of friction at room temperature for several filled PTFE materials when subjected to loads of 1.415 Mpa, or 205 psi, and of 7.074 Mpa, or 1026 psi, are shown in Figure 2. They indicate that the coefficients of friction for these PTFE specimens with various kinds and sizes of fillers are all fairly independent of sliding speed, especially at greater loads, when sliding against a mild steel surface with a roughness of s.c.a. 0.03 μm c.l.a. [4]

Nominal coefficients of friction given by a particular manufacturer may, as noted earlier, differ from those shown in Figure 2 because of the amount, size, orientation, and kind of filler material used. Their static coefficient of

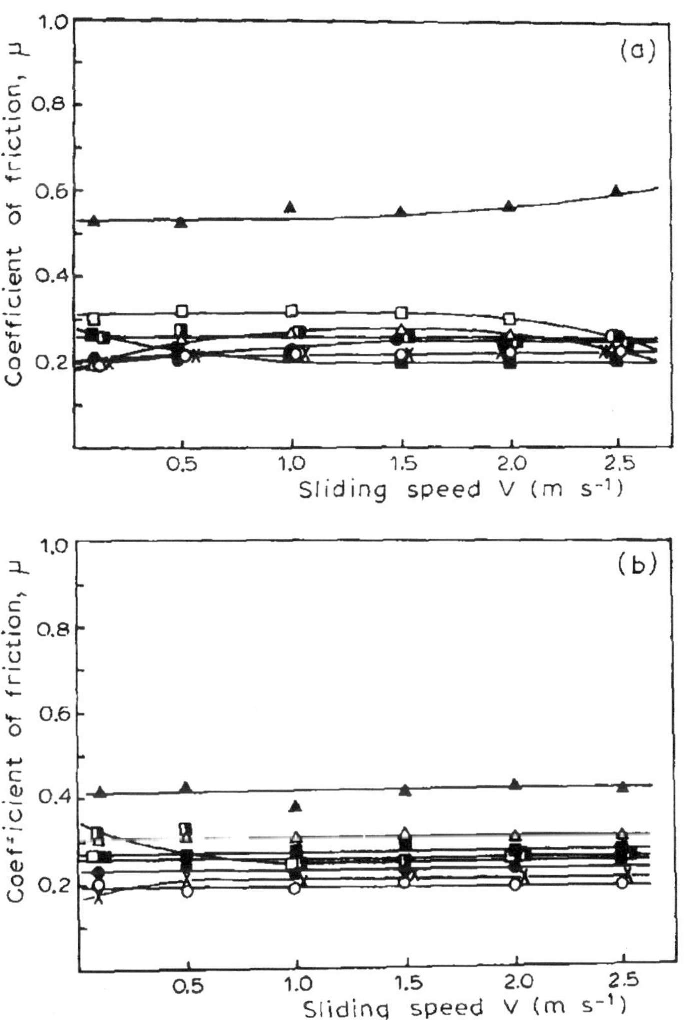

FIGURE 2 Coefficient of friction versus sliding speed at an average pressures of 1.415 Mpa, or 205 psi (a), and 7.074 Mpa, or 1,026 psi (b), for the fillers as indicated; Open triangle: TiO_2; filled triangle: ZrO_2; open square: glass; filled square: bronze; open circle: graphite; filled circle: MoS_2; X: unfilled, half-filled rectangle: Turcite (proprietary material, probably PTFE with bronze filler). (Courtesy Elsevier Science Publishers, New York.)

friction may from 0.089 to 0.108 and their dynamic coefficient may vary from 0.078 to 0.117 [3]. Under light loads of 1.0 Mpa (145 psi) and sliding speeds of 0.03 m/s (1.22 in./s) it has been reported that PTFE filled with bronze mesh displayed friction coefficients ranging from approximately 0.03 to 0.25 [5].

It may be of interest to note that unfilled Teflon has the property that its coefficient of friction, μ, is not given by $\mu = F_n/F_t$, but rather by $\mu = F_n^{0.85}/F_t$, where F_n denotes the force normal to the contact surface and F_t denotes the force tangential to the surface [6]. Fillers may modify this property by an amount that depends upon the kind, amount, or orientation of the filler.

Obviously, wear is also an important consideration in the selection of lining and facing materials because it determines the cost of the lining per hour of use in terms of main tenance time to replace the lining or facing in addition to the cost of the material itself, which is often the lesser of the two. Fortunately, experimental data, as shown in Figure 3, indicates that these fillers,

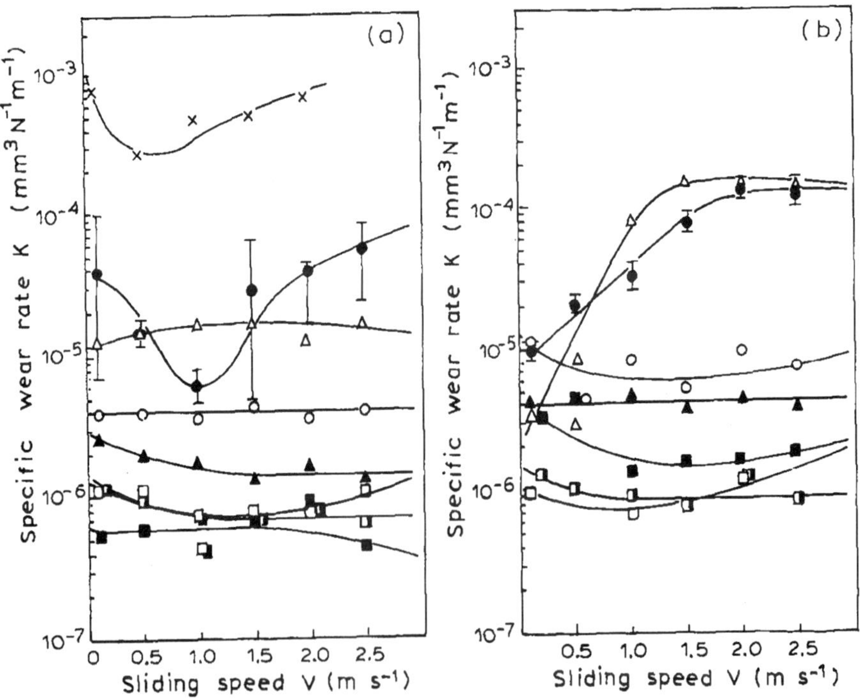

FIGURE 3 Specific wear of PTFE as a function of sliding speed for.

such as bronze, glass, and graphite, significantly add to the wear resistance of PTFE [3]. Glass appears to be the most commonly used.

B. Kevlar

Kevlar is the Du Pont trade name for an aramid (aromatic polyamide fiber) that has a tensile strength greater than some steels, i.e., some of these fibers have a modulus up to 27×10^6 psi (1.86×10^5 Mpa). Nevertheless they are flexible enough to be woven and processed as textiles, so Kevlar brake linings and clutch facings are available in either woven or nonwoven forms. They are used along with proprietary polymer binders in the manufacture of brake linings and clutch facings for both wet (oil bath) and dry clutch applications.

In dry brake and clutch applications, a flexible, nonwoven form can withstand dynamic pressure up to 3100 kPa (450 psi), are nonabrasive to iron, steel, and copper surfaces, and display and nominal coefficient of friction of 0.36 ± 0.1, as stated by one manufacturer. This manufacturer also states that in a dry environment these brake linings show significant fade at 260°C (500°F) that becomes greater at 370°C (700°F) [7]. Hence, they may be used in those industrial, marine, and off-road applications where fade is not a limiting factor; applications can include agricultural, industrial, marine, and off-road equipment.

In wet applications this nonwoven form of facing material is said to withstand dynamic pressure up to 2760 kPa (400 psi) with a nominal coefficient of friction in the 0.10–0.15 range when dissipating 23–290 W/cm^2 (0.2–2.5 hp/in^2) [7]. Ambient operating temperatures are replaced by power per unit area at the lining face in wet applications because the enveloping fluid bath cools the lining as it transfers the heat to cooling fins or to an oil cooler. Clutch facings and brake linings that contain no metal reinforcing wires or segments provide low wear on mating surfaces and eliminate the possibility of metal fragments in cooling system filters.

Kevlar has also been used in a proprietary solid form to obtain higher coefficients of friction in a woven material in which Kevlar fibers are mixed with other organic and inorganic fibers that enclose brass wire yarns to produce a lining that may be used as a direct replacement for older linings that contain asbestos [8]. Because of the brass wire and inorganic fibers, these lining may be more abrasive than those without these materials. This is, of course, a natural consequence of having higher friction coefficients on the orders of 0.40 dynamic and 0.42 static.

Representative second fade and second recovery curves of the friction coefficient vs. temperature of a representative of such linings are shown in Figure 4, as determined according to the SAE J661 standard. Field perform-

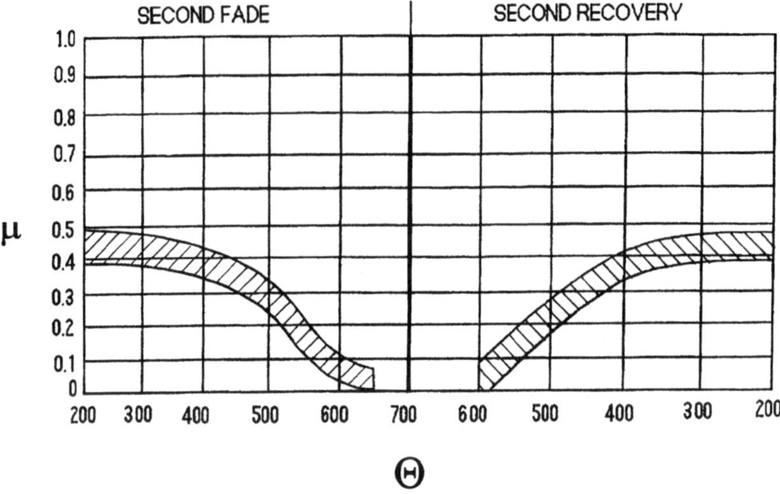

FIGURE 4 Second fade and second recovery vs. drum temperature (°F) for a proprietary lining containing Kevlar with a nominal coefficient of friction of 0.40. (Adapted from Reddaway Manufacturing Co., Inc., Newark, NJ.)

ance may be different from that shown in these graphs because of drum conditions, contamination, and other factors that depend upon the particular application.

The material whose may fall within the cross-hatched regions in Figure 4 may operate at a pressure no greater that 1379 kPa (200 psi) and a temperature no greater than 260°C (500°F) when in either a wet (oil) or a dry environment. This lining material may be used for band brakes and band clutches that work against steel or cast iron surfaces, as recommended by the manufacturer [8].

C. Mineral Enhanced

Mineral-based linings and facings are generally in the form of castings that can provide nominal friction coefficients ranging from 0.1 to 0.61. These friction materials may operate either dry or wet (oil) and find applications from tension control in manufacturing processes through overhead cranes, hoists, and industrial brakes and clutches and in farm and garden tractors.

Some or all of the following materials, and others, that are now necessary to produce a lining having a high friction coefficient may be embedded in the resin binders used [9].

Cyanamid	Barite	Wollastonite
Mica	Rubber dust	Collan
Glass fibers	Cellulose	Alumina
Vermiculite	Petroleum coke	Acrylic fibers
Graphite		

Because some of these materials have been classified as hazardous by one or both of the Occupational Safety and Health Administration (OSHA) or the American Conference of Governmental Industrial Hygienists (ACGIH) organizations, dust in the vicinity of their use and storage must be removed by vacuuming or by a dust suppressant according to the time schedules of one or both of OSHA and ACGIH.

Second fade and second recovery for such a material that has a nominal friction coefficient of 0.61, that may be subjected to a pressure of 350 psi (2.41 MPa), and that has a flash point above 600°C (1112°F) is shown in Figure 5. It has been used as a snubber for rail cars and is suitable for applications where high torque at low lining pressure is required. Test curves shown for this lining material hold for a test pressure of 1.034 MPa (1.50 psi) and a sliding speed of 6.1 m/s (20 ft/s).

D. Sintered

Proprietary sintered lining material, designated DM81, was available as an option on all Chevrolets, including Corvette, Corvair, and Chevy II brakes, in 1962 [10]. In normal driving they wore about twice as long as conventional linings of that time, with a larger ratio in favor of sintered linings for more severe service, as experienced by taxicabs.

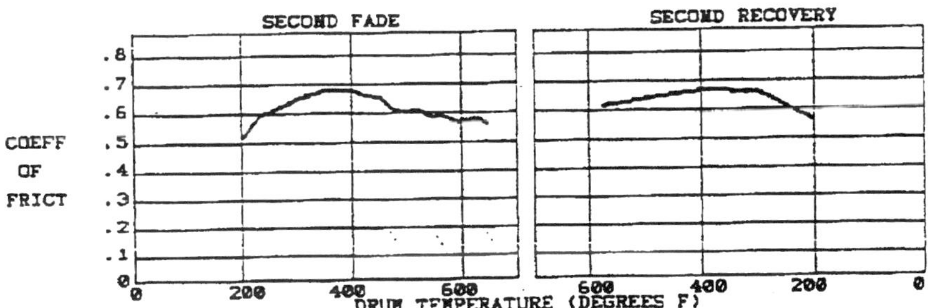

FIGURE 5 Second fade and second recovery vs. drum temperature (°F) for a mineral-enhanced lining (HF-61) with a nominal coefficient of friction of 0.61. (Courtesy Hibbing International Friction, New Castle, IN.)

Sintered metal friction material and carbon–carbon composites are widely used in brakes for large aircraft, such as long-range commercial jets, and in military aircraft. Their typical construction is shown in Figure 6. Brakes using sintered metal linings that press against steel plates are known as steel brakes, and those that press carbon lining material against carbon plates are known as carbon brakes in Figure 6. Stator plates are keyed to the brake housing, and rotor plates are keyed to the torque tube that rotates with the wheel to which it is attached.

Wear is greater in the lower-cost steel brake. The lining material in the steel brake is usually either a base of copper with additions of iron, graphite, and silicon as an abrasive and a high-temperature lubricant, such as molybdenum disulfide, or a base of iron with additions of copper and the other

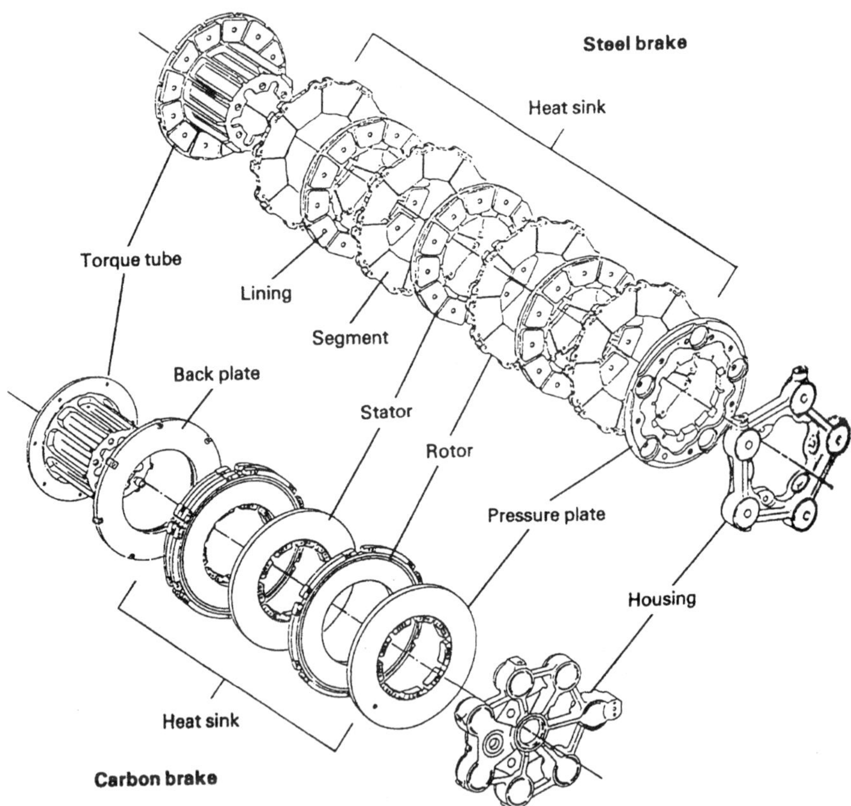

FIGURE 6 Construction of brakes having sintered linings working against steel plates (upper) and having carbon linings working against carbon plates (lower). (Courtesy *ASM Handbook*, ASM International, Materials Park, OH.)

additives just listed. The iron-based lining tends to provide a larger friction coefficient but may be more difficult to bond to its carrier plate.

Temperature dependence of the friction coefficient is indirectly indicated in Figure 7, where the energy per unit mass is the energy dissipated per unit mass for a series of alternate stator and rotor plates stacked along the axis of the brake [11].

E. Carbon–Carbon

Carbon–carbon brakes are made from manufactured carbon that is a composite of coke aggregate and carbon binders. It has been thermally stabilized to temperatures as high as 3000°C, and it has no melting point at atmospheric pressure. It sublimes at 3850°C. A useful characteristic for clutch and brake linings is that its strength increases with temperature up to anywhere from 2200°C to 2500°C. Beyond these temperatures it becomes viscoelastic and will, therefore, creep when stressed. Graphite crystals themselves are anisotropic because of their layered structure with their greater strength in the basal plane. In the basal plane a single crystal may have a tensile strength of approximately 1 × 105 MPa (14.5 × 06 psi), and graphite fibers have a tensile strength of the order of 2 × 104 MPa (2.9 × 104 psi).

So-called conventional graphites may have a tensile strength ranging from 6.5 to about 280 MPa (approximately 940 to about 40,600 psi) [12].

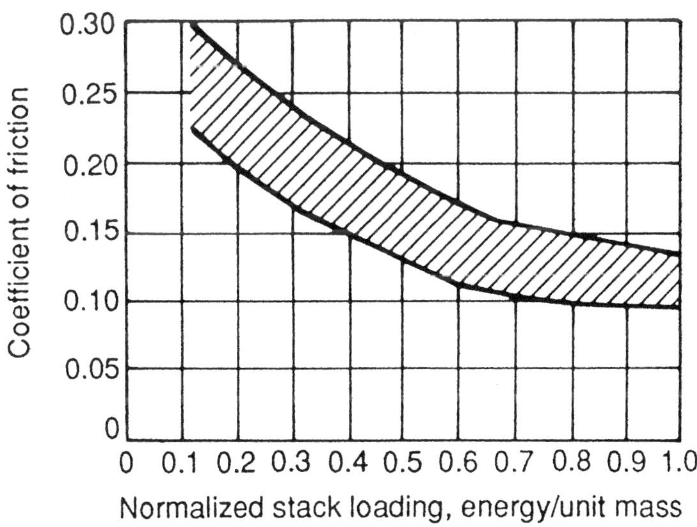

FIGURE 7 Typical range of friction coefficients for a steel brake based upon stack loading. (Courtesy *ASM Handbook*, ASM International, Materials Park, OH.)

Both carbon and graphite display porosity that varies with their grades. Blocking these pores with thermosetting resins that include phenolics, furans, and epoxies produces what is known as impervious graphite. Impervious graphite, graphite, and carbon resist corrosion by acids, alkalies, and many inorganic and organic compounds [12].

Carbon–carbon linings may display a range of friction coefficients, depending upon many factors, some of which remain proprietary with the lining manufacturers. Brake design, however, is known to have an effect in that μ increases with the number of rotors. Because carbons and graphites have an affinity for moisture, brakes that have been allowed to absorb moisture for several hours have a lower μ, sometimes known as *morning sickness*. The friction coefficient returns to its dry value because braking causes the moisture to evaporate [11].

The greatest wear on aircraft brakes occurs during a rejected takeoff (RTO) in which an aircraft taxies up to takeoff speed and then must brake to a stop. RTOs are scheduled several times during a manufacturer's ground test of prototype aircraft but rarely occur during the operation of properly maintained aircraft in service. An RTO is a spectacular display of smoke, burning rubber, and the roar of engines with the thrust reversers on. Break wear during an RTO is said to range anywhere from 100 to 1000 times greater than during a normal service stop. Wheels and brakes after an RTO are normally scrapped. Changes in μ during an RTO are shown in Figure 8, which

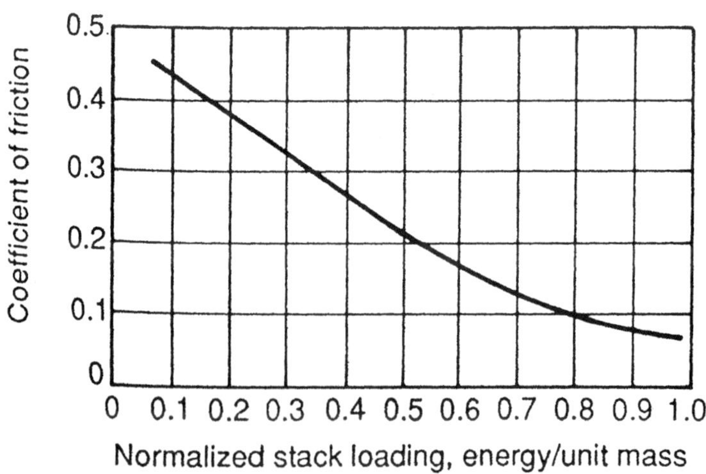

FIGURE 8 Variation of μ from taxing on the left-hand side to RTO on the right-hand side. (Courtesy *ASM Handbook*, ASM International, Materials Park, OH.)

indicates a change in μ from approximately 0.45 to approximately 0.07, for a change of 84.4%. In contrast to this, Figure 7 for steel brakes with sintered linings shows a change in μ from a midrange value of about 0.25 to a midrange value of about 0.14, for a change of 44%. However, carbon–carbon brakes are lighter than steel brakes and can be made from a single material [11].

F. Other Proprietary Materials

Friction materials produced by most manufacturers are proprietary to the extent that not all of their ingredients are disclosed. None of the ingredients may be listed for those lining materials that perform satisfactorily without components provided by others, such as Kevlar. Absence of asbestos always will be noted by U.S. suppliers.

Many manufacturers of entirely or partially proprietary linings provide data on the nominal friction coefficients, wear, and recommended temperature ranges of their products, although a few will supply data only to a manufacturing customer. This data may be in either tabular or graphical form. Typical tabular data for dry lining materials may be similar to that shown in Table 2, and typical data for wet (oil, transmission fluid) lining materials may be similar to that shown in Table 3 for two different lining materials. The first of these, GL 483-110, is described as a layered Kevlar mat with embedded carbon particles that are highly wear resistant. This layered composition is bound together with a high strength, temperature resistant phenolics resin. The second, GL 383-114, is a non-asbestos, cellulose fiber composite friction paper that is saturated with a similar phenolics resin. (See Fig. 1.)

Data in both tables were obtained from conditions that may differ from those experienced by the lining material in any particular application. Data in Table 3 especially may differ significantly from that found in any given application because of the profound effects of the finish and hardness of the mating surface along with the effects of the nature and temperature of the enveloping fluid upon the performance of the lining material.

TABLE 2 Proprietary Dry Clutch/Brake Lining Material

Product type	p_{max}	$\mu_{dynamic}$ (normal)	μ_{hot}	μ_{static}	Wear rate (hp hr)	Comment
GL 121-120	150 psi	0.48	0.47	0.66	0.009 $in.^3$	Flexible
GL 134-142	450 psi	0.42	0.40	0.56	0.011 $in.^3$	Flexible
GL 181-142	33,000 psi	0.56	0.52	0.49	0.009 $in.^3$	Rigid

Source: Web site: Great Lakes Friction Products, Milwaukee, WI.

TABLE 3 Proprietary Wet Clutch/Brake Lining Material

p_{max} (psi)	1800	μ_{static}	0.11
Energy (hp/in.^{2})	4.6	F^o_{max}	600
$\mu_{dynamic}$	0.12	F^o_{spike}	750

Source: Web site: Great Lakes Friction Products, Milwaukee, WI.

V. NOTATION

A	area (l^2)
d	distance (l)
F	force (ml/t^2)
K	specific wear (lt^2/m)
p	pressure (m/lt^2)
P	power (ml^2/t^3)
t	time (t)
Γ	wear rate (ml^2/t^2)
δ	thickness removed (l)
Θ	temperature (1)
υ	volume removed (l^3)

REFERENCES

1. SAE Handbook, 2003.
2. Ludema, K. C. (1996). *Friction, Wear, Lubrication.* Boca Raton, FL: CRC Press.
3. Engineering Plastics. 190 Turnpike Rd., Westboro, MA.
4. Tanaka, K., Kawakami S. (1982). Effects of various fillers on the friction and wear of Polytetrafluoroethylene-based composites. *Wear* 79: 221–234.
5. Anderson, J. C. (1986). The wear and friction of commercial polymers and composites. In: Friedrich, K., ed. *Friction and Wear of Polymer Composites. Composite Materials, Series 1.* New York: Elsevier, pp. 329–362.
6. Rabinowicz, Ernest. (1955). *Friction and Wear of Materials.* 2nd ed. New York: Wiley.
7. Tribco, Inc., 1700 London Rd., Cleveland, OH.
8. Reddaway Manufacturing Co., Inc., 32 Euclid Ave., Newark, NJ.
9. Hibbing International Friction, 2001 Troy Ave., New Castle, IN.
10. Reinsch, E. W. (1970). Friction and Antifriction Materials. In: Hausner, H. H., Roll, K. H., Johnson, P. K., eds. *Perspective in Powder Metallurgy.* New York: Plenum Press. Reprinted from a paper by the same name and author in *Progress* in Powder Metallurgy, 1962, pp. 131–138.
11. Tatarrzycki, E. M., Webb, R. T. (1992). Friction and Wear of Aircraft Brakes. Vol. 18. 10th ed. *ASM Handbook.* Metals Park, OH: ASM International, pp. 527–582.
12. Grayson, M. ed. (1983). *Encyclopedia of Composite Materials and Components.* New York: Wiley, pp. 188–221.

2

Band Brakes

Band brakes are simpler and less expensive than most other braking devices, with shoe brakes, as perhaps their nearest rival. Because of their simplicity, they may be produced easily by most equipment manufacturers without having to purchase special equipment and without having to use foundry or forging facilities. Only the lining must be purchased from outside sources.

Band brakes are used in many applications such as in automatic transmissions (Figure 1) and as backstops (Figure 5—devices designed to prevent reversal of rotation), for bucket conveyors, hoists, and similar equipment. They are especially desirable in the last-mentioned application because their action can be made automatic without additional controls.

I. DERIVATION OF EQUATIONS

Figure 2 shows the quantities involved in the derivation of the force relations used in the design of a band brake. Consistent with the direction of rotation of the drum, indicated by ω, the forces acting on an element of the band are as illustrated in the lower right section of Figure 2. In this figure, r is the outer radius of the brake drum and F_1 and F_2 are the forces applied to the ends of the brake band. Because of the direction of drum rotation, F_1 is greater than F_2. Equilibrium of forces in directions parallel and perpendicular to the tangent to a typical brake-band element at its midpoint requires that

$$(F + dF) \cos \frac{d\theta}{2} - F \cos \frac{d\theta}{2} - \mu\, pwr\, d\theta = 0 \qquad (1\text{-}1)$$

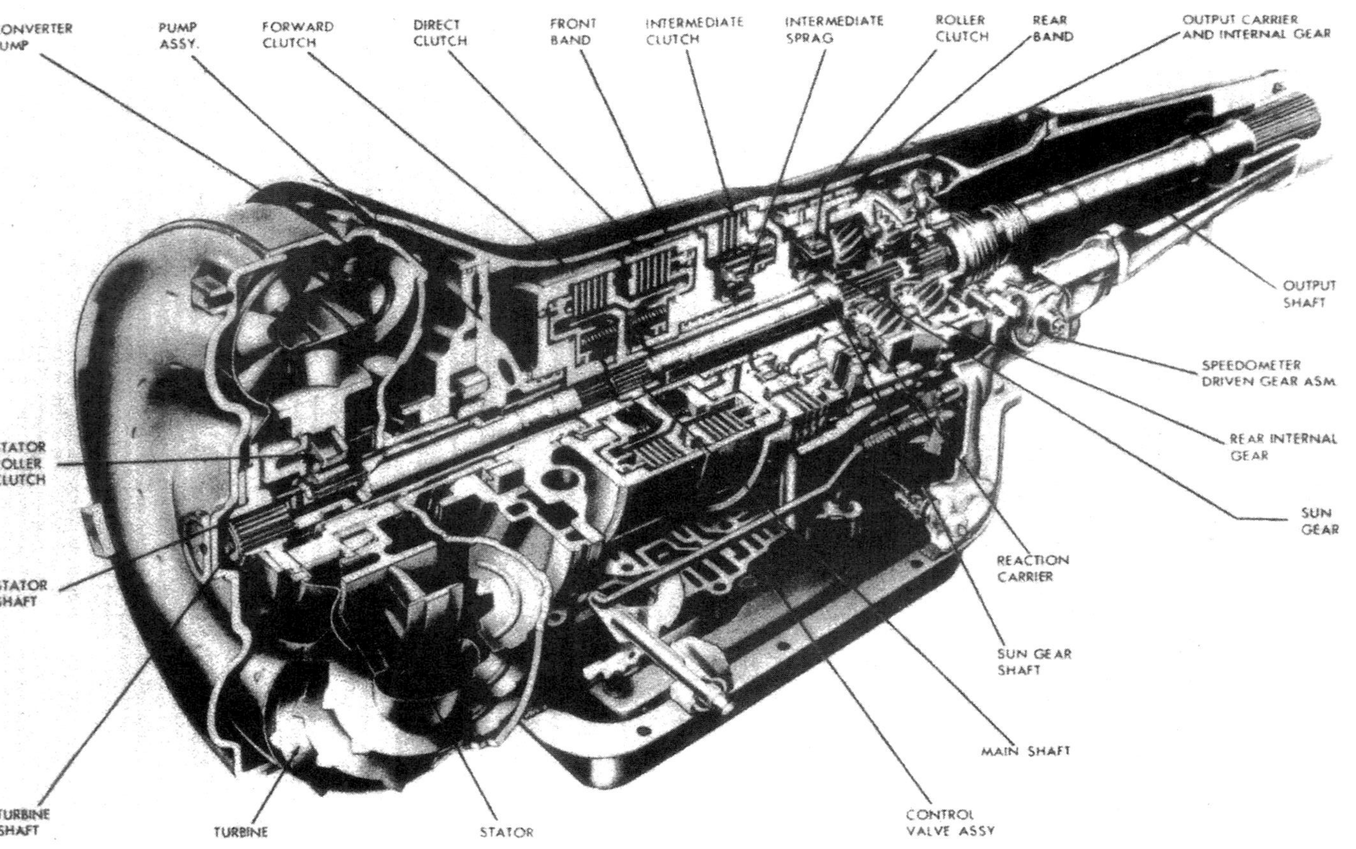

FIGURE 1 Band brakes used in an automatic transmission system.

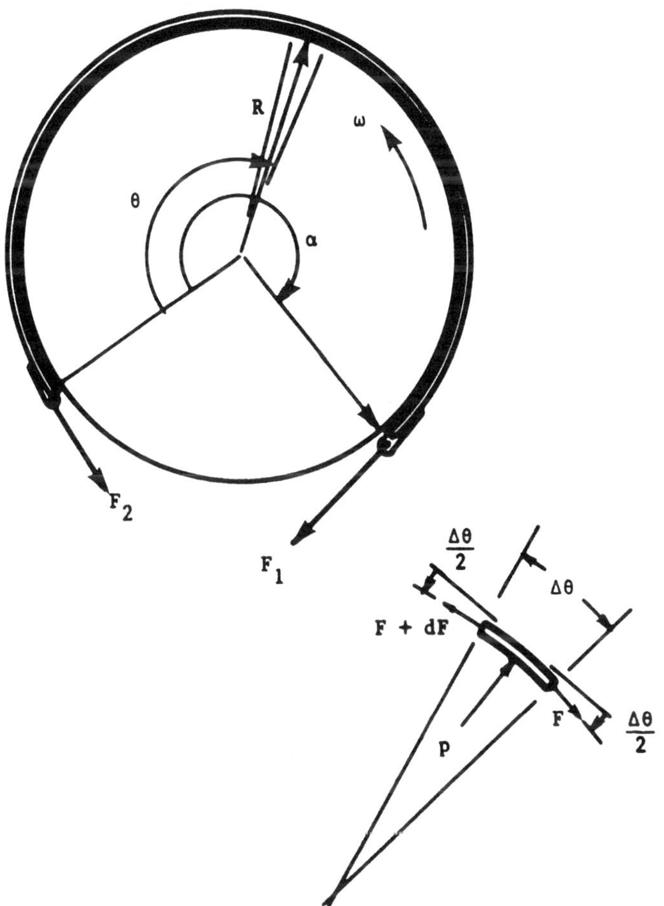

FIGURE 2 Quantities and geometry used in the derivation of the band-brake design relations.

$$(F + dF)\ \sin\ \frac{d\theta}{2} +\ F\ \sin\ \frac{d\theta}{2} - pwr\ d\theta = 0 \tag{1-2}$$

when the brake lining and the supporting brake band together are assumed to have negligible flexural rigidity, where μ represents the coefficient of friction between the lining material and the drum, p represents the pressure between the drum and the lining, and w represents the width of the band. Upon simplifying equations (1-1) and (1-2) and remembering that as the element of band length approaches zero, $\sin(d\theta/2)$ approaches $d\theta/2$, $\cos(d\theta/2)$

approaches 1, and the product $dF(d\theta/2)$ becomes negligible compared to $F\,dt$, we find that these two equations reduce to

$$dF = \mu pwr\,d\theta \tag{1-3}$$

so that

$$F = pwr \tag{1-4}$$

Substitution for pwr from equation (1-4) into equation (1-3) yields an expression that may be integrated to give

$$\ln F - \ln F_2 = \ln \frac{F}{F_2} = \mu\theta \tag{1-5}$$

where θ is taken to be zero at the end of the band where F_2 acts. It is usually more convenient to write this relation in the form

$$\frac{F}{F_2} = e^{\mu\theta} \tag{1-6}$$

which expresses the tangential force in the band brake as a function of position along the brake.

We may find F_1 from equation (1-6) by simply setting $\theta = \alpha$ to obtain

$$\frac{F_1}{F_2} = e^{\mu\alpha} \qquad (\alpha = \text{wrap angle}) \tag{1-7}$$

Since this equation shows that the maximum force occurs at $\theta = \alpha$, it follows from equation (1-4) that

$$F_1 = wrp_{max} \tag{1-8}$$

in terms of the radius r of the drum and the width w of the band. This equation points out a disadvantage of a band brake: The lining wear is greater at the high-pressure end of the band. Because of this the lining must be discarded when it is worn out at only one end, or it must be reversed approximately halfway through its life, or the brake must have two, or perhaps even three, different lining materials with different coefficients of friction so that the lining does not need to be changed as frequently.

The torque exerted by the brake is related to the band force according to

$$T = (F_1 - F_2)r \tag{1-9}$$

Upon factoring out F_1 by referring to equation (1-7) and then replacing F_1 by the right-hand side of equation (1-8), we get

$$T = F_1 r(1 - e^{-\mu\alpha}) = p_{max} wr^2(1 - e^{-\mu\alpha}) \tag{1-10}$$

which gives the brake's maximum restraining torque as a function of its dimensions and its maximum compressive pressure. This equation may be applied if the leading link can withstand the force $F_1 = rwp_{max}$ and if the band is strong enough to support the force given by equation (1-6) for $0 \leqq \theta \leqq \alpha$.

A measure of the efficiency of a band brake is the ratio of the torque applied by the brake to the torque that could be obtained if the force were applied directly to the drum itself:

$$\frac{T}{F_1 r} = 1 - e^{-\mu\alpha} \tag{1-11}$$

The maximum value of this ratio for a single-turn band brake is 0.998 when $\mu = 1.00$. From the plot of this ratio, Figure 3, it is apparent that reductions in the angle of wrap from 360° to 270° has relatively little effect on the efficiency for $\mu = 0.5$ or greater. We also see that the brake should subtend an arc of 270° or more if degradation of the friction coefficient, perhaps due to a dirty environment and infrequent maintenance, is to be expected.

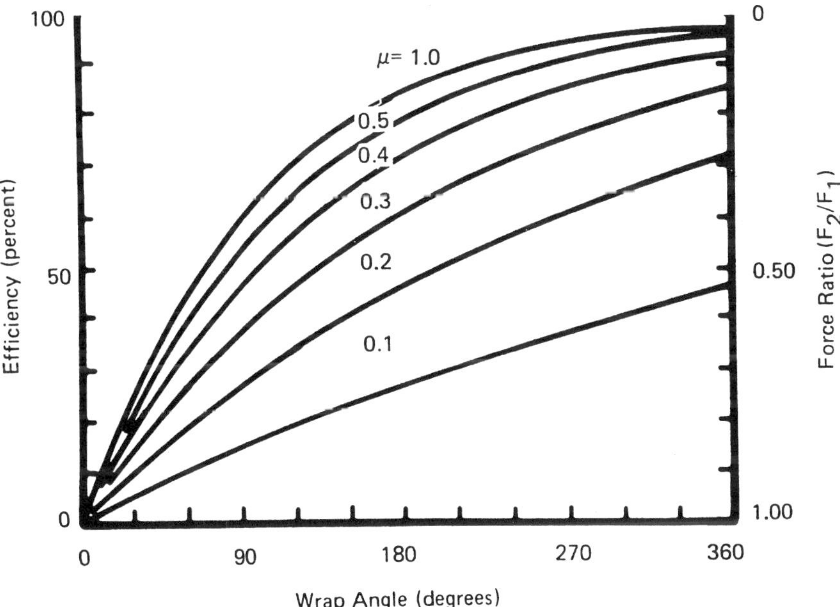

FIGURE 3 Efficiency (T/F_1r) and force ratio (F/F_1) as a function of angle from the leading end of the brake band.

Since reinforcement of the band near its leading end depends on the force decay as a function of angle along the band, it may be of interest to display how F decreases with ϕ, measured from the leading end of the band. To do this we simply replace F_2 with F and replace α with ϕ in equation (1-7) to obtain

$$\frac{F}{F_1} = e^{-\mu\phi} \qquad \phi = \alpha - \theta \tag{1-12}$$

For a brake band extending over an angle ϕ from F_1.

$$T = (F_1 - F)r = F_1 r\left(1 - \frac{F}{F_1}\right) = F_1 r(1 - e^{-\mu\phi}) \tag{1-13}$$

Thus the decay of the band force from its maximum at the leading end of the band may be found from Figure 3 using the scales shown on the right-hand ordinate and associating the abscissa with ϕ.

It is because of the low coefficient of friction for wet friction material that the brake bands in an automatic transmission are relatively thick and curved to fit the drum with only a small clearance. The thickness is required to support the large band force necessary to deliver a relatively large torque when operating at low efficiency and the small clearance is necessary to minimize the required activation force to bend the band and lining to the drum radius.

II. APPLICATION

In this section we consider the design of a band brake to exert a torque of 9800.0 N-m subject to the conditions that the drum width be no greater than 100 mm and that the drum diameter be no greater than 750 mm. To complete the design we should also specify the necessary link strength for a safety factor of 3.5 when using a steel that has a working stress of 410 MPa. Other mechanisms require that the angle of wrap not exceed 290°. Lining temperature is not expected to rise above 300°F (148°C) during the most severe conditions. Select a lining material that can sustain a maximum pressure of 1.10 MPa.

Return to Chapter 1 to find that the lining represented by Figure 4 is one of several that is flexible enough for use in a band brake and has the limiting temperature and pressure capability. Thus, use $\mu = 0.4$ and equation (1-10) to find that at the maximum radius the band width should be given by

$$w(r) = \frac{T}{p_{\max} r^2 (1 - e^{-\mu\alpha})} \tag{2-1}$$

where lining width w is written as a function of r in a numerical analysis program. Likewise, the lining area is given by

$$A(r) = \alpha r w(r) \tag{2-2}$$

where α is in radians. Similarly, substitution for $w(r)$ from equation (2-1) into equation (1-8) gives

$$F(r) = p_{\max} w(r) r \tag{2-3}$$

which enables calculation of the link diameter for a safety factor ζ and maximum operating stress σ from the relation.

$$d_1(r) = 2\sqrt{\frac{F(r)}{\pi\sigma}\zeta} \tag{2-4}$$

For the largest drum diameter, which is 375 mm, turn to equation (2-1) to find that for this drum the lining width should be

$$w(375) = 72.992 \text{ mm}$$

which is within the width limits. The corresponding lining area and link diameter $d_1(r)$ as given by equations (2-2) and (2-4) are

$$A(375) = 1.385 \times 10^5 \text{ mm}^2 = 1385 \text{ cm}^2 \qquad d_1(r) = 2r_1(r) = 18.09 \text{ mm}$$

For the largest lining width, solve equation (2-1) for the drum radius and find the drum diameter as a function of the lining width from

$$d(w) = 2\left[\frac{T}{p_{\max} w(1 - e^{-\mu\alpha})}\right]^{1/2} \tag{2-5}$$

which yields that the drum diameter for a 100-mm lining width should be

$$d(100) = 640.77 \text{ mm}$$

According to equations (2.2) and (2.4), the corresponding lining area and link diameter are

$$A(320.38) = 1622 \text{ cm}^2 \qquad \text{and} \qquad d_1 = 2r_1 = 19.57 \text{ mm}$$

Select the design with the larger lining area in order to reduce the energy dissipation per unit area, lower the operating temperature, and thereby decrease lining wear. Selecting a convenient drum diameter slightly larger than 640.77mm, namely, 641 mm, while retaining the lining width of 100 mm will only increase the brake's torque capability for a negligibly smaller link force while reducing the pressure upon the lining.

III. LEVER-ACTUATED BAND BRAKE: BACKSTOP DESIGN

This type of brake may be represented as shown in Figure 4(a). Moment equilibrium about the pivot point of the lever requires that

$$F_1 a - F_2 b + P(b + c) = 0 \qquad (3\text{-}1)$$

so that substitution for F_2 from equation (1-7) yields

$$-F_1(a - be^{-\mu\alpha}) = P(b + c) \qquad (3\text{-}2)$$

as the force P required to activate the brake. Substitution for F_1 in equation (3-2) from relation (1-11) yields

$$P = \frac{be^{-\mu\alpha} - a}{r(1 - e^{-\mu\alpha})} \frac{T}{b + c} \qquad (3\text{-}3)$$

Note that not only is the force related to the lever arm length, as is to be expected from elementary statics, but a braking torque may be exerted with no activating force if

$$a = be^{-\mu\alpha} \qquad (3\text{-}4)$$

In other words, the lever portion of length c could be removed and the mechanism would stop rotation in the direction shown [Figure 4(b)]. The brake is then termed "self-locking in one direction."

Mechanisms of this sort, illustrated in Figure 4(c), are known as backstops. Their function is to permit rotation is one direction and prevent rotation in the other direction.

If the direction of rotation is reversed, the brake will loosen because a slight rotation in the counterclockwise direction of the lever will cause a larger motion at B than at A. Brake-band sag should be sufficient to provide enough friction force to activate the brake whenever the rotation reverses direction.

A backstop using the linkage shown in Figure 4(c) is shown in Figure 5. The two small tabs on the brake band are to prevent it from slipping off the drum. A relatively close fit (with a slight increase in power dissipation) is intended between the band and the drum to maintain sufficient frictional force to assure quick response whenever the direction of rotation is reversed.

IV. EXAMPLE: DESIGN OF A BACKSTOP

Design a backstop similar to that shown in Figure 2.4(c) to prevent gravity unloading of a bucket elevator similar to that shown in Figure 6 that has 41 buckets on each side. For design purposes assume that all buckets on the downward-moving side are empty and that all of the buckets on the upward-

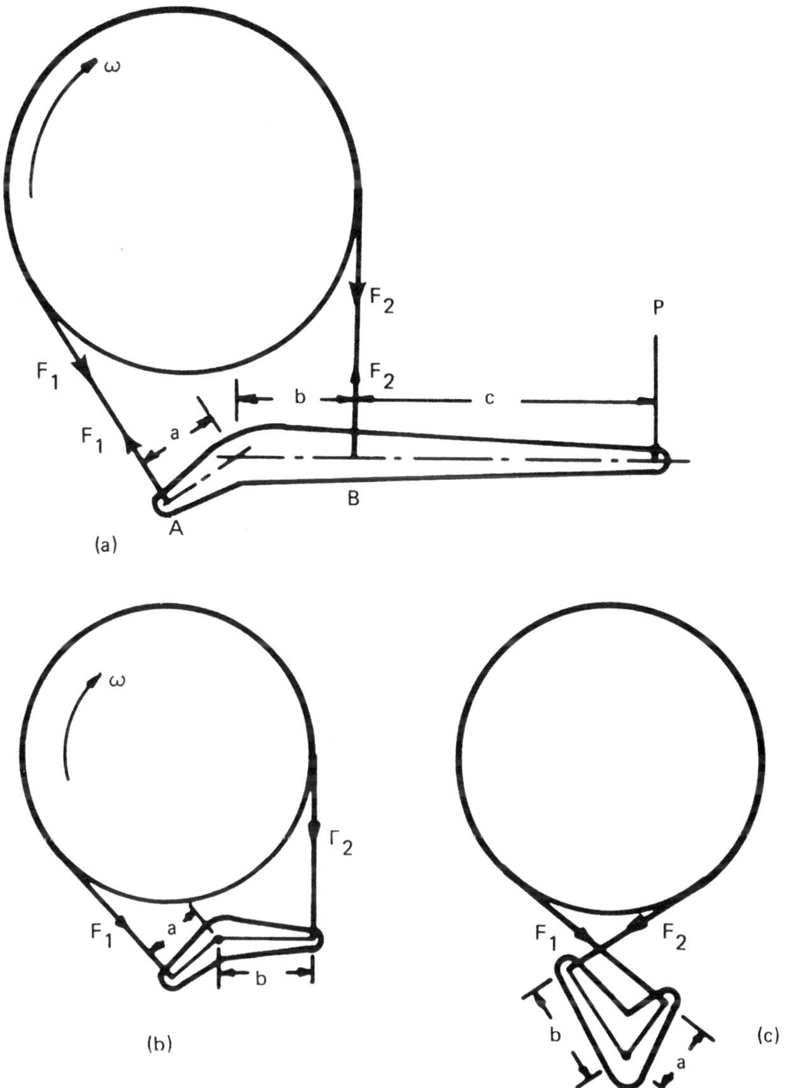

Figure 4 (a) Lever-activated band brake; (b) backstop configuration with $a = be^{-\mu\alpha}$; (c) backstop with levers a and b rearranged to provide a greater wrap angle.

FIGURE 5 Backstop.

moving side are filled when the power is turned off, with each bucket containing 129 lb of material. The pitch diameter d_s of the sprocket is 34 inches.

Assume that the friction coefficient of the lining will always be 0.4 and that the minimum value of p_{max} is 275 psi. Housing requirements demand that the backstop drum diameter be no larger than 33 in. Use a safety factor of 1.5 in sizing the drum band, which is to be made from spring steel having a yield stress of 102,000 psi.

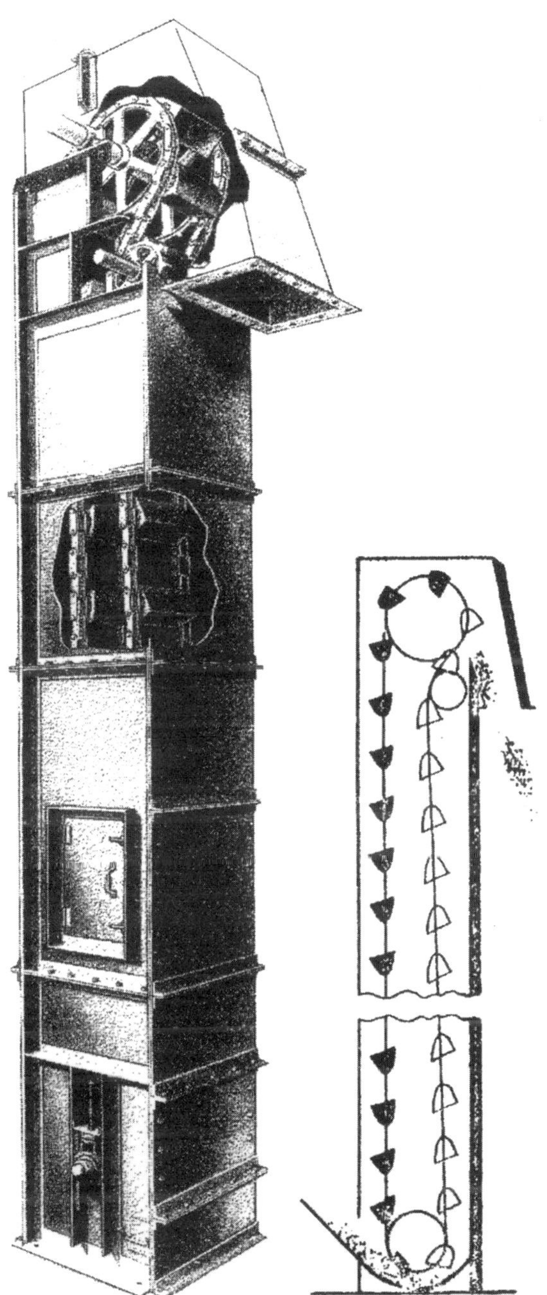

Figure 6 Positive discharge bucket conveyor cutaway and cross section. (Courtesy American Chain Association, Washington, DC.)

To ensure clearance, let the drum diameter be 32 in., and design for a wrap angle, α, of 300°. From the sprocket pitch diameter and the bucket weights, find

$$T = (d_s/2)WN = (34/2)129(41) = 89,913 \text{ in.-lb}$$

as the maximum expected value of the torque. Here W denotes the expected maximum weight of material in a bucket and N denotes the number of buckets one each side. Chain and empty bucket weights were ignored because the chain and empty bucket assembly is in equilibrium due to the symmetry of the conveyor system about its vertical axis.

After solving equation (1-10) for w, we have

$$w = \frac{T}{p_{\max} r^2 (1 - e^{-\mu\alpha})} \tag{4-1}$$

so substitution of $\alpha = 300\pi/180 = 5.2360$ radians along with the given values into this expression yields

$$w = 1.457 \text{ in.}$$

Force F_1 may be calculated for this width from equation (1-8), to get the maximum force as

$$F_1 = 6409 \text{ lb}$$

The thickness of the spring steel band to which the lining is attached may be calculated from

$$t = \frac{\zeta F}{w\sigma} \tag{4-2}$$

in which ζ represents the safety factor and σ represents the yield stress of the steel band. Substitution of these values along with w and F into equation (4-2) yields $t = 0.065$ in. Finally, from equation (3-4), we have

$$b/a = e^{\mu\alpha} = e^{0.4(5.236)} = 8.121$$

Although relation (3-4) may be derived from the backstop configuration using the equilibrium equation for the backstop lever, which is

$$F_1 a = F_2 b$$

together with equation (1-7), use of equation (3-3) has the advantage of showing that when b/a is less than $e^{\mu\alpha}$, the direction of force P on the lever reverses. This implies that the backstop lever proportions should obey the inequality

$$a/b = e^{-\mu\alpha} \tag{4-3}$$

to function properly. In particular, the equality follows by setting $P = 0$ and the inequality follows by setting $P < 0$ in equation (3-3).

Return to equation (4-1) and substitute that expression for w into equation (1-8) to find that F_1 may be written as

$$F_1 = \frac{T}{r(1 - e^{-\mu\alpha})} \tag{4-4}$$

to show that F_1 is independent of p_{max}. Therefore w may be increased to 1.50 in. without affecting F_1. It merely slightly reduces the p_{max} required to provide the design torque capability.

V. NOTATION

a	lever length (l)
b	lever length (l)
c	lever length (l)
d	drum diameter (l)
d_l	link diameter (l)
F	band force (mlt^{-2})
F_1	maximum band force (mlt^{-2})
F_2	minimum band force (mlt^{-2})
P	force applied to brake lever (mlt^{-2})
p	lining pressure ($ml^{-1}t^{-2}$)
p_{max}	maximum lining pressure ($ml^{-1}t^{-2}$)
r	drum radius (l)
T	torque (ml^2t^{-2})
t	band thickness (l)
w	band width (l)
α	brake wrap angle (1) $0 \leqq \theta \leqq \alpha$
θ, ϕ	Angle subtended at the center of the drum (1)
μ	Friction coefficient (1)
σ	yield stress ($ml^{-1}t^{-2}$)
ζ	safety factor (1)

VI. FORMULA COLLECTION

Torque:

$$T = F_1 r(1 - e^{-\mu\alpha}) = (F_1 - F_2)r = p_{max} w r^2 (1 - e^{-\mu\alpha})$$

Activation force in terms of torque:

$$F_1 = \frac{T}{r(1 - e^{-\mu\alpha})}$$

Activation force in terms of maximum pressure:

$$F_1 = p_{\max} w r$$

Link diameter:

$$d_1 = 2\sqrt{\frac{F}{\pi\sigma}\zeta}$$

Maximum pressure in terms of torque:

$$p_{\max} = \frac{T}{r^2 w(1 - e^{-\mu\alpha})}$$

Band thickness:

$$t = \frac{F\zeta}{w\sigma}$$

Minimum wrap angle:

$$\alpha = \frac{1}{\mu}\ln\frac{1}{1 - T/(F_1 r)}$$

Drum diameter in terms of activation force:

$$d = \frac{2F_1}{w p_{\max}}$$

Drum diameter in term of torque:

$$d = \frac{2T}{F_1(1 - e^{-\mu\alpha})} = 2\left[\frac{T}{p_{\max} w(1 - e^{-\mu\alpha})}\right]^{1/2}$$

Band brake lever force:

$$P = \frac{be^{-\mu\alpha} - a}{r(1 - e^{-\mu\alpha})}\frac{T}{b + c}$$

Band width in terms of torque:

$$w = \frac{T}{p_{\max} r^2(1 - e^{-\mu\alpha})}$$

REFERENCES

1. Roark, R. J., Young, W. C. (1975). *Formulas for Stress and Strain.* 5th ed. New York: McGraw-Hill.
2. Popov, E. P. (1976). *Mechanics of Materials.* 2nd ed. Englewood Cliffs, NJ: Prentice-Hall.

3

Externally and Internally Pivoted Shoe Brakes

Typical externally and internally pivoted shoe brakes are shown in Figures 1 and 2. In all but extremely rare designs, equal forces act upon both shoes to produce equal applied moments about their pivots. External shoe brake control is usually through a lever system that may be driven by electromechanical, pneumatic, or hydraulic means. Internal shoe brake control is usually by means of a double-ended cylinder or a symmetrical cam.

Calculation of the moments and shoe lengths to achieve a specified braking torque cannot be carried out directly when the two shoes are pivoted as shown in either of these figures and when the opposing shoes are subjected to equal moments. The tedious task of manually iterating these formulas to get a satisfactory design under these conditions may be eliminated with the use of computer programs, such as those mentioned in the following sections, that can quickly produce either graphical or numerical design solutions.

I. PIVOTED EXTERNAL DRUM BRAKES

A. Long Shoe Brakes

Externally pivoted, long shoe brakes similar to that shown in Figure 1 are often used as holding brakes. As its name implies, a holding brake is to hold a shaft stationary until the brake is released. The compression spring on the left-hand side of the brake in Figure 1 applies a clamping force to the brake shoes

Figure 1 Externally pivoted shoe brake. (Courtesy Automation & Process Technology Div., Ametek Paoli, PA.)

on either side of the brake drum to hold it stationary without the need for external power. Electrical current through the solenoid on the left side of the assembly releases the brake and holds it open for as long as voltage is applied to the solenoid. Other holding brakes may use slightly different mechanical arrangements and may use either a hydraulic or a pneumatic cylinder to release the brake.

One of the applications of a holding brake is in the design of an overhead crane. The value of a holding brake is that it allows a load that has been raised to be held in position without external power. This is also a safety feature because the crane will not allow its load to be raised or lowered until the brake is intentionally released. Likewise, these brakes are also used in hoists, in punch and forming presses, and in some conveyor systems, for similar reasons.

We begin the derivation of the governing equations for the braking torque by considering only one shoe (Figure 3) and then extending those results to more than one shoe. Under the assumption that the shoe, lever, and drum are all rigid, and that the stress–strain relations of the lining are linear,

FIGURE 2 Internal pivoted shoe brake. (Courtesy of Dyneer Mercury Products, Canton, Ohio.)

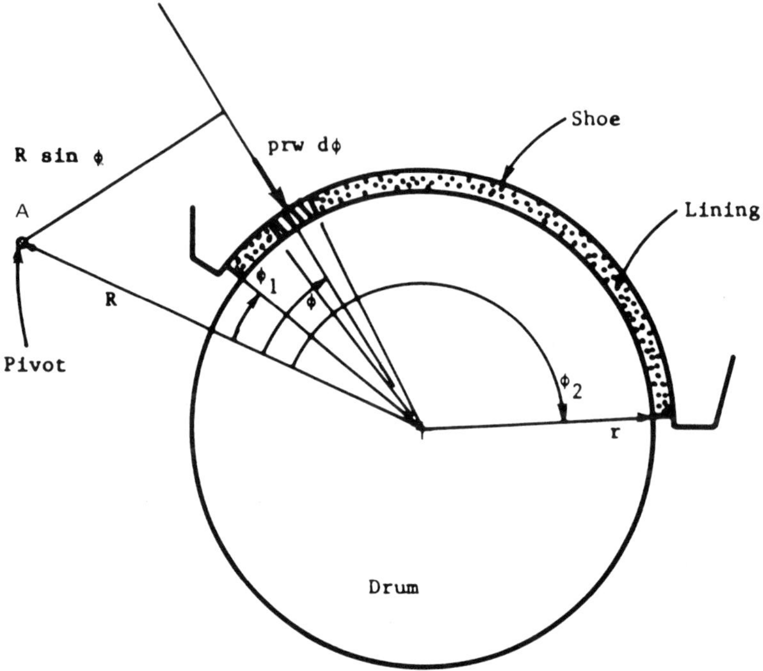

FIGURE 3 Geometry involved in calculating moment M_p about pivot point *A*.

the pressure p at any position along the lining due to infinitesimal rotation $\delta\beta$ of the shoe about pivot A will be given by

$$p = kR\delta\beta \sin \varphi \tag{1-1}$$

in terms of the quantities shown in Figure 3. Upon the introduction of the maximum pressure, written as

$$p_{\max} = kR\delta\beta \, (\sin \phi)_{\max} \tag{1-2}$$

we find that

$$kR\delta\beta = \frac{p_{\max}}{(\sin \phi)_{\max}} \tag{1-3}$$

so that after substitution from equation (1-3) into equation (1-1), we find that the pressure may be written in terms of the maximum pressure as

$$p = \frac{p_{\max}}{(\sin \phi)_{\max}} \sin \phi \tag{1-4}$$

With this expression for pressure as a function of position, the torque on the drum will be the integral over the shoe length of the incremental friction force $\mu(prw - d\phi)$ acting on the surface of a drum of radius r. Thus

$$T = \mu r^2 w \frac{p_{\max}}{(\sin\phi)_{\max}} \int_{\phi_1}^{\phi_2} \sin\phi \, d\phi \tag{1-5}$$

in which $(\sin\phi)_{\max}$ denotes the maximum value of $\sin\phi$ within the range $\phi_1 \le \phi \le \phi_2$. Integration of equation (1-5) yields

$$T = \frac{\mu p_{\max} r^2 w}{(\sin\phi)_{\max}} (\cos\phi_1 - \cos\phi_2) \tag{1-6}$$

in which ϕ_1 is the angle from radius R between the drum axis and pivot A to the near edge of the drum sector subtended by the brake lining. As drawn in Figure 3, angle ϕ_2 is measured from radius R toward the far edge of the brake lining. Hence the angle subtended by the shoe is given by

$$\phi_0 = \phi_2 - \phi_1 \tag{1-7}$$

To calculate the moment that must be applied about pivot A in Figure 3 to obtain the torque found by equation (1-6), we first calculate the moment reaction at the pivot due to both the incremental normal forces and the incremental friction forces acting on the lining. An equal and opposite moment must, of course, be supplied to activate the brake.

Radial force $prw\, d\phi$ on each incremental area also contributes to a pressure moment M_p about pivot A. Relative to the geometry in Figure 3, and with the aid of equation (1-4), this moment may be written as

$$M_p = \int_{\phi_2}^{\phi_2} (pwr\, d\phi) R \sin\phi = \frac{p_{\max}\, wrR}{(\sin\phi)_{\max}} \int_{\phi_1}^{\phi_2} \sin^2\phi \, d\phi \tag{1-8}$$

which integrates to

$$M_p = \frac{p_{\max}\, wrR}{4(\sin\phi)_{\max}} (2\phi_0 - \sin 2\phi_2 + \sin 2\phi_1) \tag{1-9}$$

where ϕ_0 is given by equation (1-7). This moment is positive in the counter-clockwise direction, and its algebraic sign is independent of the direction of drum rotation relative to the brake lever's pivot point.

Reactive moment M_f at pivot A due to the friction force acting on the shoe may be calculated using the geometry sketched in Figure 4. Thus,

$$\begin{aligned} M_f &= \int_{\phi_1}^{\phi_2} (\mu pwr\, d\phi)(R\cos\phi - r) \\ &= \frac{\mu p_{\max}\, wr}{(\sin\phi)_{\max}} \int_{\phi_1}^{\phi_2} (R\cos\phi \sin\phi - r\sin\phi) d\phi \end{aligned} \tag{1-10}$$

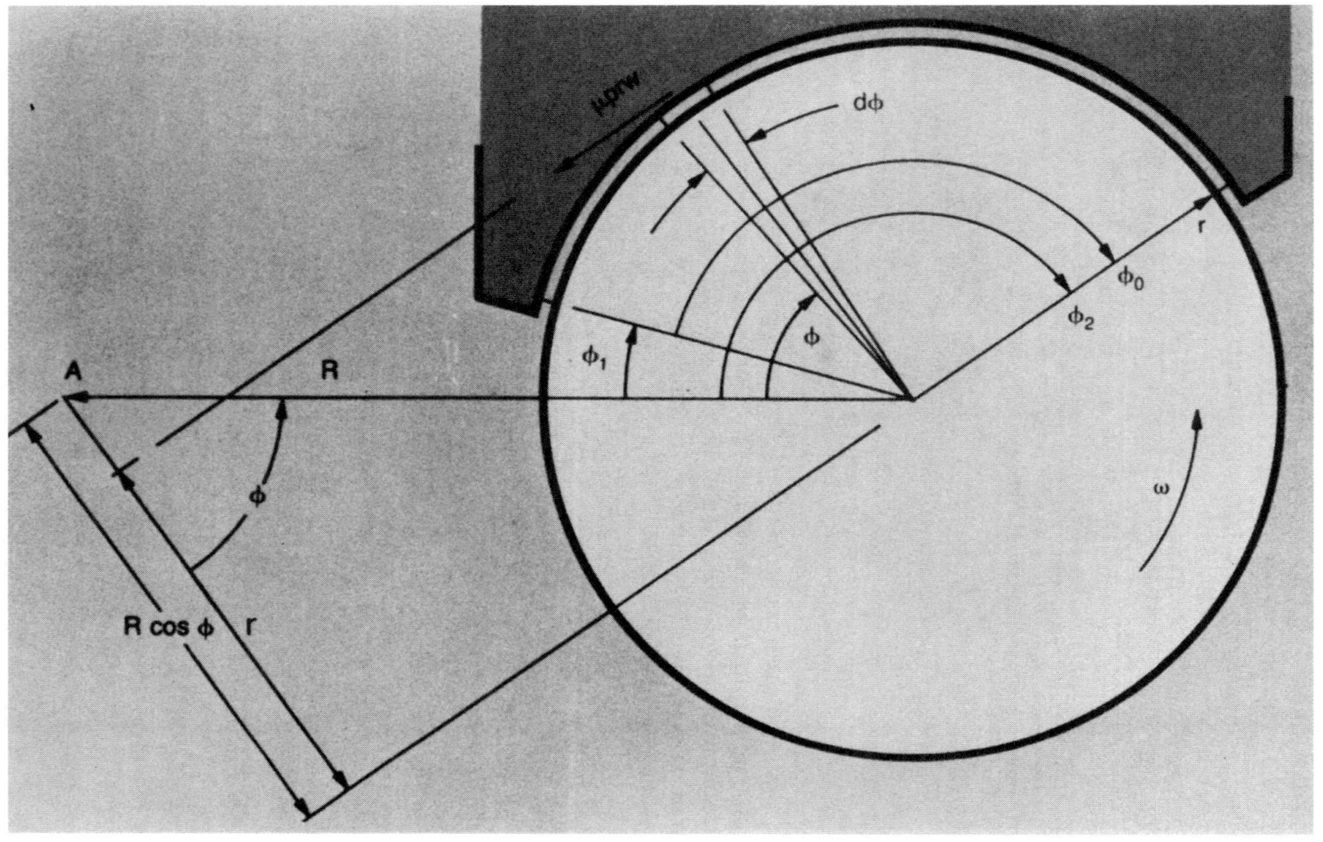

FIGURE 4 Geometry involved in calculating moment M_f about pivot point A.

Integration of equation (1-10) yields

$$M_f = \frac{\mu p_{max} wr}{4(\sin\phi)_{max}}[R(\cos 2\phi_1 - \cos 2\phi_2) - 4r(\cos\phi_1 - \cos\phi_2)] \quad (1\text{-}11)$$

Here the quantity enclosed by the square brackets determines the algebraic sign of M_f and may cause it to be zero. The physical significance of the algebraic sign for nonzero values of moment M_f depends upon the direction of rotation ω of the drum.

If the rotation is toward the pivot, as in Figure 4, a positive value of M_f signifies a clockwise moment about the pivot that applies the brake by forcing the shoe against the drum, which would cause self-locking. Therefore, a negative or zero value for M_f from equation (1-11) is required to produce either a counterclockwise or a zero moment, respectively, about the pivot point.

The interpretation is reversed if the drum rotation ω is away from the pivot. In this case a positive value from equation (3.11) indicates a counterclockwise rotation of the shoe about the pivot that tends to release the brake. Obviously, a negative value in this situation indicates a clockwise moment about the pivot that tends to rotate the shoe toward the drum.

From these observations it follows that brake activation requires an applied moment M_e about the pivot point A such that

$$\begin{aligned} M_p + M_f &= M_e > 0 \qquad \omega \text{ away from the pivot} \\ M_p - M_f &= M_e > 0 \qquad \omega \text{ toward the pivot} \end{aligned} \quad (1\text{-}12)$$

where M_f itself, as calculated from equation (1-11), must be negative or zero when rotation ω is toward the pivot and positive or zero when it is away from the pivot—hence the minus sign in the second of equations (1-12).

Self-locking is of use only when the brake is to serve as a backstop or as an emergency brake during control failure. Otherwise, self-locking is generally to be avoided because it does not allow the braking torque to be controlled by the control of M_e.

B. Short Shoe Brakes

Short shoe brakes are generally defined as those for which the angular dimension of the brake, ϕ_0, is small enough (generally less than 20°) that $\sin\phi \cong (\sin\phi)_{max}$ and $p \cong p_{max}$ so that with these restrictions equation (1-5) may be approximated by

$$T = \mu p w r^2 \phi_0 = \mu r F \quad (1\text{-}13)$$

where

$$F = p w r \phi_0 \quad (1\text{-}14)$$

is the force exerted on the short shoe. Application of these approximations to equation (1-9) before integration yields

$$M_f = \mu F(R \cos \phi_1 - r) \tag{1-15}$$

Similarly, application of these approximations to equation (1-10) before integration yields

$$M_p = FR \sin \phi_1 \tag{1-16}$$

so that substitution into equation (1-12) with the minus sign in effect reveals that the short shoe will not be self-locking if

$$\sin \phi_1 - \mu\left(\cos \phi_1 - \frac{r}{R}\right) > 0 \tag{1-17}$$

II. PIVOTED INTERNAL DRUM BRAKES

The equations derived in Section IA dealing with long external shoe brakes apply equally well to internal shoe drum brakes. There is one essential difference, however, that does not appear explicitly in the equations themselves: The physical significance of positive values of moments M_p and M_f is different. The geometry used to obtain these relations for internal shoe brakes is shown in Figures 5 and 6; the different interpretations for the various combinations of direction of rotation and internal or external shoes are listed in Table 1. In that table rotation of the drum from the far end of the shoes to the end near the pivot (termed rotation from the toe of the brake to the heel) is indicated by an arrow pointing toward the letter p; rotation in the opposite direction is indicated by an arrow pointing away from the letter p. The acronym cw indicates clockwise rotation (or the direction of rotation of an advancing right-hand screw), and ccw indicates counter-clockwise rotation.

From Figure 5 it follows that

$$dM_f = (\mu wrp \, d\phi)(r - R \cos \phi) \tag{2-1}$$

This is the negative of the integrand in equation (1-10). The rotation indicated causes the shoe to pivot in the counterclockwise direction about A; but because equation (1-10) used the negative of the integrand above, the rotation shown corresponds to a negative M_f value as calculated using either equation (1-10) or equation (1-11). Hence, negative M_f from these formulas implies counterclockwise rotation and positive M_f corresponds to clockwise rotation of the shoe about its pivot.

Braking requires a moment M_a applied to the shoe as given by

$$M_p - M_f = M_a \qquad \omega \text{ away from the pivot}$$

$$M_p + M_f = M_a \qquad \omega \text{ toward the pivot}$$

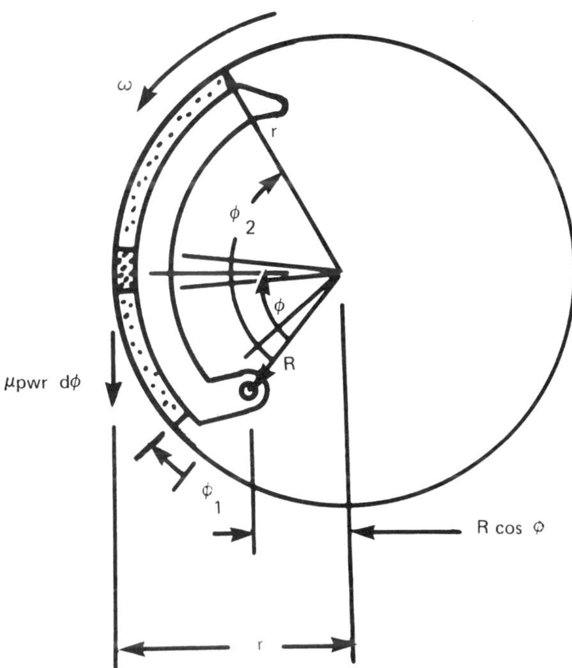

FIGURE 5 Geometry for calculating the moment due to friction about point *A* for an internal shoe brake.

for internal shoes. The physical significance of the algebraic signs associated with the moment expressions derived in the preceding sections as applied to external and internal brakes is displayed in Table 1. It is may be helpful the rewrite the equations for either internal or external brakes in terms of different symbols if the use of a single set of equations for two different cases becomes too confusing. After using these equations enough to become familiar with them, the reader may find that analysis is easier if they are again combined into a single set, as has been done here.

Drum brake efficiency may be measured in terms of the ratio of the torque produced by the brake itself to the torque required to activate the brake, also known as the shoe factor; namely,

$$\frac{T}{M_a} = \frac{T}{M_p \pm M_f} \tag{2-2}$$

Brake efficiency is generally not a design factor in the analysis of drum brakes because it is dependent on too many factors [ϕ_1, ϕ_2, r/R, μ, w, and $(\sin \phi)_{max}$]

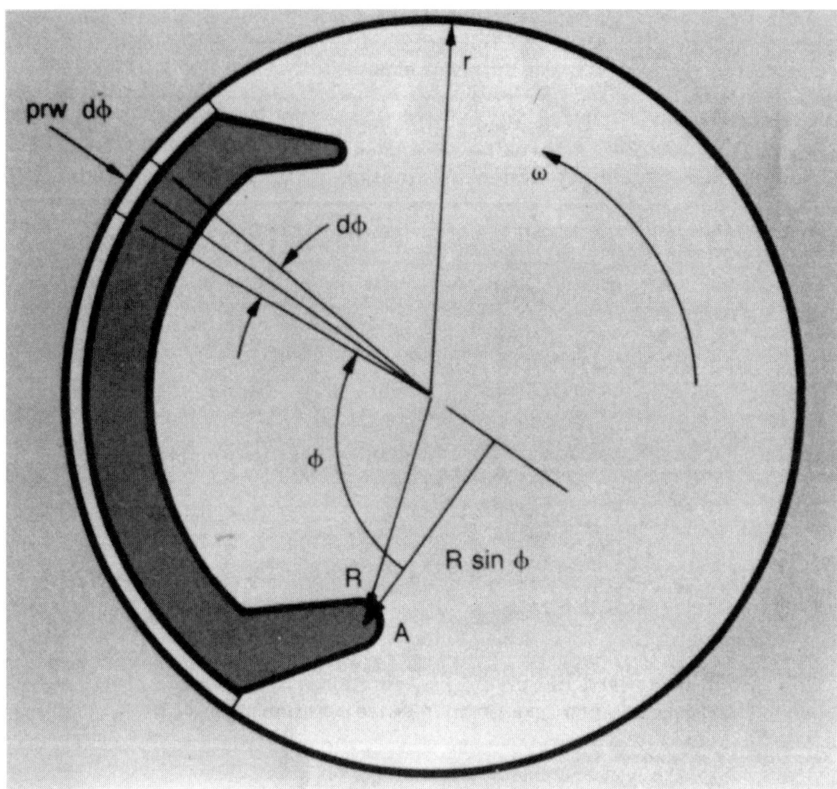

FIGURE 6 Geometry for calculating the moment due to pressure about point A for an internal shoe brake.

to make it useful. More significance is usually associated with brake life, heat dissipation, fading, and braking torque capability.

III. DESIGN OF DUAL-ANCHOR TWIN-SHOE DRUM BRAKES

For both external and internal shoes and for either direction of rotation a positive M_e value indicates that an external moment of that magnitude must be applied to activate the brake. The formulas also clearly indicate that the extent of the braking action may be controlled by controlling this activation moment. The role of M_f, the moment due to friction, in determining the required activation moment M_e may be seen by returning to equation (1-11)

TABLE 1 Moment Relations for Internal and External Drum Brakes

		External shoe		Internal shoe	
Rotation[a]	Moment	Implied braking action	Implied shoe rotation	Implied braking action	Implied shoe rotation
$p\rightarrow$	$M_p > 0$	Open	ccw	Open	cw
$p\leftarrow$	$M_p > 0$	Open	ccw	Open	cw
$p\rightarrow$	$M_f > 0$	Open	ccw	Close	ccw
$p\leftarrow$	$M_f > 0$	Close	cw	Open	cw
		Applied Moment Relations			
		External Shoe		Internal Shoe	
$p\rightarrow$	–	$M_p + M_f = M_a$		$M_p - M_f = M_a$	
$p\leftarrow$	–	$M_p - M_f = M_a$		$M_p + M_f = M_a$	

[a] $p\rightarrow$, Rotation toward the pivot; $p\leftarrow$, rotation away from the pivot; cw, clockwise rotation; ccw, counterclockwise rotation.

and observing that this moment may be either positive or negative, depending on the choices for the quantities appearing in brackets.

One measure of the contribution of the friction moment to the entire amount acting to force the shoe against the drum is the actuation factor, defined by

$$\frac{M_f}{M_a} \quad \left(\text{sometimes defined as } \frac{M_f}{M_p}\right) \tag{3-1}$$

which is independent of the torque produced by the brake.

If the quantities in brackets in equation (1-11) are chosen such that the bracket becomes both negative and relatively large, M_f may dominate M_p and M_a becomes negatfive. This means that the brake has become self-locking: contact between the shoe and the drum causes uncontrolled motion of the shoe toward the drum. Since the resulting braking action is beyond the control of the usual single-direction activation mechanism, self-locking is generally to be avoided.

Return to relations (2-2), equate the denominators, and then divide both sides by M_p, which is always positive, to obtain

$$\frac{M_a}{M_p} = 1 \pm \frac{M_f}{M_p} \tag{3-2}$$

Hence self-locking of external brakes in which the drum rotates toward the pivot can be avoided if the relation M_f/M_p is always less than $+1$; if the drum rotates away from the pivot, self-locking can be avoided if M_f/M_p is always greater than -1. Similar criteria hold for internal brakes except that the directions of rotation are reversed for the same algebraic signs. Since most brakes are designed for rotation in both directions, it is generally convenient to combine these criteria into a single criterion, which is that self-locking of both internal and external drum brakes may be avoided if

$$-1 \leq \frac{M_f}{M_p} \leq +1 \tag{3-3}$$

Selection of shoe and drum angles and dimensions in accordance with this criterion may be aided by construction of design curves such as illustrated in Figures 7 and 8, in which the ratio $M_f/\mu M_p$ is plotted against angle ϕ_2 for selected values of the coefficient of friction. External shoes are characterized by R/r ratios greater than unity and internal shoes by r/R ratios less than unity. The ratio $M_f/\mu M_p$ has been plotted instead of M_f/M_p in Figures 7 and 8 because it itself is independent of the coefficient of friction and thus must be

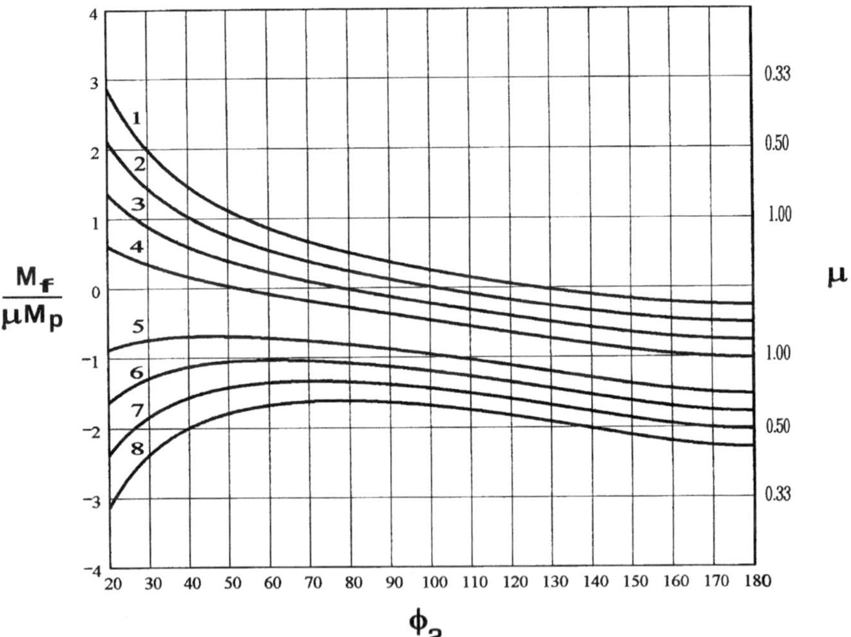

FIGURE 7 Design curves for $M_f/(\mu M_p)$ for $\varphi_1 = 10°$. r/R ratios for the upper, external brake, curves are: 1—$r/R = 0.2$; 2—$r/R = 0.4$; 3—$r/R = 0.6$; 4—$r/R = 0.8$. r/R ratios for the lower, internal brake, curves are: 5—$r/R = 1.2$; 6—$r/R = 1.4$; 7—$r/R = 1.6$; 8—$r/R = 1.8$.

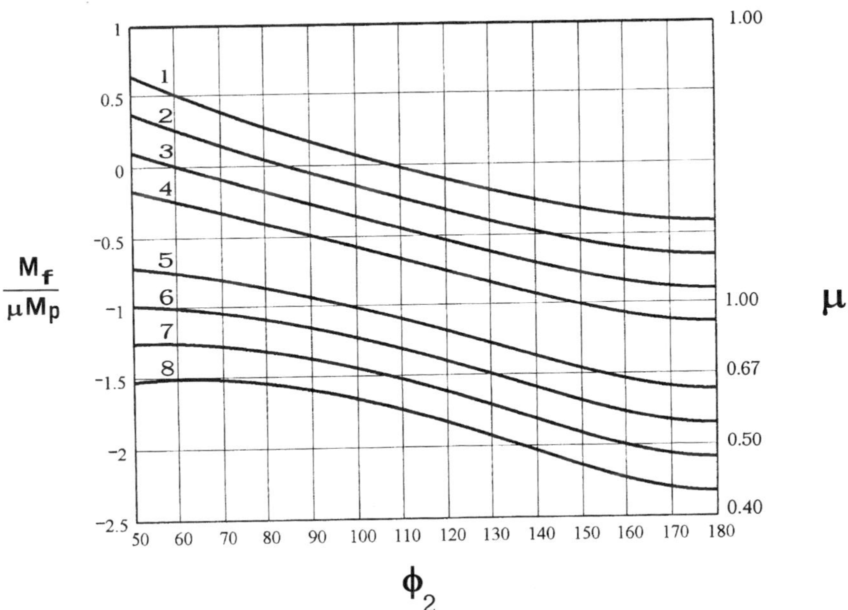

Figure 8 Design curves for $M_f/(\mu M_p)$ for $\varphi_1 = 45°$. r/R ratios for the upper, external brake, curves are: 1—$r/R = 0.2$; 2—$r/R = 0.4$; 3—$r/R = 0.6$; 4—$r/R = 0.8$. r/R ratios for the lower, internal brake, curves are: 5—$r/R = 1.2$; 6—$r/R = 1.4$; 7—$r/R = 1.6$; 8—$r/R = 1.8$.

plotted only once. To use it for any coefficient of friction within the range shown, it is only necessary to note that the requirement that the ratio M_f/M_p lie between -1 and $+1$ is equivalent to

$$-\frac{1}{\mu} \leq \frac{M_f}{\mu M_p} \leq \frac{1}{\mu} \tag{3-4}$$

Since $p_{\max}$, $(\sin\phi)_{\max}$, and μ cancel out when equation (1-11) is divided by the product of μ and equation (1-9), the ratio $M_f/(\mu M_p)$ is a function of only three quantities: r/R, ϕ_1, and ϕ_2. Thus, $M_p/(\mu M_p)$ may be plotted as a function of ϕ_2 for fixed values of r/R and ϕ_1, as in Figures 7 and 8. Criterion (3.4) also can be included in these graphs by noting that $1/\mu > 0$ pertains to external drum brakes and $1/\mu < 0$ pertains to internal drum brakes, so these values may be shown on the left-hand ordinate of these graphs by relating them to the limiting values of $M_f/(\mu M_p)$ according to relation (3.4), namely, that at the lower limit,

$$-1/\mu = M_f/(\mu M_p)$$

and that at the upper limit,

$$1/\mu = M_f/(\mu M_p)$$

Consequently, the ordinates on the right-hand sides of the graphs in Figures 7 and 8 are the reciprocals of the ordinates on the left-hand sides. Thus, we may read directly from these graphs that to be non-self-locking, the $M_f/(\mu M_p)$ ratio must fall below the $1/\mu$ value for external drum brakes, and it must fall above the $-1/\mu$ value for internal drum brakes.

Note that these curves show that the range of possible values for $M_p/(\mu M_p)$ that ensure that a dual-shoe brake will be free of self-locking decreases as the lining coefficient of friction increases, as should be expected.

The length of a single shoe for a desired torque may be found algebraically from equation (1-6). However, selection of the shoe length to provide a specified braking torque cannot be accomplished directly if two external or two internal shoes operating about fixed pivot points, or anchor pins, are to be used for greater braking torque.

Whenever two shoes are required and the arc length of the lining, $r\phi_0 = r(\phi_2 - \phi_1)$, is to be selected, it is necessary to select ϕ_1, say, and then find a value of ϕ_2 such that the total torque T is the sum of T_a and T_b, where T_a represents the braking torque contribution from the shoe with the larger peak pressure and T_b represents the braking torque from the shoe with the smaller peak pressure, p_b. Torque T_a, as given by the equation

$$T_a = \frac{\mu p_a r^2 w}{(\sin \phi)_{\max}} (\cos \phi_1 - \cos \phi_2) \tag{3-5}$$

will be the reference torque for both shoes. For simplicity in writing the remaining equations it is convenient to introduce the quantities

$$\begin{aligned} A &= R(2\phi_2 - 2\phi_1 - \sin 2\phi_2 + \sin 2\phi_1) \\ B &= \mu[R(\cos 2\phi_1 - \cos 2\phi_2) - 4r(\cos \phi_1 - \cos \phi_2)] \\ b_a &= p_a b \qquad b_b = p_b b \qquad b = \frac{rw}{4(\sin \phi)_{\max}} \end{aligned} \tag{3-6}$$

so that moments M_f and M_p may be written as

$$\begin{aligned} M_{p_a} &= b_a A \qquad M_{f_a} = b_a B \\ M_{p_b} &= b_b A \qquad M_{f_b} = b_b B \end{aligned} \tag{3-7}$$

In these terms the applied moment to one of the shoes may be written as

$$M_a = b_a \min \begin{cases} (A + B) \\ (A - B) \end{cases} \tag{3-8}$$

and the relation

$$M_a = b_b \max \begin{cases} (A + B) \\ (A - B) \end{cases} \tag{3-9}$$

then determines the maximum pressure on the other shoe. Recall that braking torque T is linearly dependent on the maximum pressure, in this case p_a, so

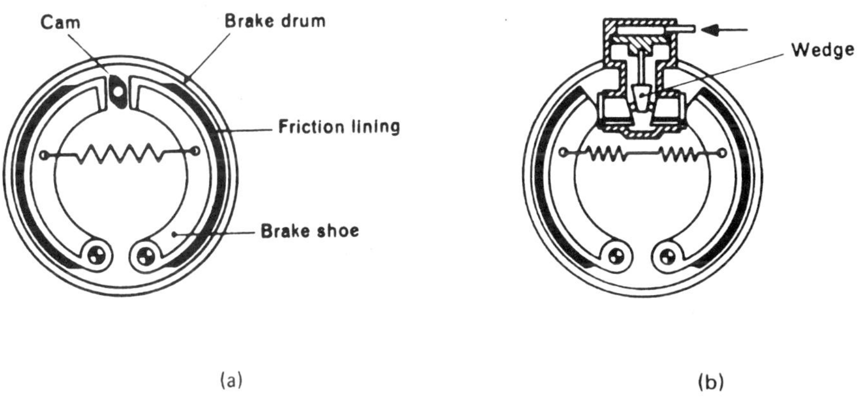

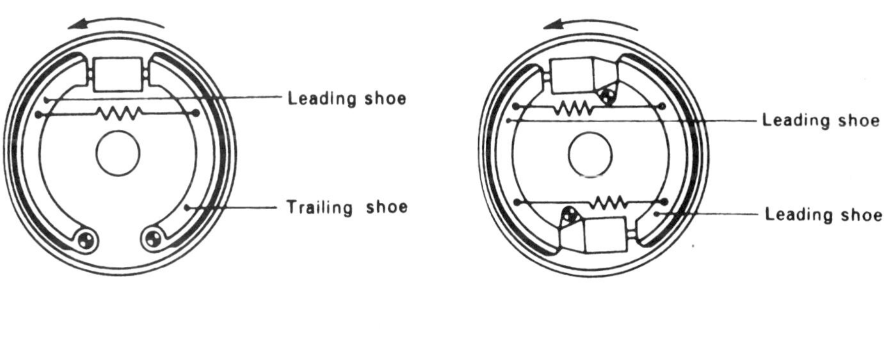

FIGURE 9 Four styles of dual-anchor brakes: (a) cam brake; (b) wedge brake; (c) simplex brake; (d) duplex brake.

that reevaluation of the integral in equation (1-5) is not necessary whenever the two shoes are of the same size; T_b is simply given by

$$T_b = \frac{p_b}{p_a} T_a = \frac{b_b}{b_a} T_a \tag{3-10}$$

Consequently, the total braking torque expression becomes

$$T = \left(1 + \frac{b_b}{b_a}\right) T_a \tag{3-11}$$

Designing a double-shoe brake, either internally pivoted (automotive type) or externally pivoted, to provide a specified braking torque consists of finding values that satisfy equations (2.2) and (3.5) through (3.11).

Moment M_a is most commonly applied by forces supplied by a cam at the toe of each brake, as shown in Figure 9(a), which is known as a cam brake; by an integral hydraulic system that drives a wedge between two pistons, which in turn act against the toe of each shoe, as shown in Figure 9(b), which is known as a wedge brake; or by a hydraulic cylinder between the two shoes, as shown in Figure 9(c), which is known as a simplex brake. In all of these the force necessary to provide moment M_a is given by

$$F = \frac{M_a}{2r \sin(\phi_0/2)} \tag{3-12}$$

The iteration process may be eliminated using the duplex brake shown in Figure 9(d), but at the expense of a brake that is more effective for one direction of rotation (rotation from toe to heel) than for the reverse rotation. This brake style is therefore usually limited to machines where rotation is in one direction, such as conveyor belts.

IV. DUAL-ANCHOR TWIN-SHOE DRUM BRAKE DESIGN EXAMPLES

Example 4.1

Design an external dual-anchor twin-shoe drum brake to provide a torque of 6050 N-m. The drum diameter should not exceed 400 mm and the drum thickness should not exceed 90 mm, based on interference with other machine components. If possible, select angle ϕ_1 to be 25° in order to use stock hydraulic components already under contract for other products. Heating during braking may occasionally be large.

Comparison of Figures 7 and 8 shows that increasing ϕ_1 from 10° to 45° has relatively little effect on these curves, so that we may refer to Figure 7 for $\phi_1 = 25°$. Hence we find that an external brake will be free of self-locking as long as the brake shoes subtend an angle of 70° or more.

Guided by the design limitations in the problem statement, let the drum diameter be 350 mm and the width be 80 mm to ensure extra clearance if needed. Both dimensions can be increased if no satisfactory brake can be designed within these smaller dimensions. Since we plan to design shoes having ϕ_2 of the order of 140° use this as an initial value of ϕ_2 along with $p_{\max} = 3.00$ MPa (435 MPa), which may be had using a proprietary material from Chapter 1 that has $\mu_{\text{bot}} = 0.40$. Take $\mu = 0.35$ to find if this will yield a satisfactory shoe. If it does, the longer band will aid in cooling and may have a longer life The required activation moment M_a may be found from equation (3.9) after A and B have been calculated.

Upon entering the selected values

$T = 6{,}050{,}000$ N–m	$R = 230$ mm
$w = 80$ mm	$r = 175$ mm
$\mu = 0.35$	$\phi_1 = 25°$
$p_{\max} = 3.00$ MPa	

into a Mathcad work page as shown later we can use the graphics capability of Mathcad to produce the graph in Figure 10 to show the torque as a function of angle φ.

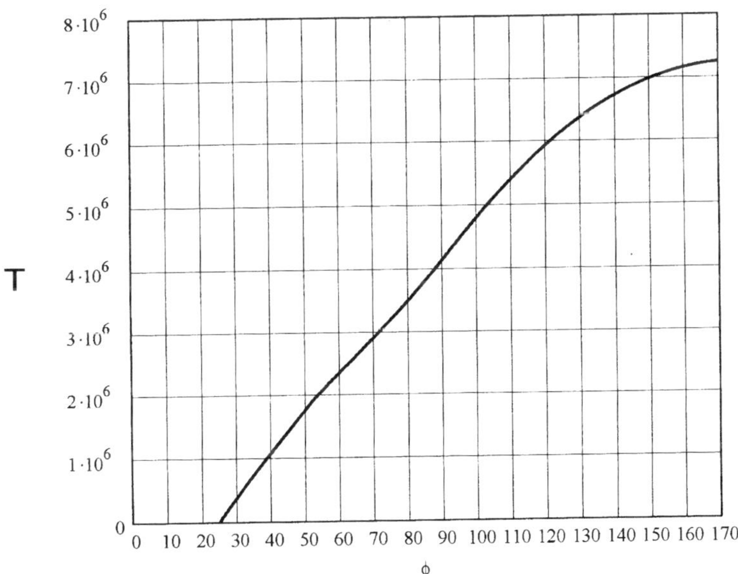

FIGURE 10 Torque vs. φ for an external drum brake.

The functional notation $\phi \cdot \text{deg}$ in the trigonometric functions is a Mathcad requirement when ϕ is in degrees. It enters the corresponding radian measure as the argument of the trigonometric function involved.

No value for M_a is given because examination of equations (3.6) and (3.10) reveals that the torque is independent of M_a. It is dependent upon $p_{\max}$, the maximum pressure, which provides the activating forces.

$$w = 80 \qquad \mu = 0.35 \qquad R = 230 \qquad r = 175 \qquad \phi_1 = 25$$

$$p_{\max} = 3.00 \qquad T_0 = 605000$$

$$A(\phi) = R(2\phi \cdot \text{deg} - 2\phi_1 \cdot \text{deg} - \sin(2\phi \cdot \text{deg}) + \sin(2\phi_1 \cdot \text{deg}))$$

$$B(\phi) = \mu[R(\cos(2\phi \cdot \text{deg}) - \cos(2\phi \cdot \text{deg})) - 4r(\cos(\phi_1 \cdot \text{deg}) - \cos(\phi \cdot \text{deg}))]$$

$$C(\phi) = A(\phi) + B(\phi) \qquad D(\phi) = A(\phi) - B(\phi)$$

$$M_1(\phi) = \begin{cases} C(\phi) & \text{if } C(\phi) \le D(\phi) \\ D(\phi) & \text{otherwise} \end{cases}$$

$$M_2(\phi) = \begin{cases} C(\phi) & \text{if } C(\phi) \ge D(\phi) \\ D(\phi) & \text{otherwise} \end{cases}$$

$$b_a(\phi) = \frac{1}{M_1(\phi)} \qquad b_b(\phi) = \frac{1}{M_2(\phi)}$$

$$\sin\phi_m(\phi) = \begin{cases} 1 & \text{if } \phi \ge 90 \\ \sin(\phi \cdot \text{deg}) & \text{otherwise} \end{cases}$$

$$T_a(\phi) = \frac{(\mu p_{\max} r^2 w)}{\sin\phi_m(\phi)}(\cos(\phi_1 \cdot \text{deg}) - \cos(\phi \cdot \text{deg}))$$

$$T(\phi) = \left(1 + \frac{b_b(\phi)}{b_a(\phi)}\right) T_a(\phi)$$

The Track feature in Mathcad prints the coordinates of the points where the screen crosshairs lie upon a curve. Smooth transition from point to point along a curve may not be possible, however, because of the difficulty of producing very small motion of the tracking ball on the mouse being used. Nevertheless, we can come sufficiently close to be within most manufacturing and design tolerances. In this example we can read from Figure 10 that

$$T = 6{,}044{,}200 \text{ N–m} \qquad \text{at } \phi_2 = 122.57^\circ$$

$$T = 6{,}052{,}200 \text{ N–m} \qquad \text{at } \phi_2 = 122.74^\circ$$

Alternatively, the bisection procedure provided by TK Solver, which is more accurate, yielded $\phi_2 = 122.693^\circ$.

Example 4.2

Design an internal twin-shoe drum brake to provide a braking torque of 5800 N-m with an outside diameter of 350 mm and a shoe width of 80 mm using a lining material with a friction coefficient of 0.35. Use $\phi_1 = 25°$ and let the pivot radius be 120 mm, if possible, because of heat sensors to be included within the drum.

Substitution of the following values into the previous worksheet and graphing the resulting $T(\phi)$ as a function of ϕ yields the graph shown in Figure 11.

$w = 80$ mm $\qquad R = 120$ mm

$r = 175$ mm $\qquad \mu = 0.35$

$\varphi = 25°$

From it we read that a torque of 5,798,700 N-m requires $\phi_2 = 155.38°$ and that a torque of 5,801,000 corresponds to an angle $\phi_2 = 155.55°$. The bisection value found from the TK Solver program was $\phi_2 = 156.4749°$. They serve as a check upon one another because the same formulas must be entered into each program to get agreement of the order shown. As before, the values read from Figure 11 are sufficiently precise for many brake applications.

Notice in both Figures 10 and 11 that increasing angle φ beyond about 120° yields diminishing returns; the torque no longer increases nearly linearly relative to φ.

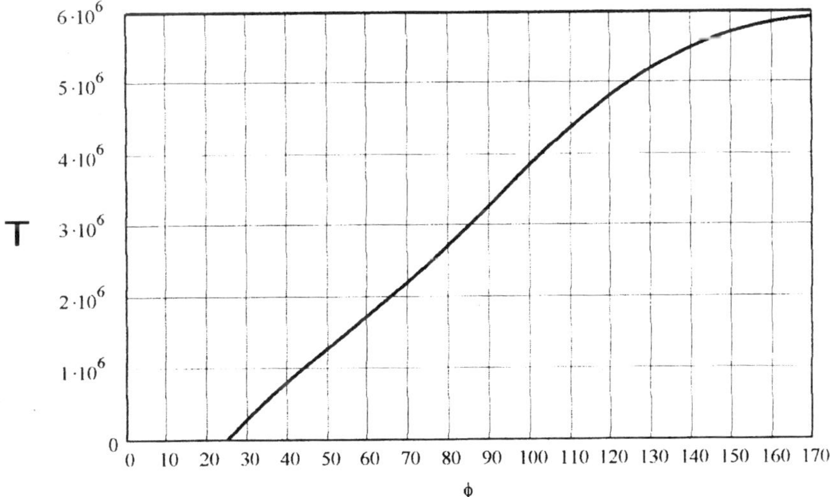

FIGURE 11 Torque vs. φ for in internal drum brake.

V. DESIGN OF SINGLE-ANCHOR TWIN-SHOE DRUM BRAKES

Internal shoe drum brakes of this type, as illustrated in Figure 12, which are also known as Bendix type, or servobrakes, have neither shoe permanently attached to an anchor pin. Each is free to shift position slightly as the direction of the drum reverses, so that for either direction of rotation one shoe pivots about the anchor pin and the othe other shoe pivots about its end of the adjusting link between shoes. Consequently, both shoes see rotation from toe to heel regardless of the direction of rotation. Although this construction facilitates the design of a self-adjusting mechanism for automotive use, it does not entirely eliminate the difference in wear between the two shoes, and it introduces additional labor to calculate brake torque and lining pressure. A

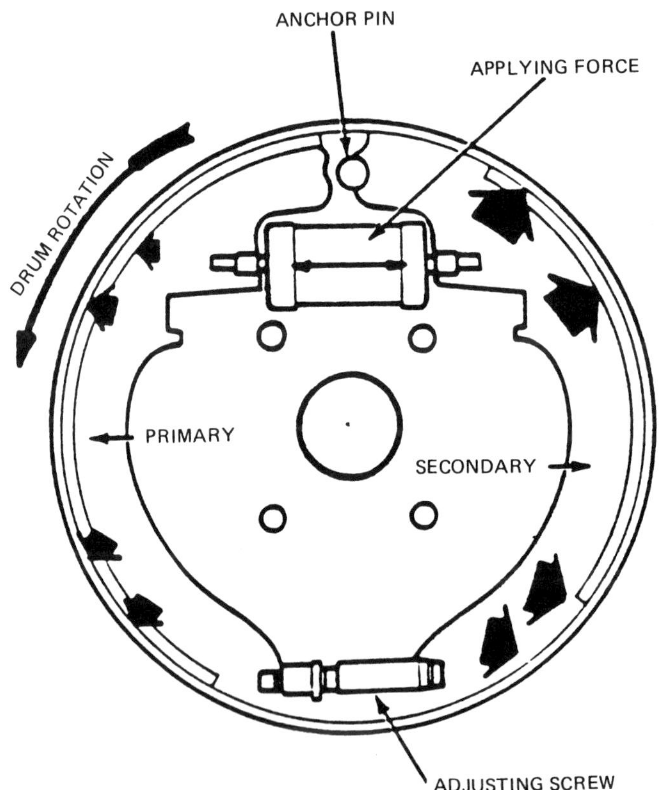

FIGURE 12 Schematic of a Bendix, or servo, single-anchor brake.

program to ease the latter tasks is described and demonstrated in the following paragraphs.

With this program it is easy to show that relatively small changes in the pressure distribution along either shoe may produce large changes in the braking torque.

Although calculation of the braking torque and consideration of the design of brakes of this type appears to be omitted from most of the machine design texts now in print, two of them do contain a brief narrative reference to their construction [1,2]. These brakes may be analyzed by the graphical method introduced by Fazekas [3] in 1957 or by the numerical method described in reference 4. In the first method the pressure is described only in terms of its center of pressure (due to the lack of easy computational facilities in 1957), while in the second method the pressure distribution is represented by either an approximating function or by a measured pressure distribution. The second method displays the marked effect the pressure distribution has on brake performance.

The governing equations for the primary shoe, which is the shoe not pivoted at the anchor pin (Figures 12 and 13), are the same as those given derived in Section 1, which are that the moments (positive in the clockwise

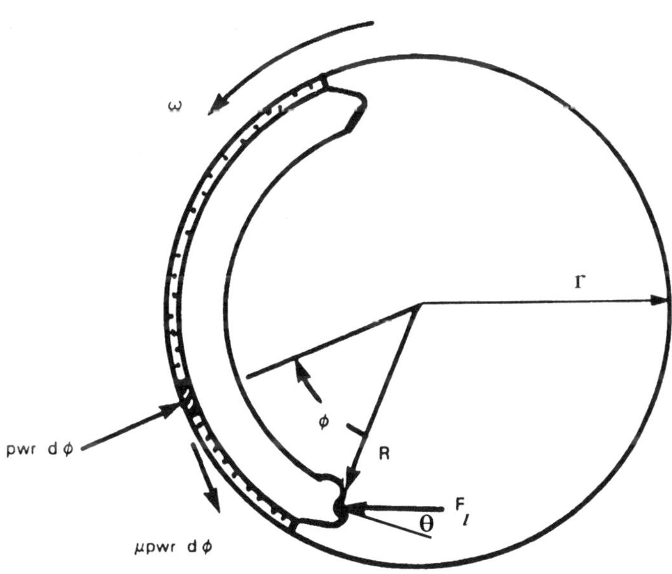

FIGURE 13 Primary link force and incremental pressure and friction forces.

direction) about the pivot due to the pressure and the frictional forces are given by

$$M_p = rwR\int_{\phi_1}^{\phi_2} p \sin\phi \, d\phi \tag{5-1}$$

and

$$M_f = \mu wr\int_{\phi_1}^{\phi_2} p(r - R\cos\phi)d\phi \tag{5-2}$$

which are repeated here for convenience. In the following discussion we shall need an expression for the radial and tangential force components acting on the pivot. These relations may be derived by taking components of the pressure and friction forces acting on the lining in directions parallel and perpendicular to R (Figure 13) from the center of the drum to the pivot point of the shoe. The result is that the force component in the radial direction is given by

$$F_{r_1} = -rw\int_{\phi_1}^{\phi_2} p\cos\phi \, d\phi + \mu rw\int_{\phi_1}^{\phi_2} p\sin\phi \, d\phi \tag{5-3}$$

and the force perpendicular to radius R is given by

$$F_{\theta_1} = rw\int_{\phi_1}^{\phi_2} p\sin\phi \, d\phi - \mu rw\int_{\phi_1}^{\phi_2} p\cos\phi \, d\phi \tag{5-4}$$

where θ is the angle between vector F_1 and a perpendicular to vector R.

Analysis based on these equations differs, however, from that associated with shoes having fixed pivot points. In particular, neither the sinusoidal-dependent pressure associated with a rigid shoe and drum discussed in Section 2 nor the constant pressure associated with a deformable shoe and drum [4] may be used in this instance because the primary shoe is to pivot about the end of the adjusting link, which can only exert a force in the direction of the chord coincident with its centerline. Consequently, the lining pressure must be such that the primary shoe is in equilibrium when acted on by the pressure, the activating moment (due to the force from the brake cylinder), and the reaction of the adjusting link along its longitudinal axis.

From this last observation and from the geometry shown in Figure 13 we see that the radial and tangential forces must satisfy the relation

$$\tan\frac{\beta}{2} + \frac{F_r}{F_\theta} = 0 \tag{5-5}$$

where β is the angle at the center of the drum subtended by the adjusting link The magnitude of the link force is given by

$$F_l^2 = F_\theta^2 + F_r^2 \tag{5-6}$$

This link force and the activating moment both act on the secondary shoe, so that the force and moment equations of equilibrium for the secondary shoe become

$$F_{r_2} = -rw\int_{\phi_1}^{\phi_2} p \cos\phi \, d\phi + \mu rw\int_{\phi_2}^{\phi_1} p \sin\phi \, d\phi + F_l \sin\left(\phi_2 + \frac{\beta}{2}\right) \quad (5\text{-}7)$$

$$F_{\theta_2} = -rw\int_{\phi_1}^{\phi_2} p \sin\phi \, d\phi + \mu rw\int_{\phi_1}^{\phi_2} p \cos\phi \, d\phi + F_l \cos\left(\phi_2 + \frac{\beta}{2}\right) \quad (5\text{-}8)$$

where 2β is the angle subtended by the adjusting link. Next,

$$M_{a_2} = M_{p_2} + M_{f_2} - 2RF_l \sin\frac{\phi_2}{2}\sin\left(\frac{\phi_2}{2} + \frac{\beta}{2}\right) \quad (5\text{-}9)$$

where M_{p_2} and M_{f_2} are again given by equations (5.1) and (5.2) in terms of the pressure distribution on the secondary shoe. Torque for either shoe may be calculated from

$$T = \mu rw\int_{\phi_1}^{\phi_2} p \, d\phi \quad (5\text{-}10)$$

Even though these equations do not uniquely determine the pressure distribution over the brake lining, they are still of use because they allow the design engineer to compare the effects of different realistic pressure distribution and to design drums and shoes whose rigidity will induce particular pressure distributions over the primary and secondary shoes.

The first of the two pressure distributions considered is a synthesis of (1) the sinusoidal distribution associated with a nondeforming drum and shoe, generally associated with a lightly loaded brake, and (2) the force peaks that occur at the ends of load-bearing members in contact. The second of the two is a systhesis of (1) the constant distribution said to be associated with more heavily loaded brakes, and (2) the previously noted force peaks. Thus the pressure distributions will be represented either by

$$\frac{p}{p_0} = e^{-c_1(\psi/\phi_0)}\left|\cos\frac{\psi}{\phi_0}\pi\right|^{c_2} + c_3 \qquad \psi = \phi - \phi_1 \quad (5\text{-}11)$$

or by

$$\frac{p}{p_0} = e^{-c_1(\psi/\phi_0)}\left|\cos\frac{\psi}{\phi_0}\pi\right|^{c_2} + c_3\sin(\psi + \phi_1) \quad (5\text{-}12)$$

for the primary shoes, where the peak pressure on each shoe occurs at the heel because its pivot cannot sustain a radial force in the outward direction. These relations produce the pressure distributions shown in Figures 14 and 15 for the values of c_1, c_2, and c_3 indicated. Similarly, peak pressure is assumed to occur at the toe of the secondary shoe, where it is subjected to the link force from the primary shoe. Since this shoe pivots at the anchor pin, it is assumed

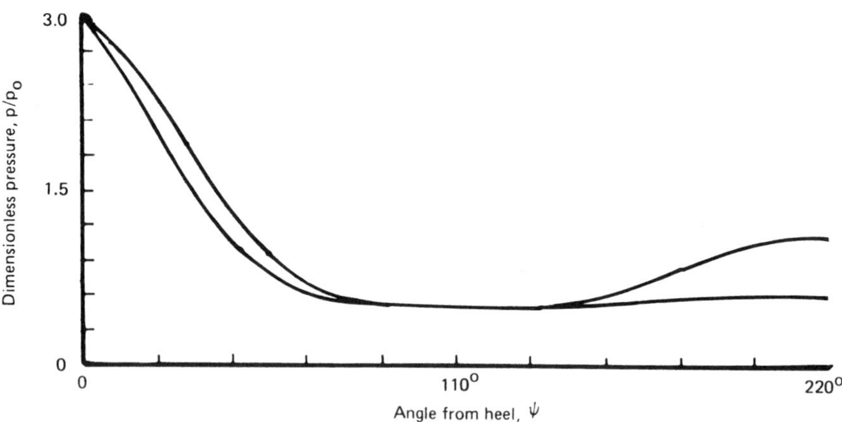

FIGURE 14 Primary pressure distributions.

that little or no increase in lining pressure is found at the heel of the secondary shoe. Thus the secondary shoe pressure distributions are represented by either

$$\frac{p}{p_0} = e^{-c_1(1 - \psi/\phi_0)} \left| \cos \frac{\psi}{\phi_0} \pi \right|^{c_2} + c_3 \tag{5-13}$$

or

$$\frac{p}{p_0} = e^{-c_1(1-\psi/\phi_0)} \left| \cos \frac{\psi}{\phi_0} \pi \right|^{c_2} + c_3 \sin(\psi + \phi_1) \tag{5-14}$$

which produce the distributions shown in Figures 16 and 17.

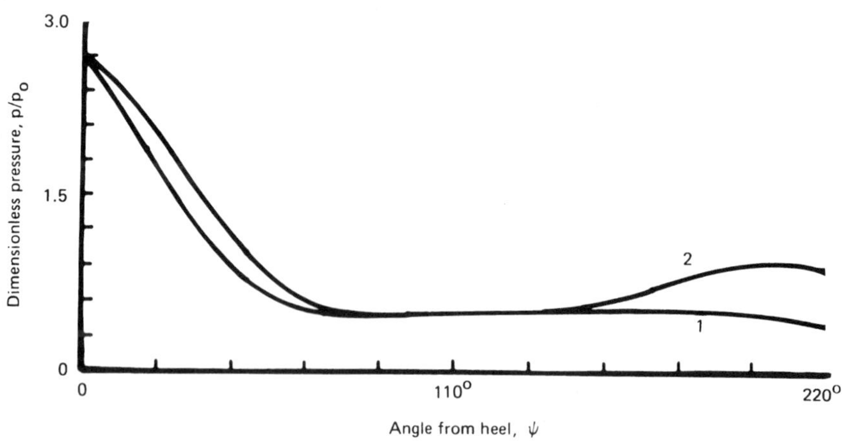

FIGURE 15 Primary pressure distributions.

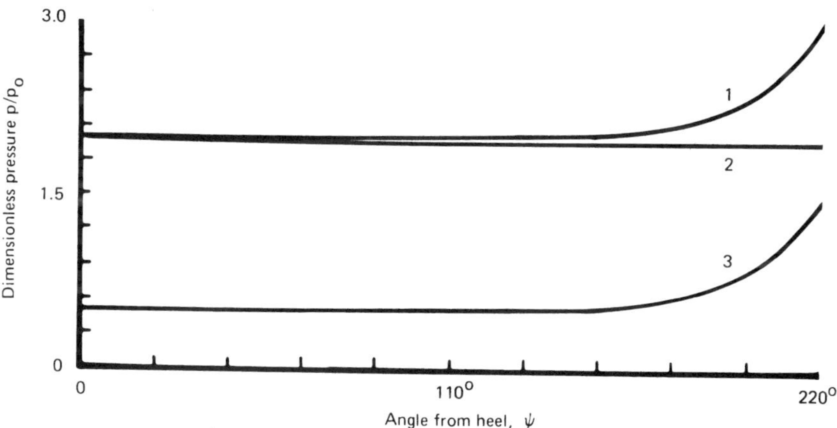

FIGURE 16 Secondary pressure distributions.

Since the secondary shoe is pivoted at the anchor pin, there are no restrictions on the direction of the resultant force and no particular mathematical restrictions on the pressure distribution itself other than that it not be infinite at any point along the shoe. The physical restrictions that these quantities be realistic motivated the use of pressure distributions similar to those on the primary shoes, based on the assumption that the shoe characteristics are similar, that the linings are very similar, if not identical, and that the force peak at the end of the shoe in contact with the link will be similar to

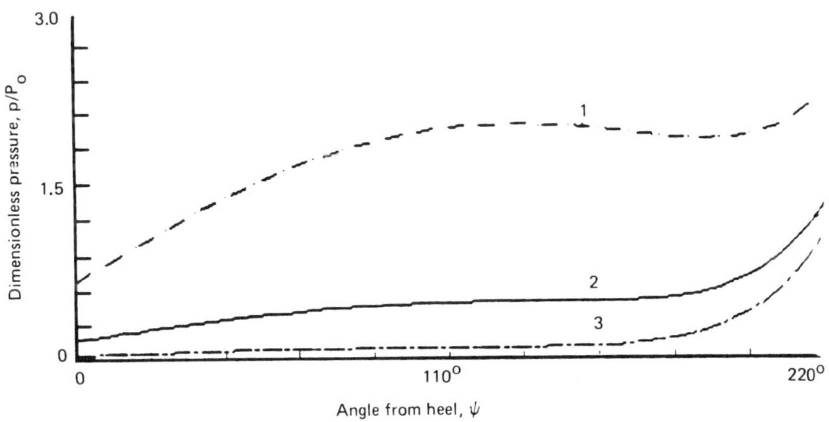

FIGURE 17 Secondary pressure distributions.

the force peak in the primary shoe at the other end of the link. Depending on the rigidity of the anchor pin and the rigidity of the brake shoe in the vicinity of the anchor pin, the force peak at this end of the lining may either be noticeably smaller than that at the free end of the primary shoe, or may vanish entirely.

The method of solution is to first solve for the unknown parameters in the expressions for the particular pressure distribution selected from either equations (5.11) or (5.12) on the primary shoe such that the condition

$$2\tan^{-1}\left(\frac{F_r}{F_\theta}\right) - \beta = 0 \tag{5-15}$$

is satisfied. Since this condition is independent of the lining pressure, that quantity may be found from the relation

$$M_1 = M_{p_1} + M_{f_1} \tag{5-16}$$

in which the external moment M_a is specified and where subscript 1 denotes the corresponding moment for the primary shoe.

With the maximum pressure on the primary shoe known, the pressure distribution over the primary shoe may be evaluated from equation (5-11) or (5-12), as appropriate, and the braking torque contributed by the primary shoe may then be calculated from equation (5-10). Link force may be found from equations (5-3), (5-4), and (5-6).

After selecting those values of c_1, c_2, and c_3 that provide a reasonable pressure distribution over the secondary shoe, the reference pressure may be found from equation (5-10) and the total braking torque becomes the sum of the torques contributed by the primary and secondary shoes.

VI. SINGLE-ANCHOR TWIN-SHOE DRUM BRAKE DESIGN EXAMPLES

Since it is the asymmetric term in the pressure distribution that is the major contributor to the control of the radial component of the force on the pivot point for a given coefficient of friction, satisfaction of equation (5-5) may be accomplished by adjusting the c_1 term in the pressure distribution. Once this is accomplished, the moment equilibrium conditions on the brake shoe may be satisfied by an appropriate choice of the pressure term, p_0. Equation (5-6) may then be evaluated to find the force transmitted to the primary shoe through the adjusting link. Straightforward calculation then yields the lining pressure for the secondary shoe and the torque contributed by both shoes.

Adjustment of constant c_1 may be carried out by finding the zero of the relation

$$\tan\frac{\beta}{2} + \frac{F_r}{F_\theta} = 0 \tag{6-1}$$

considered as a function of c_1. Location of a single zero in a given interval may be obtained using the bisection routine, such as in the TK Solver program.

It may appear that the search time may be reduced by restricting the search range for c_1 to a small neighborhood in the vicinity of the zeros found by displaying plots of equation (5-15) as a function of c_1 over an interval selected by the user for a range of values of μ, ϕ_1, and ϕ_2 suggested by the problem at hand. Plotting such a curve has proven to be unsatisfactory in practice because of the small intervals that are at times necessary to locate paired zeros. It may be faster to search for zeros by using a program that displays the integrands of F_r and F_t, their integrals, and their arctangents. This is because associating the asymmetry of the pressure distribution with the angle of the reaction at the adjusting link enables one quickly to see whether changes in the values of c_1 tend to bring the reaction into coincidence with the axis of the link. It also has the advantage of showing whether the constants chosen continue to represent adequately a physically reasonable pressure distribution.

A program for the numerical evaluation of the integrals involved in expressions (5-7) and (5-8) may easily be written using Simpson's rule. There appears to be only negligible improvement in accuracy obtained by dividing the interval into more than 50 segments.

The following four examples show the effect of changes in pressure distributions on brake performance and they also demonstrate the use of the method outlined. For this comparison all of the brake shoes subtend 120° at the center of the drum, they all have a 20° angle between the pivot and the heel of the shoe, and they all have an adjusting link which subtends 15° at the center of the drum. In addition, they are all subjected to an activating moment of 100 in.-lb and they all act on a drum having an inside diameter of 5.1 in. and a pivot at a radius of 4.2194 in. Results are summarized in Tables 2 and 3.

Example 6.1

Consider the dimensionless pressure distribution corresponding to curve 1 in Figure 14 and given by equation (5-11) with

$$c_1 = 3.15679 \qquad c_2 = 4.0 \qquad c_3 = 0.20$$

For this pressure distribution the maximum lining pressure at the heel is found to be 25.32 psi, the pressure at the toe of the shoe is 5.12 psi. the torque contribution from the primary shoe is 524.80 in.-lb, and the adjusting link is

TABLE 2 Lining Pressure and Shoe Braking Torque Associated with Primary Shoe Pressure Given by Equation (4-11) (Unit Width Shoe)

Shoe	Distribution number[a]	Heel pressure (psi)	Toe pressure (psi)	Link force (lb)	Torque (in.-lb)	Friction coefficient
P	1	25.320	5.320	1131.57	514.800	0.4
S	1	500.642	750.963	1131.57	2099.040	0.4
S	2	336.002	336.002	1131.57	2026.710	0.4
S	3	490.806	1472.418	1131.57	2310.270	0.4
P	2	25.320	9.390	908.55	338.200	0.3
S	1	314.318	417.477	908.55	988.383	0.3
S	2	321.610	321.610	908.55	969.967	0.3
S	3	294.334	883.001	908.55	1039.090	0.3

[a] *P* distributions are shown in Figure 14; *S* distributions are shown in Figure 16.

subjected to an axial force of 1131.57 lb. Upon imposing each of the three different pressure distributions shown in Figure 16 on the secondary shoe, it is found that although the maximum pressure distribution at the toe varies from about 336 to 1472 psi, the torque contribution from the secondary shoe only varies from 2027 to 2310 in.-lb.

Example 6.2

Reduction of the friction coefficient from 0.4 to 0.3 in equation (6.1) changes c_1 to 1.40460, and this change in turn modifies the pressure distribution acting over the primary shoe to that represented by curve 2 in Figure 14. The toe pressure remains unchanged, the central pressure drops, the heel pressure

TABLE 3 Lining Pressure and Shoe Braking Torque Associated with Primary Shoe Pressure Given by Equation (4-12) (Unit Width Shoe)

Shoe	Distribution number[a]	Heel pressure (psi)	Toe pressure (psi)	Link force (lb)	Torque (in.-lb)	Friction coefficient
P	1	21.140	3.380	1119.09	514.360	0.4
S	1	200.019	388.271	1119.09	2040.110	0.4
S	2	198.909	1544.466	1119.09	2277.790	0.4
S	3	170.460	6022.909	1119.09	3502.060	0.4
P	2	21.910	7.350	903.65	334.100	0.3
S	1	126.430	422.430	903.65	961.431	0.3
S	2	118.716	921.797	903.65	1019.600	0.3
S	3	92.300	2868.490	903.65	1245.890	0.3

[a] *P* distributions are shown in Figure 15; *S* distributions are shown in Figure 17.

increases to 9.39 psi, the link force (servo action) drops to 908.55 lb, and the torque contribution from the primary shoe falls to 338.20 in.-lb as a result of this change in the pressure distribution.

Secondary shoe pressures at the toe now range from 321.60 to 883.001 psi and the torque contributions from the secondary shoe now range from about 970 to 1039 in.-lb. Thus a 25% reduction in the friction coefficient has produced about a 58% reduction in the maximum torque capability of the secondary shoe, based on the 1039 in.-lb valve derived from distribution 3, Figure 16. Total torque capacity has dropped about 51%.

Example 6.3

Use of the primary distribution represented by curve 1 in Figure 15 and given by relation (5.15) with a friction coefficient of 0.4 leads to

$$c_1 = 3.45218 \qquad c_2 = 4.00 \qquad c_3 = 0.20$$

which produces a pressure distribution having lining pressures at the heel and toe of 21.14 and 3.38 psi, respectively, a torque contribution of 514.36 in.-lb, and a lining force (serve action) of 1119.09 lb. As shown in Figure 17, the pressure distributions on the secondary shoe produce maximum pressures ranging from about 388 to 6023 psi and torque contributions ranging from about 2040 to 3502 in.-lb. Maximum pressure and torque are both obtained from curve 3 in Figure 17.

Example 6.4

Reduction of the friction coefficient from 0.4 to 0.3 causes exponent c_1 to decrease to 1.51392, which corresponds to the dimensionless pressure distribution shown by curve 2 in Figure 15, wherein the heel and toe pressure become 7.35 and 21.91 psi, respectively. Braking torque from the primary shoe is calculated to be 334.100 in.-lb. and the link force is calculated to be 903.650 lb.

Secondary pressures at the toe of the lining for the pressure distributions shown in Figure 17 range from approximately 922 to 1246 in.-lb. In this case the 25% reduction in the coefficient of friction between drum and lining has produced almost a 61% reduction in the total braking torque.

Together these examples show that percentage changes in the torque capacity of the brake are a magnification of the percentage change in the friction coefficient. In large part this magnification appears to be due to the pressure maximum near the adjusting link which is necessary if the primary shoe is to be held in equilibrium by the link force and the lining pressure. Pressure distributions satisfying these equilibrium conditions may be of the form given by equations (5-11) and (5-12). Based on these pressure distributions, we have found that although the maximum pressure on the secondary shoe is strongly dependent on the width of the pressure maximum in the vicinity of the adjusting link, the magnitude of the braking torque contributed by the secondary shoe does not change quite as rapidly as the change in the lining pressure at the toe.

VII. ELECTRIC BRAKES

Common usage has associated the term electric brakes with friction brakes which are electrically activated, rather than with those brakes that rely upon electrical and magnetic forces rather than friction to provide the braking torque. Typical electric brakes are pictured in Figures 18 and 20. Both are single-anchor drum brakes that use the servo action associated with these brakes to obtain the required braking torque in response to an activating force indirectly related to the applied magnetic field.

FIGURE 18 Electric drum brake activated by a cam attached to a magnet arm. (Courtesy of Warner Electric Brake & Clutch Co., South Beloit, IL.)

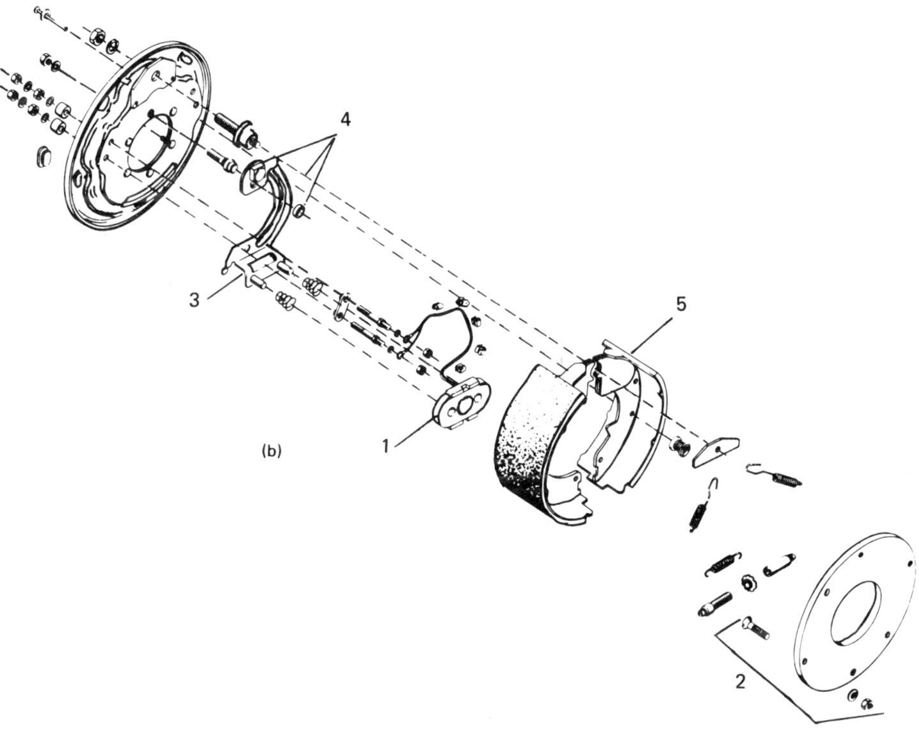

FIGURE 19 Exploded view of the components of the electric brake shown in Figure 18. (Courtesy of Warner Electric Brake & Clutch Co., South Beloit, IL.)

In the first type, shown in Figure 18, the small spot magnet 1 in Figure 19 is attracted to armature 2, which rotates with the drum. Friction between the spot magnet and the armature cause lever arm 3 to rotate and to actuate lever mechanism 4 to bring shoes 5 into contact with the drum. Depending upon the direction of rotation, one of these shoes will be the leading shoe, which by servo action will drive the trailing shoe against the drum.

Greater braking torque may be had from the model shown in Figure 20. In that design the annular electromagnet 1 in Figure 21 is attached to non-rotating backing plate by means of a pilot ring, which allows it to rotate slightly in the plane of the backing plate. When the electromagnet is energized it attracts armature 2, which rotates with the drum (not shown) but is allowed sufficient axial motion to contact the friction material on the face of the electromagnet. Friction between the electromagnet and the armature causes the electomagnet to rotate just enough to activate cam pair 3 (only one is shown)

Figure 20 Electric drum brake activated by a cam and pin attached to the slightly rotating electromagnet. (Courtesy of Warner Electric Brake & Clutch Co., South Beloit, IL.)

and force shoes 4 against the drum. Again, servo action is relied upon to drive the trailing shoe against the drum so that together they provide a relatively large braking torque.

Simplified schematics of these two brakes which emphasize their means of operation are given in Figure 21.

These brakes were designed for use with highway trailers where a quick response time may be important. They have both fewer total parts and fewer exposed parts than either hydraulic or air brakes, but do not have as great a braking torque for a given size of drum and shoes.

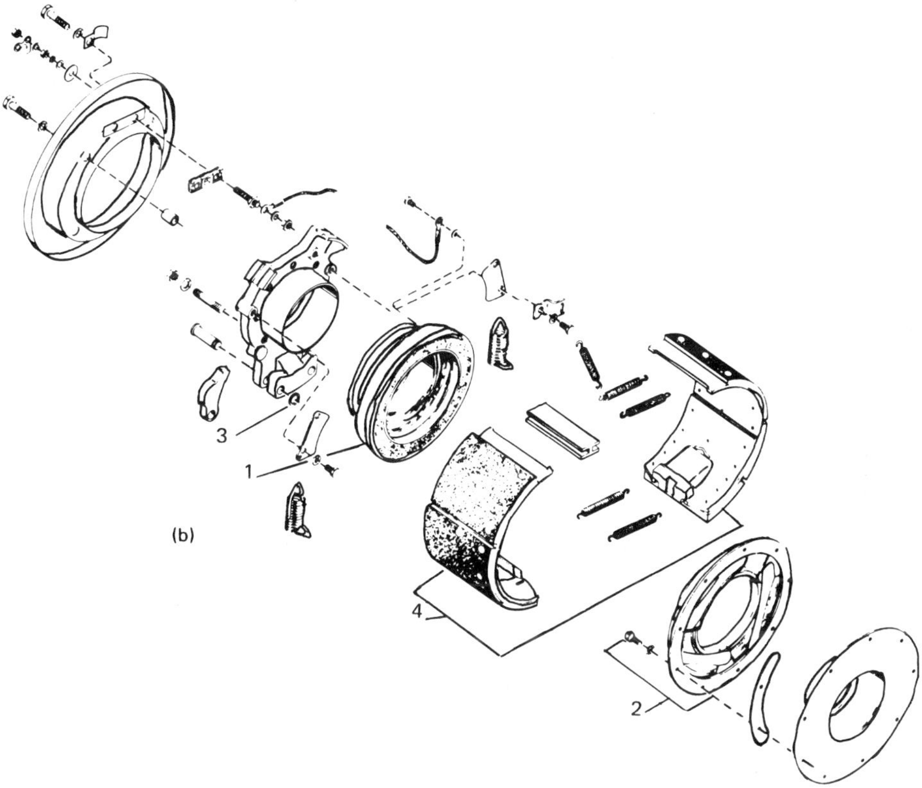

FIGURE 21 Exploded view of the components of the electric brake shown in Figure 20. (Courtesy of Warner Electric Brake & Clutch Co., South Beloit, IL.)

VIII. NOTATION

A	dummy variable (l)
B	dummy variable (l)
b	dummy variable (l^2)
b_a, b_b	dummy variables (mt^{-2})
c_1, c_2, c_3	pressure distribution coefficients (l)
F	force (mlt^{-2})
F_l	force along the axis of the adjusting link (mlt^{-2})
F_r	radial force (mlt^{-2})
F_θ	circumferential force (mlt^{-2})
k	effective spring constant of the lining (mt^{-2})
M_e	externally applied moment (ml^2t^{-2})

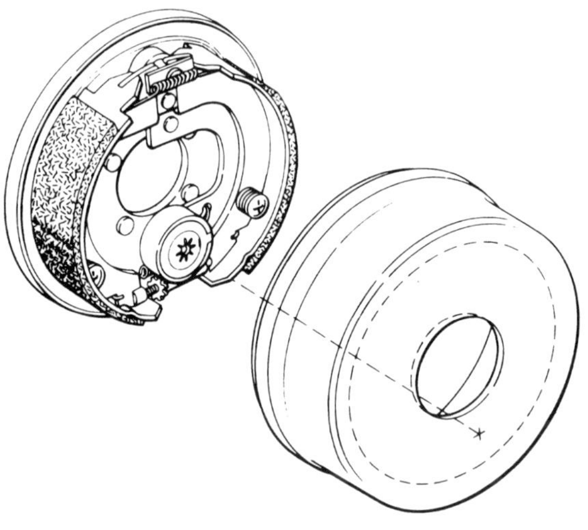

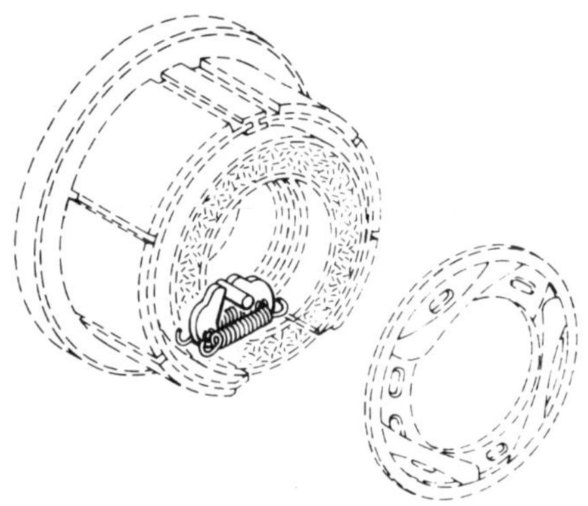

Figure 22 Schematic of brakes shown in Figures 18 and 20 to show method of operation. (Courtesy of Warner Electric Brake & Clutch Co., South Beloit, IL.)

M_f	moment on the shoe due to friction (ml^2t^{-2})
M_p	moment on the shoe due to lining pressure (ml^2t^{-2})
p	lining pressure $(ml^{-1}t^{-2})$
p_0	reference pressure $(ml^{-1}t^{-2})$
R	radius from drum center to shoe pivot point (l)
r	drum radius $(r = d/2)$ (l)
T	torque (ml^2t^{-2})
T_a	torque contributed by primary shoe (ml^2t^{-2})
T_b	torque contributed by secondary shoe (ml^2t^{-2})
w	lining and shoe width (l)
β	half of the angle subtended by the adjusting link at the center of the drum (1)
$\delta\alpha$	angular motion of the shoe during braking (1)
μ	friction coefficient (1)
ϕ	angle subtended at the center of the drum (1)
ϕ_0	angle subtended by the lining at the center of the drum (1)
ψ	distribution parameter (1)

IX. FORMULA COLLECTION

A. Long Shoe–External and Internal

Angle subtended by the shoe:

$$\phi_0 = \phi_2 - \phi_1$$

Pressure:

$$p = \frac{p_{\max}}{(\sin\phi)_{\max}}$$

Torque:

$$T = \frac{\mu p_{\max} r w^2}{(\sin\phi)_{\max}}(\cos\phi_1 - \cos\phi_2)$$

Moment due to friction:

$$M_f = \frac{\mu p_{\max} r w}{4(\sin\phi)_{\max}}[R(\cos 2\phi_1 - \cos 2\phi_2) - 4r(\cos\phi_1 - \cos\phi_2)]$$

Moment due to pressure:

$$M_p = \frac{p_{\max} w r R}{4(\sin\phi)_{\max}}(2\phi_0 - \sin 2\phi_2 + \sin 2\phi_1)$$

External (activation) moment:

$$M_e = M_p \pm M_f \qquad \text{(see Table 1 to select proper sign)}$$

Radial force on a fixed anchor pin:

$$F_r = -r_w \int_{\phi_1}^{\phi_2} p \cos \phi \, d\phi + \mu r w \int_{\phi_1}^{\phi_2} p \sin \phi \, d\phi$$

Tangential force of a fixed anchor pin:

$$F_\theta = r_w \int_{\phi_1}^{\phi_2} p \sin \phi \, d\phi + \mu r w \int_{\phi_1}^{\phi_2} p \cos \phi \, d\phi$$

Figures 3–6 show the quantities involved in the foregoing formulas. Quantity ϕ_1 in the short-shoe formulas is identical to the same quantity defined for long shoes.

B. Short Shoe

Torque:

$$T = \mu p w r^2 \phi_0 = \mu r F$$

Pressure:

$$p = p_{\max}$$

Force:

$$F = p r w \phi_0$$

Moment due to friction:

$$M_f = \mu F(R \cos \phi_1 - r)$$

Moment due to pressure:

$$M_p = FR \sin \phi_1$$

REFERENCES

1. Burr, A. H. (1981). *Mechanical Analysis and Design*. New York: Elsevier.
2. Juvinal, R. C. (1983). *Fundamentals of Machine Component Design*. New York: Wiley.
3. Fazekas, G. A. G. (1958). Some basic properties of shoe brakes. *Journal of Applied Mechanics* 25:7–10.
4. Orthwein, W. C. (1985). Estimating torque and lining pressure for bendix type drum brakes. *SAE Paper 841234, SAE Transactions* 86:5.617–5.622.

4

Linearly Acting External and Internal Drum Brakes

Linearly acting drum brakes are those fitted with shoes which, when activated, approach the drum by moving parallel to a radius through the center of the shoe. Typical linearly acting drum brakes are illustrated in Figures 1–3.

Analysis of linearly acting brakes includes those in which the centrally pivoted shoes are attached to pivoted levers, as in Figure 1. Including brakes of this design within the category of linearly acting brakes is justified if they are designed so that the applied forces on the shoes and linings act along the radii of the shafts that they grip when the brakes are applied. Brakes of this type may act either upon brake drums or directly upon rotating shafts and are suitable for use in heavy-duty applications, such as found in mining and construction equipment and in materials-handling machinery.

Internal linearly acting drum brakes, such as used on trucks in Europe, may be designed as in Figure 2. Either pneumatic or hydraulic cylinders or cams may be used to force the shoes outwardly against the drum. The cylinders or cams also serve as anchors to prevent rotation and react against the vehicle frame. The springs shown are to retract the shoes when the brake is released.

A collection of segmented brake pads (backing plate plus lining) along the entire circumference of the drum may be arranged as in Figure 3 to move outwardly against a drum, as in Figure 3a, or inwardly against a drum, as in Figure 3b. The brake pads, or shoes, are themselves constrained against

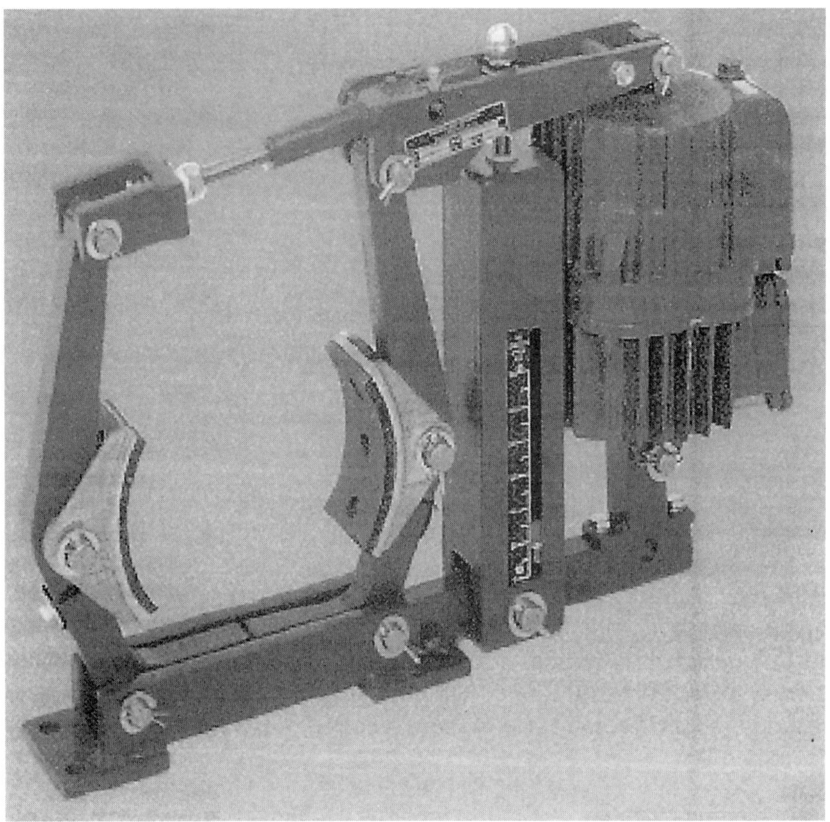

FIGURE 1 Linearly acting, centrally pivoted shoe brake. (Courtesy of the Hindon Corp., Charleston, SC)

rotation by anchor pins that fit into short radial slots between the shoes and are attached to the rim of the circular frame, as shown in Figure 3a. Brake actuation is accomplished by using air to expand the normally flat elastomeric-fabric annular tube shown in that figure between the brake pads and the circular frame. When designed to move inwardly against a drum, as in Figure 3b, the brake lining is riveted to a differently contoured backing plate which has shoulders at each end to resist a twisting torque and which is fitted with a central slot that accepts the anchor pin to the outer frame at each side of the shoe. This radial slot allows the pad to move rapidly inward but prevents tangential motion.

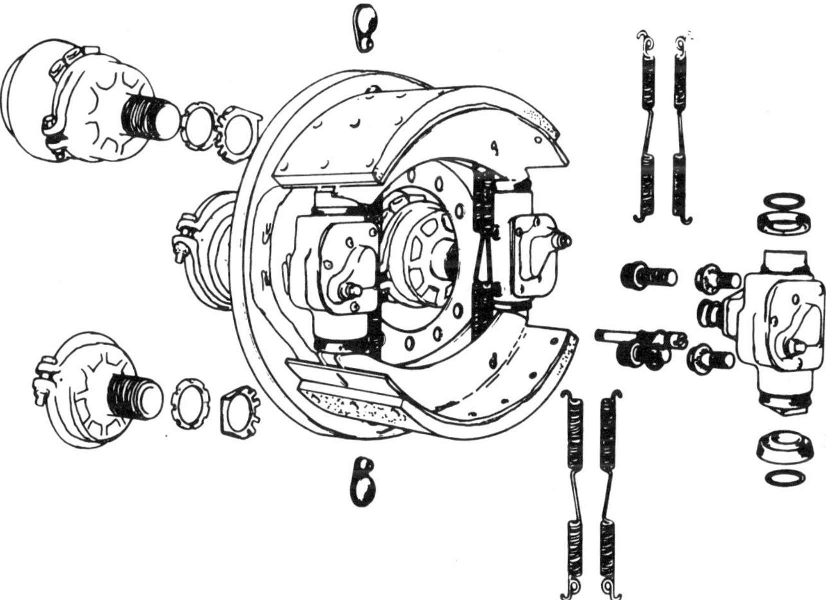

FIGURE 2 Linearly acting, twin-shoe, internal drum brake with pneumatic activation (Girling Twinstop). (Reprinted with permission. ©1977 Society of Automotive Engineers, Inc.)

I. BRAKING TORQUE AND MOMENTS FOR CENTRALLY PIVOTED EXTERNAL SHOES

To calculate the torque, we must first find an expression for the lining pressure. Guided by the geometry shown in Figure 4, we see that the lining pressure will be given by

$$p = k\Delta \cos\theta \tag{1-1}$$

in terms of the lining deformation Δ if the shoe and drum are assumed to be absolutely rigid. Maximum pressure occurs when $\theta \simeq 0$, so that $p_{\max} = k\Delta$. Thus equation (1-1) becomes

$$p = p_{\max} \cos\theta \tag{1-2}$$

and the incremental tangential friction force is given by

$$dF = \mu p_{\max} \cos\theta \, rw \, d\theta \tag{1-3}$$

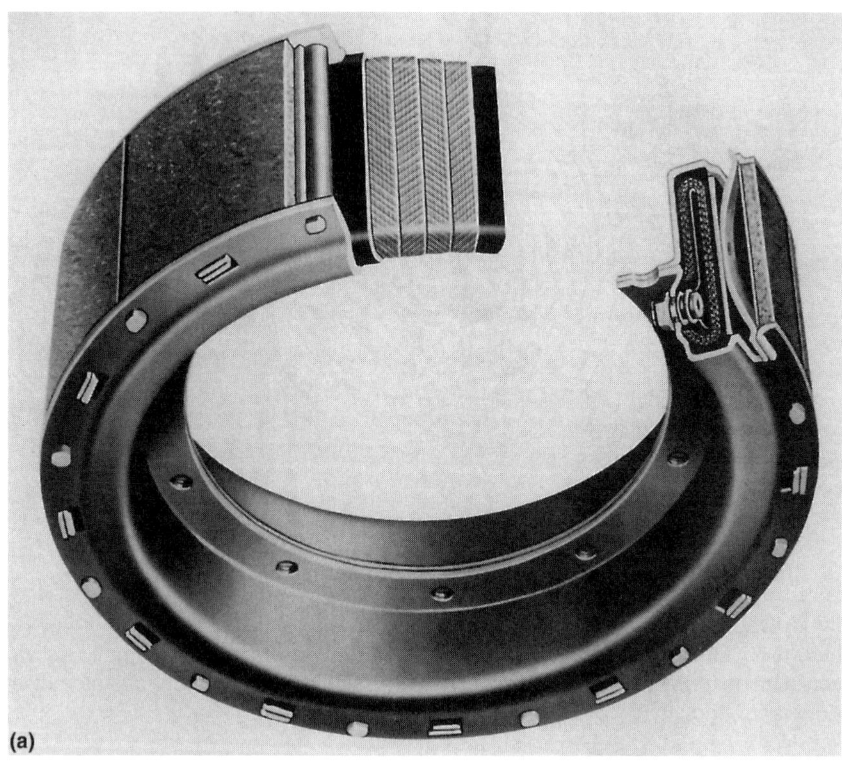

FIGURE 3 Rim brakes with pneumatic activation (also used as rim clutches). (Courtesy of Eaton Corp., Airflex Division, Cleveland, Ohio.)

so the braking torque becomes

$$T = \mu p_{\max} r^2 w \int_{\theta_1}^{\theta_2} \cos\theta \, d\theta = \mu p_{\max} r^2 w (\sin\theta_2 - \sin\theta_1) \tag{1-4}$$

In designs different from those shown in Figure 1 it may prove convenient to have the shoe pivoted about a point at a radial distance R on the axis of symmetry, such as point A in Figure 4. The moment M_p due to the pressure on the lining is zero about point A because of the symmetry of the shoe about this point. No such symmetry exists for the friction moment M_f, however, so from the incremental moment due to friction

$$dM_f = (\mu p_{\max} r w \cos\theta \, d\,\theta)(R\cos\theta - r)$$

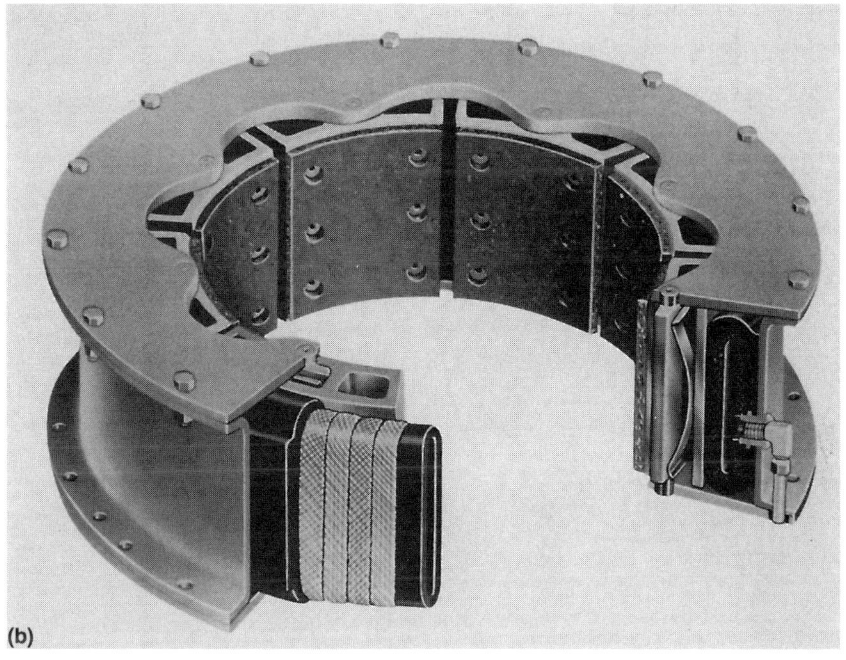

FIGURE 3 Continued.

it follows that

$$M_f = \mu p_{\max} r w \int_{\theta_1}^{\theta_2} \left(R \cos^2\theta - r \cos\theta\right) d\theta$$

$$= \mu p_{\max} r w \left\{ R \left[\frac{\phi_0}{2} + \frac{1}{4}(\sin 2\theta_2 - \sin 2\theta_1) \right] - r(\sin\theta_2 - \sin\theta_1) \right\} \tag{1-5}$$

where $\phi_0 = \theta_2 - \theta_1$. The expression in equation (1-5) may be simplified by observing that the symmetry of the shoe about A requires that

$$-\theta_1 = \theta_2 = \frac{\varphi_0}{2} \tag{1-6}$$

where φ_0 is the angle subtended by the lining. Substitution of these values into equation (1-5) leads to

$$M_f = \mu p_{\max} r w \left[\frac{R}{2}(\phi_0 + \sin\phi_0) - 2r \sin\frac{\phi_0}{2} \right] \tag{1-5a}$$

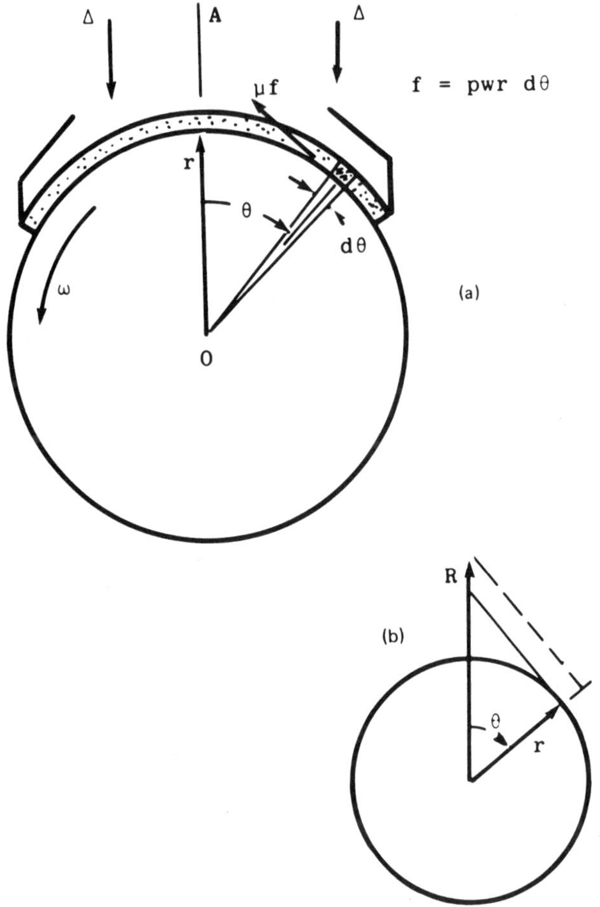

FIGURE 4 Geometry used for the analysis of a linearly acting external shoe.

which suggests that moment M_f will vanish if the shoe is pivoted at

$$R = r\frac{4\sin(\phi_0/2)}{\phi_0 + \sin\phi_0} \tag{1-7}$$

Upon plotting R/r we obtain Figure 5, wherein the ratio increases smoothly from 1.0 at $\phi_0 = 0$ to 1.273 at $\phi_0 = \pi$ rad. $= 180°$. This clearly indicates that it is impossible to find a pivot point for which $M_f = 0$ for an internal linearly acting shoe. This conclusion is, of course, unaffected by the

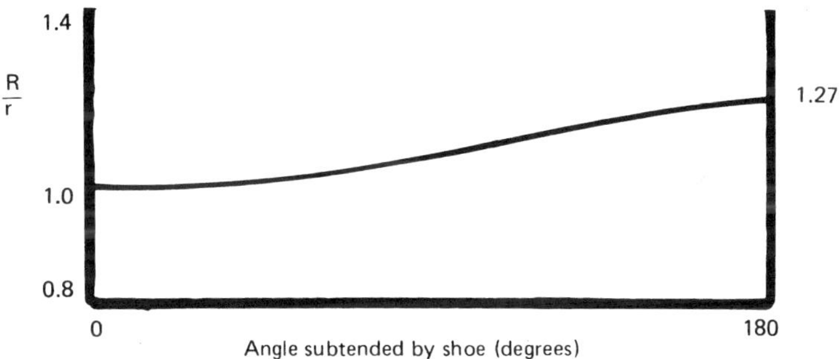

FIGURE 5 Variation of R/r with angle ϕ_0.

sign reversal found in the expression $(R \cos\theta - r)$ when equation (1-5a) is applied to an internal shoe. The sign reversal simply changes the direction of rotation implied by a positive value of M_f, as was discussed in an earlier section.

The nearly horizontal portion from 0° to about 30° implies that for external shoes which subtend an angle less than 30°, any changes in the length of the shoe that do not increase the subtended angle beyond 30° will have a negligible effect on the R/r ratio. This correlates with the short-shoe segments used in the brakes shown in Figure 3. Moreover, the value of M_f caused by a deviation from the R/r ratio that yields a zero value of M_f will be small if ϕ_0 remains small. In particular, if the R value that yields $M_f = 0$ is replaced by $R + \delta R$ in equation (1-5a), the moment due to friction will increase to only

$$\mu p_{\max} rw \frac{\delta R}{2}(\phi_0 + \sin\phi_0)$$

which is small enough to be easily resisted by the shoulders shown on the shoes in Figure 3.

Activation force F_s and tangential force F_t on a symmetrically placed pivot are given by the relations

$$F_s = 2p_{\max} rw \int_0^{\phi_0/2} \cos^2\theta \, d\theta = \frac{1}{2} p_{\max} rw(\phi_0 + \sin\phi_0) \tag{1-8}$$

and

$$F_t = 2\mu p_{\max} rw \int_0^{\phi_0/2} \cos^2\theta \, d\theta = \mu F_s \tag{1-9}$$

Let us define the efficiency of a brake as the ratio of the torque provided by the brake to the torque that could be had by applying the force directly to the drum, or shaft. According to this definition, the efficiency becomes

$$\frac{T}{F_s r} = \frac{\mu p_{\max} r^2 w(\sin \theta_2 - \sin \theta_1)}{(1/2) p_{\max} r^2 w(\phi_0 + \sin \phi_0)} = 2\mu \frac{\sin \theta_2 - \sin \theta_1}{\phi_0 + \sin \phi_0} \tag{1-10}$$

Upon substituting for θ_1 and θ_2 in equation (1-8) and recalling equations (1-6) and (1-7) we find that

$$\frac{T}{F_s r} = \frac{4\mu \sin(\phi_0/2)}{\phi_0 + \sin \phi_0} = \mu \frac{R}{r} \tag{1-11}$$

where the right-hand side has already been plotted in Figure 5. From that figure we find that although maximum efficiency may be achieved only if each shoe and lining extend over half of the drum, or shaft, relatively little efficiency is lost if the lining extends over only 160° instead of 180°. This, together with the near impossibility of maintaining good contact between the lining and the drum near the ends of a shoe subtending 180° at the center of the drum, accounts for the angular dimensions of the brake linings shown in Figure 1.

Finally, it follows from equation (1-11) that if the shoe is symmetrically pivoted and if equation (1-7) holds, the applied torque is given by

$$T = \mu R F_s \tag{1-12}$$

II. BRAKING TORQUE AND MOMENTS FOR SYMMETRICALLY SUPPORTED INTERNAL SHOES

Pressure p and braking torque are again given by equations (1-2) and (1-4), respectively, for an internal shoe moved against a rotating drum along a line parallel to its axis of symmetry, line OB, in Figure 6. In the following analysis it may be more descriptive to measure the angle along the shoe from the end rather than from the middle because the activation forces are now applied at the ends. Denote this angle by ϕ. Since the expression for the torque is unaffected by this choice of angle, substitution of equation (1-6) into equation (1-4) shows it can be given by

$$T = 2\mu p_{\max} r^2 w \sin \frac{\phi_0}{2} \tag{2-1}$$

The pressure distribution described by equation (1-2) may be rewritten in terms of ϕ according to

$$p = p_{\max} \cos(\phi - \alpha) \tag{2-2}$$

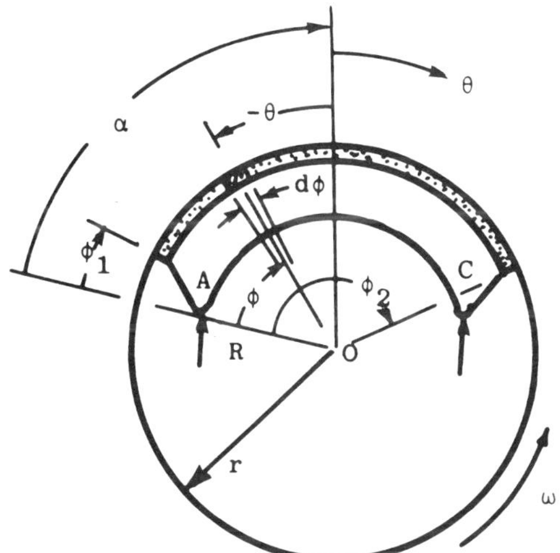

FIGURE 6 Geometry used in the analysis of a linearly acting internal shoe drum brake.

Let the shoe be restrained at A to prevent it from rotating with the drum, with the restraint moving with the shoe. This may be accomplished using guide pins and/or plates which may also serve as anchors to transfer braking torque from the shoes to the appropriate structure.

The moment M_p about A due to pressure p is given by integration of

$$dM_p = (prw\ d\phi)R \sin \phi = p_{max} Rrw \cos(\phi - \alpha) \sin \phi\ d\phi \tag{2-3}$$

to obtain

$$M_p = p_{max} wRr \int_{\phi_1}^{\phi_2} (\cos \alpha \cos \phi + \sin \alpha \sin \phi) \sin \phi\ d\phi \tag{2-4}$$

which may be integrated directly to give

$$M_p = \frac{p_{max}}{4} wRr[2 \cos \alpha\ (\sin^2\phi_2 - \sin^2\phi_1) + \sin \alpha(2\phi_0 - \sin 2\phi_2 + \sin 2\phi_1)] \tag{2-5}$$

Let α represent the central angle from the R vector to the middle of the shoe (i.e., from R to the transverse plane of symmetry through radius OB in Figure 6), so that

$$\phi_1 = \alpha - \frac{\phi_0}{2} \qquad \phi_2 = \alpha + \frac{\phi_0}{2} \tag{2-6}$$

After substitution for ϕ_1 and ϕ_2 from equations (2-6) and using common trigonometric identities, M_p may be written as

$$M_p = \frac{p_{\max}}{2} rRw(\phi_0 + \sin \phi_0)\sin \alpha \tag{2-7}$$

Similarly, the moment about A due to friction may be found from

$$dM_f = (aprw\ d\phi)(r - R \cos \phi) \tag{2-8}$$

which with the aid of equation (2-2) leads to the integral

$$M_f = p_{\max} rw\mu \int_{\phi_1}^{\phi_2} (\cos \alpha \cos \phi + \sin \alpha \sin \phi)(r - R \cos \phi)\ d\phi \tag{2-9}$$

which, upon integration, produces

$$\begin{aligned} M_f = \frac{\mu}{4} rwp_{\max} [&4r(\sin \phi_2 - \sin \phi_1)\cos \alpha \\ &+ 4r(\cos \phi_1 - \cos \phi_2)\sin \alpha - R(2\phi_0 + \sin 2\phi_2 \\ &- \sin 2\phi_1)\cos \alpha - 2R(\sin^2\phi_2 \sim \sin^2\phi_1)\sin \alpha] \end{aligned} \tag{2-10}$$

Substitution for ϕ_1 and ϕ_2 from equations (2-6) into equation (2-10) and use of common trigonometric identities enables equation (2-10) to be simplied to read

$$M_f = \frac{p_{\max}}{2} \mu rw \left[4r \sin \frac{\phi_0}{2} - R(\phi_0 + \sin \phi_0)\cos \alpha \right] \tag{2-11}$$

According to the geometry shown in Figure 6, a positive M_p corresponds to clockwise rotation of the shoe about point A and positive M_f corresponds to counterclockwise rotation when the drum rotation is from the opposite end of the shoe toward point A. Reversing the direction of drum rotation will not affect the implied direction of shoe rotation due to M_p but will reverse the direction by a positive M_f; that is, positive M_f will then imply clockwise rotation about point A. This last observation is of academic interest only, however, if the shoes are supported at each end, because in that case each shoe will tend to pivot about the end toward which the drum rotates, regardless of the direction of rotation of the drum.

For symmetrically supported symmetric shoes it follows that the shoes will be free of self-locking if

$$M_p - M_f > 0 \tag{2-12}$$

Substitution for M_p and M_f from equations (2-7) and (2-11) into equation (2-12) yields

$$M_p - M_f = \frac{p_{\max}}{2} Rrw\left[(\phi_0 + \sin\phi_0)\sin\alpha - 4\mu\frac{r}{R}\sin\frac{\phi_0}{2} + \mu(\phi_0 + \sin\phi_0)\cos\alpha\right] > 0 \tag{2-13}$$

Condition (2-12) is, according to equation (2-13), equivalent to the condition

$$4\mu\frac{r}{R}\sin\frac{\phi_0}{2} < (\phi_0 + \sin\phi_0)(\sin\alpha + \mu\cos\alpha) \tag{2-14}$$

for internal, linearly acting brakes. Consequently, the r/R ratio must satisfy

$$1 \geq \frac{R}{r} > \frac{4\sin(\phi_0/2)}{(\phi_0 + \sin\phi_0)[\cos\alpha + (1/\mu)\sin\alpha]} \tag{2-15}$$

to ensure that the brake will not become self-locking when it is applied.

III. DESIGN EXAMPLES

Example 4.1

Design an external, linearly acting, twin-shoe brake to provide a braking torque of 2700 N-m when acting on a flywheel hub 260 mm in diameter. The lining material to be used here has a design maximum pressure of 3.41 MPa and $\mu = 0.41$.

Since the torque on either an external or an internal shoe is given by equation (2-1), it follows from the $\sin(\phi_0/2)$ term that 90% of the maximum theoretical torque (i.e., for $\phi_0 = 180°$) may be obtained from $\phi_0 = 128.3°$, that 95% may be had from $\phi_0 = 143.6°$, and 98% may be had from $\phi_0 = 157.0°$. It we select $\phi_0 = 145°$ for each shoe, assume that each shoe will supply half of the design braking torque, and solve equation (2-1) for w, we find that

$$w = \frac{T}{2\mu p_{\max} r^2 \sin(\phi_0/2)} = \frac{1350 \times 10^3}{0.82(3.41)(130)^2 \sin 72.5°} = 29.95 \text{ mm} \rightarrow 30 \text{ mm}$$

The required vertical force on each shoe as calculated from equation (1-8) becomes

$$F_s = \frac{p_{\max}}{2} rw(\phi_0 + \sin\phi_0) = \frac{3.41}{2} 130(29.95)(2.531 + \sin 145°) = 20{,}609\,N$$

So if the brake is to be pneumatically activated, as shown in Figure 1, the pressure and diaphragm diameter are related according to

$$F_s = \pi r^2 p_{\text{dia}}$$

Upon solving this relation for the diaphragm pressure pdia, and using an active diaphragm diameter of 250 mm, we find that the line pressure to the diaphragm must be 4.20 atm.

Finally, if the shoes are to be pivoted about an axis in their planes of symmetry, the radial distance to the pins may be calculated from equation (1-7), which yields

$$\frac{R}{r} = \frac{4\sin(\phi_0/2)}{\phi_0 + \sin\phi_0}$$

$$= \frac{4\sin 72.5^\circ}{2.531 + \sin 145^\circ}$$

So $R = 159.74$ mm from the center of the drum, or 29.74 mm from the drum surface.

Example 4.2

Design an internal, linearly acting, twin-shoe drum brake to provide a braking torque of 413,000 in. - 1b acting on a drum whose maximum inside diameter may be 26.0 in. The lining material to be used has a maximum design pressure of 450 psi and a friction coefficient of 0.50 or greater over the design temperature range.

Substitution into the expression obtained by solving equation (2-1) for w yields, for $\phi_0 = 130^\circ$,

$$w = \frac{206,500}{2(0.5)(450)(13)^2(\sin 65^\circ)} = 2.996 \text{ in.} + 3.00 \text{ in.}$$

If self-locking is to be avoided, the pivot point for each shoe should obey the inequality (2-15), which in this case becomes, for $\alpha = 70^\circ$,

$$R > \frac{4\mu r \sin(\phi_0/2)}{(\phi_0 + \sin\phi_0)(\sin\alpha + \mu\cos\alpha)}$$

$$= \frac{4(0.5)(13)\sin 65^\circ}{(2.2689 + \sin 130^\circ)(\sin 70^\circ + 0.5\cos 70^\circ)} = 6.990 \text{ in.}$$

Equal forces that must be applied at points A and C in Figure 6 to achieve the 450 psi maximum pressure may be found from equation (1-8) after replacing F_s with $2F_s$, where F_s in equation (3-1) represents the force at A and at C. Thus

$$F_s = \frac{p_{\max}}{4} \mathrm{rw}(\phi_0 + \sin(\phi_0))$$

$$= \frac{450}{4} 13(3)\left(130\frac{\pi}{180} + \sin\left(130\frac{\pi}{180}\right)\right) \tag{3-1}$$

$$= 13,315.9 \text{ lbs}$$

If an axial piston hydraulic pump capable of a continuous pressure of 4500 psi is used, this force may be had from a hydraulic cylinder whose piston diameter is equal to, or greater than

$$\mathrm{d(p)} = 2\sqrt{\frac{F_s}{\pi \cdot \mathrm{p}}} = 1.941 \text{ inches.}$$

Space available for such a cylinder may be found from the geometry in Figure 6 by finding the distance from a plane P through the center of the drum and perpendicular to the $\theta = 0$ line. The available distance will be twice this value.

Angle β between R and plane P may be found from

$$2\beta - 180^\circ - \phi_0 - 2\phi_1 \tag{3-2}$$

so that the distance h_0 from point A to the corresponding point on the opposite shoe becomes

$$h_0 = 2\mathrm{R} \sin \beta. \tag{3-3}$$

If we let $\mathrm{R} = 10$ inches the result is that since $\phi_1 = 5^\circ$

$$\beta = (180^\circ - 130^\circ - 10^\circ)/2 = 20^\circ$$

so that

$$h_0 = 2(10) \sin (20) = 6.840 \text{ inches.}$$

Distance h_1 from point A to the drum surface may be found from the relation that

$$h_1(\phi_1, R) = r \sin\left(\operatorname{acos}\left(\frac{R}{r} \cdot \cos(\beta(\phi_1) \cdot \deg)\right)\right) - h_0(\phi_1, R) \tag{3-4}$$

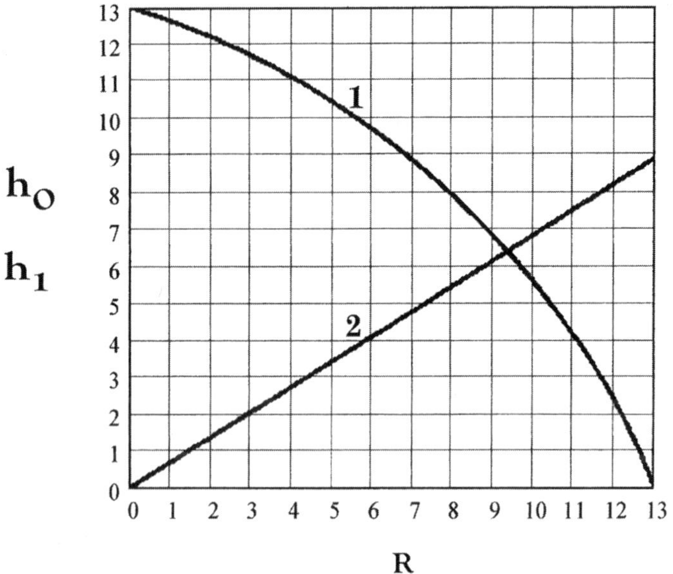

Figure 7 Curve 1: Distance h_1 between points A, or C, and the drum surface as a function of radius R; Curve 2: Height h_0 available for a hydraulic cylinder as a function of R. All dimensions in inches.

which for R = 10 inches yields

$$13 \sin\left(\operatorname{acos}\left(\frac{10}{13}\cos\left(20\frac{\pi}{180}\right)\right)\right) - 3.4202 = 5.56299$$

which may suggest that selecting a larger value for R would give more space for the hydraulic cylinders that force the shoes against the drum and would also require less material in each shoe. Plotting h_0 and h_1 for other values of R results in the curves shown in Figure 7. Distances are measured along chords that pass through points A on opposing shoes and through points C on opposing shoes.

IV. NOTATION

F_s	Force in the transverse plane of symmetry of the shoe (mlt^{-2})
F_t	force tangential to the shoe at the transverse plane of symmetry (mlt^{-2})
h_0	Length available for an activation mechanism

h_1	Length available for the shoe structure
k	equivalent spring constant for lining material (mt^{-2})
M_f	moment due to friction (ml^2t^{-2})
M_p	moment due to pressure (ml^2t^{-2})
p	lining pressure ($ml^{-1}t^{-2}$)
p_{max}	maximum lining pressure ($ml^{-1}t^{-2}$)
R	radius to effective pivot point from the drum center (l)
r	drum radius (l)
T	braking torque (ml^2t^{-2})
w	shoe and lining width (l)
α	Lining half-angle (1)
Δ	Lining deflection in compression (l)
θ	Angle (1)
μ	Friction coefficient (1)
ϕ	Angle (1)

V. FORMULA COLLECTION

Pressure distribution:

$$p = p_{max}\cos\theta = p_{max}\cos(\phi - \alpha)$$

Lining pressure in terms of torque for external and internal shoes:

$$p_{max} = \frac{T}{2\mu r^2 w \sin(\phi_0/2)}$$

Lining width in terms of torque for external and internal shoes:

$$w = \frac{T}{2\mu p_{max} r^2 \sin(\phi_0/2)}$$

Moment due to friction for a symmetrically pivoted external shoe:

$$M_f = \mu p_{max} r w \left[\frac{R}{2}(\phi_0 + \sin\phi_0) - 2r\sin\frac{\phi_0}{2}\right]$$

Moment due to friction about the trailing end of an internal shoe:

$$M_f = \frac{p_{max}}{2}\mu r w\left[4r\sin\frac{\phi_0}{2} - R(\phi_0 + \sin\phi_0)\cos\alpha\right]$$

Moment due to pressure about the trailing end of an internal shoe:

$$M_p = \frac{p_{max}}{2} r R w(\phi_0 + \sin\phi_0)\sin\alpha$$

Activation force

$$F_s = \frac{1}{2} p_{\max} r w(\phi_0 + \sin \phi_0)$$

Tangential force for a centrally pivoted external shoe:

$$F_t = \mu F_s$$

Anchor pin location for symmetrically pivoted external shoes:

$$\frac{R}{r} = \frac{4 \sin(\phi_0/2)}{\phi_0 + \sin \phi_0}$$

Support point location for a linearly acting internal shoe:

$$\frac{R}{r} > \frac{4 \sin(\phi_0/2)}{(\phi_0 + \sin \phi_0)[\cos \alpha + (1/\mu)\sin \alpha]}$$

Length available for an activation mechanism

$$h_0 = R \sin(\beta)$$

Length available for the shoe structure

$$h_1 = \mathrm{r} \sin\left(\operatorname{acos}\left(\frac{R}{r}\cos(\beta)\right)\right) - h_0$$

5

Dry and Wet Disk Brakes and Clutches

This chapter on disk brakes and clutches will consider annular contact disk clutches and both caliper and annular contact disk brakes, as illustrated in Figures 1, 2, and 3.

Caliper disk brakes, as shown in Figure 1, are used on aircraft, automotive, industrial, and mining equipment. Their two main advantages compared to drum brakes are greater heat dissipation, and hence less fading, because of their open construction, and a more uniform braking action, due to self-cleaning by brake pad abrasion. Their main disadvantage is that they require a larger activation force than is required for drum brakes because they have neither a friction moment nor servo action to aid in brake application.

Annular contact disk brakes and clutches are available as either dry or wet brakes, as shown in Figure 2 and 3. These units may be used as either a brake or as a clutch because the only differences between the two are whether one side of the unit is fastened to a stationary frame or to a rotating shaft and whether the unit has the necessary fittings for it to be controlled while in rotation. For example, both of these functions are combined in Minster combination dry clutch and brake units, illustrated in Figure 2, which are pneumatically controlled using air passages in the shaft to the combination unit.

Wet multiple-disk brakes and clutches, illustrated in Figure 3, have similar multiple-disk construction, but operate in an oil bath. Thus these brakes are isolated from dirt and water, and the circulation of the oil through a heat exchanger usually provides greater heat dissipation than can be had from direct air cooling. Because of these advantages, wet brakes have been

Figure 1 Floating, or sliding, caliper disk brake. (Courtesy of Misco, Inc., North Mankato, MN.)

used on large earth-moving equipment, on mine shuttle cars, and similar equipment which may require large braking torque and which may be designed to operate in a dirty environment.

I. CALIPER DISK BRAKES

From the moment of contact until the disk is stopped, the velocity of the disk relative to the brake pads will vary linearly with the disk radius. If the thickness of the lining material removed is denoted by δ and if δ is dependent on the relative velocity and the pressure, as is commonly assumed, then according to the uniform wear assumption,

$$\delta = kpr \tag{1-1}$$

where k is a constant of proportionality. Since the caliper brake pads are usually small enough for their supports to be considered rigid, we shall assume

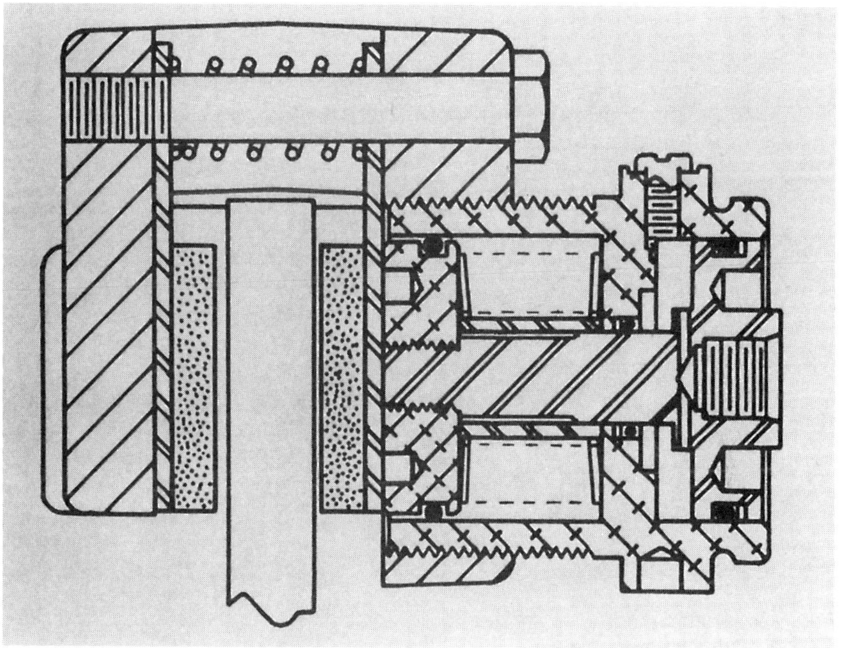

FIGURE 1 Continued.

that δ is constant over the brake pad (i.e., the wear is uniform). Whenever these conditions hold, equation (1-1) implies that the pressure increases as the radius decreases, so the maximum pressure is found at the inner radius, r_i. Thus

$$\delta = kp_{\max} r_i \tag{1-2}$$

Elimination of k and δ from equations (1-1) and (1-2) yields

$$p = p_{\max} \frac{r_j}{r} \tag{1-3}$$

With the lining pressure known, we may now calculate the required axial force from

$$F = \int_A p \, da \tag{1-4}$$

and the resulting braking torque from

$$T = \mu \int_A pr \, da \tag{1-5}$$

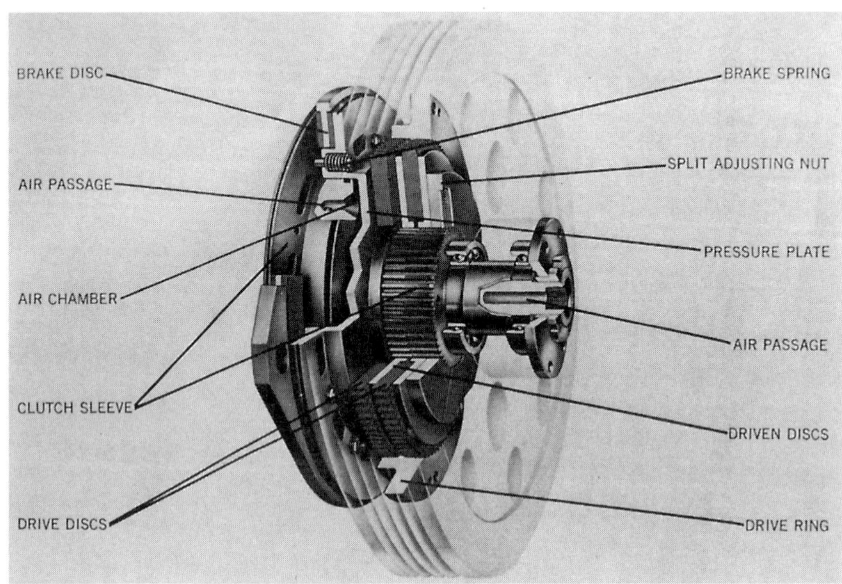

FIGURE 2 Combination disk brake and disk clutch, both dry. (Courtesy of Minster Machine Co., Minster, OH.)

Evaluation of these integrals is easiest for brake pads with radial and circular boundaries, as in Figure 4, for which equation (1-4) and (1-5) may be written using a dummy variable ϕ as

$$
\begin{aligned}
F &= p_{\max} r_i \int_A \frac{1}{r}\, da = p_{\max} r_i \int_{r_i}^{r_0} \int_0^{\theta} d\phi\, dr \\
&= p_{\max} r_i \theta (r_0 - r_i)
\end{aligned}
\tag{1-6}
$$

and

$$
\begin{aligned}
T &= \mu p_{\max} r_i \int_A da = \mu p_{\max} r_i \int_{r_i}^{r_o} r\, dr \int_0^{\theta} d\phi \\
&= \mu p_{\max} r_i \frac{\theta}{2} \left(r_o^2 - r_i^2 \right)
\end{aligned}
\tag{1-7}
$$

From equation (1-7) we find that for the pressure distribution given by relation (1-3) the torque may be easily calculated for any brake pad whose area is

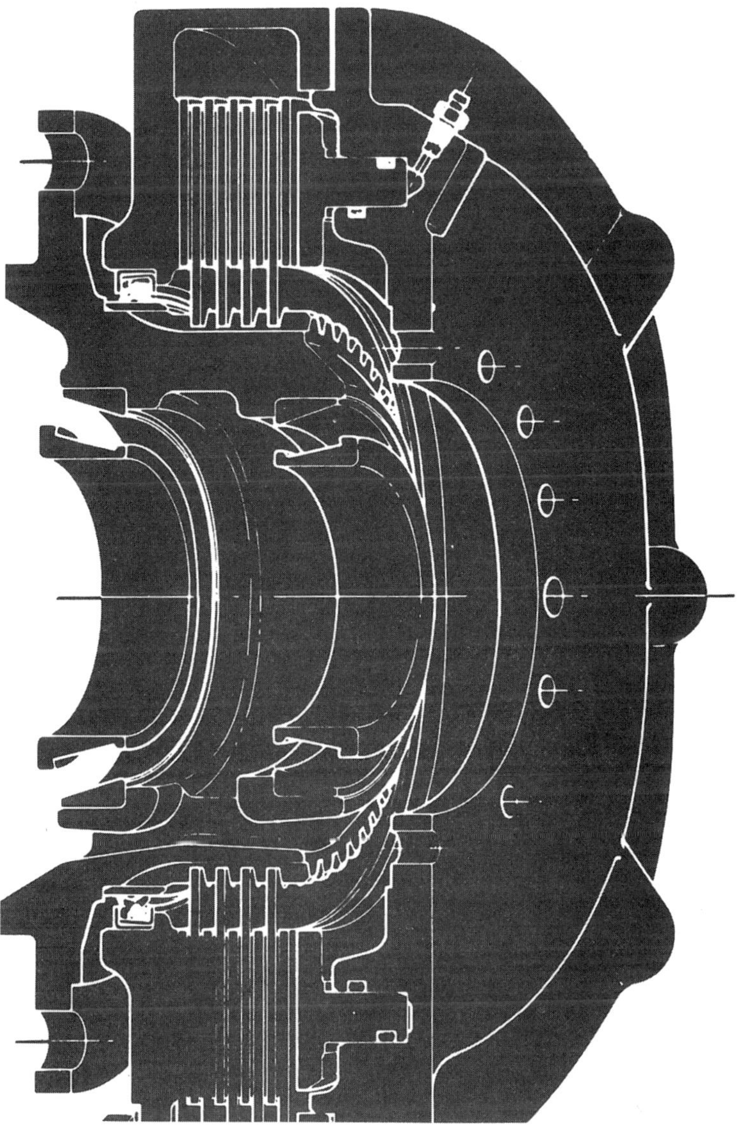

FIGURE 3 Wet multiple-disk brake. (Courtesy D. A. B. Industries, Inc., Troy, MI.)

FIGURE 4 Annular sector caliper disk brake. (Courtesy of Horton Manufacturing Co., Inc., Minneapolis, MN.)

known or simply calculated. For a circular pad of diameter d, for example, the torque is given by

$$T = \mu p_{\max} r_i \frac{\pi}{4} d^2 \tag{1-8}$$

According to equation (1-7), the torque provided by a caliper brake having pads similar to those in Figure 4 usually will be greater than that provided by circular pads of equal area, as shown in Figure 5, when acting on disks of equal outside diameter because the proportions of the pads in the brake shown in Figure 4 generally place the center of pressure at a larger radius from the center of the disk. (See also Figure 6.)

Circular pads are often used, nevertheless, in hydraulically activated caliper brakes whenever the hydraulic pressure may be increased relatively cheaply because the pads themselves are supported entirely by the piston face and are therefore cheaper to produce because no additional supporting

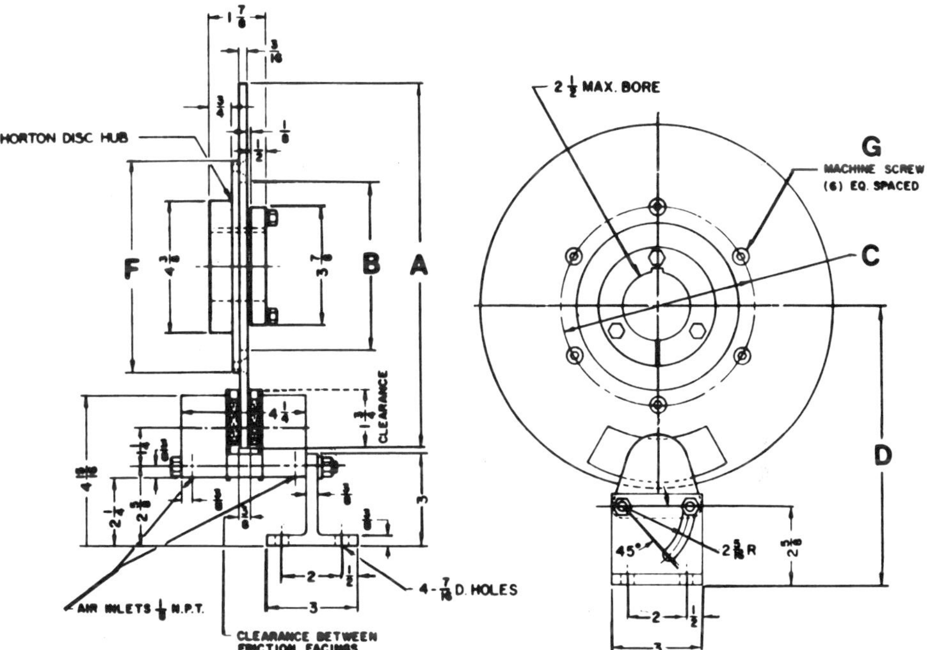

FIGURE 4 Continued.

structure is required. Noncircular pads are used where increasing the pressure may be relatively expensive and where the maximum performance is required for the pressure that is available, as in aircraft brakes.

If we replace $d^2/4$ in equation (1-8) with r_p^2, where r_p is the pad radius ($r_p = d/2$), and also replace r_i in equation (1-8) according to the relation $r_i = r_o - 2r_p$, we have

$$T = \mu\pi p_{\max}(r_o - 2r_p)r_p^2 \tag{1-9}$$

Upon differentiating equation (1-9) with respect to r_p we obtain

$$\frac{dT}{dr_p} = \mu\pi p_{\max} 2r_p(r_o - 3r_p) \tag{1-10}$$

which is equal to zero when $r_p = r_o/3$, indicating an extreme value of T for that pad radius. Since dT^2/dr_p^2 is negative at this value of r_p, it follows that T has its maximum value at $r_p = r_o/3$.

Calculating the activation force for a circular pad is more involved than it is for an annular sector pad because radius r remains in the denominator of

FIGURE 5 Caliper disk brake with circular pads, two pistons. (Courtesy of Misco, Inc., North Mankato, MN.)

the integrand. An element of the pad area may be written as $da = \rho\, d\rho\, d\theta$, where radius ρ is measured from the center of the pad, as shown in upcoming Figure 13, associated with later Example 4.1. Complexity arises from the requirement that the expression for the radius r from the center of the disk to the element of area of the circular pad must now be written in terms of ρ and θ.

From the law of cosines we have

$$r = (r_c^2 + \rho^2 - 2r_c\rho \cos \theta)^{1/2} \tag{1-11}$$

where r_c is the radius from the center of the brake pad to elemental area da, as shown in Figure 13 (see later Example 4.1).

Substitution of equation (1-11) into the first integral in equation (1-6) and writing the element of area as $\rho\, d\rho\, d\theta$ allows the activation force to be written as

$$F = p_{\max} r_i \int_0^{2\pi} \int_0^{r_p} \frac{\rho}{(\rho^2 + r_c^2 - 2\rho_i r_c \cos \theta)^{1/2}} d\rho d\theta \tag{1-12}$$

FIGURE 6 Typical caliper brake pad of sintered material for heavy aircraft brakes. (Note contour of the pad to place lining material toward the outer periphery of the disk.) (Courtesy Friction Products, Medina, OH.)

Since analytical evaluation of the integrals in equation (1-12) is somewhat tedious, it is easier to turn to numerical methods. Evaluation using a numerical program, such as Mathcad, may provide graphical data that displays the dependence of force F on the pad radius r_p, as will be demonstrated later in Example 4.1.

The Mathcad manual specifies the integration method used in its program and the references used in writing the program. They may be consulted for the details of mathematical analysis.

II. VENTILATED DISK BRAKES

Although disk brakes are less susceptible to fade than drum brakes, they will be heated by friction, which may lead to brake fade in situations requiring heavy and frequent braking. This heating may be reduced by using ventilated disk brakes, which consist of two disks separated by radial vanes, so that additional cooling surface is provided, as shown in Figure 7.

Ventilation also increases brake life, as implied by the representative brake pad life as a function of the surface temperature as given in Figure 8, where the longest life is realized for that pad and caliper combination which provides the largest heat sink, shown in Figure 9 and the shortest life for that with the smallest heat sink, shown in Figure 10.

III. ANNULAR CONTACT DISK BRAKES AND CLUTCHES

Annular contact, or face contact, disk brakes are available either as dry multiple-plate disk brakes, as shown in Figure 2, or as wet multiple-plate disk brakes, as shown in Figure 3. Their construction is similar to that of multiple disk cluthes to the extent that many manufacturers produce both multiple disk cluthes and brakes that have many components in common.

Conventional design formulas for these brakes are predicated on one of two assumptions: uniform wear or uniform pressure. Although the first of these assumptions may be a better approximation of brake behavior, it involves more calculation than the second. Following established practice, we shall consider the consequences of both of these assumptions.

A. Uniform Wear

The uniform wear assumption employed in the derivation of the force and torque relations given by equations (1-6) and (1-7) may be applied to disk brakes if the plates and the clamping structure tend to maintain uniform lining thickness. Application of equations (1-6) and (1-7) to annular contact disk brakes requires only that θ be replaced by 2π in both relations to get

$$T = \mu\pi p_{\max} r_i (r_o^2 - r_i^2) \tag{3-1}$$

and

$$F = 2\pi p_{\max} r_i (r_o - r_i) \tag{3-2}$$

So the ratio T/F of the torque to the activating force is given by

$$\frac{T}{F} = \mu \frac{r_o + r_i}{2} \tag{3-3}$$

Examination of equation (3-1) yields the somewhat surprising result that if we cover the entire face of a single-plate brake or clutch with lining material, the brake or clutch will soon become ineffective. In other words, the braking torque predicted by equation (3-1) will be zero whenever $r_i = r_o$, as reasonably expected, or whenever $r_i = 0$ and $r_o > 0$, as may not be expected.

FIGURE 7 Ventilated caliper disk brake. (Courtesy of Eaton Power Transmission Systems, Airflex Division, Cleveland, OH.)

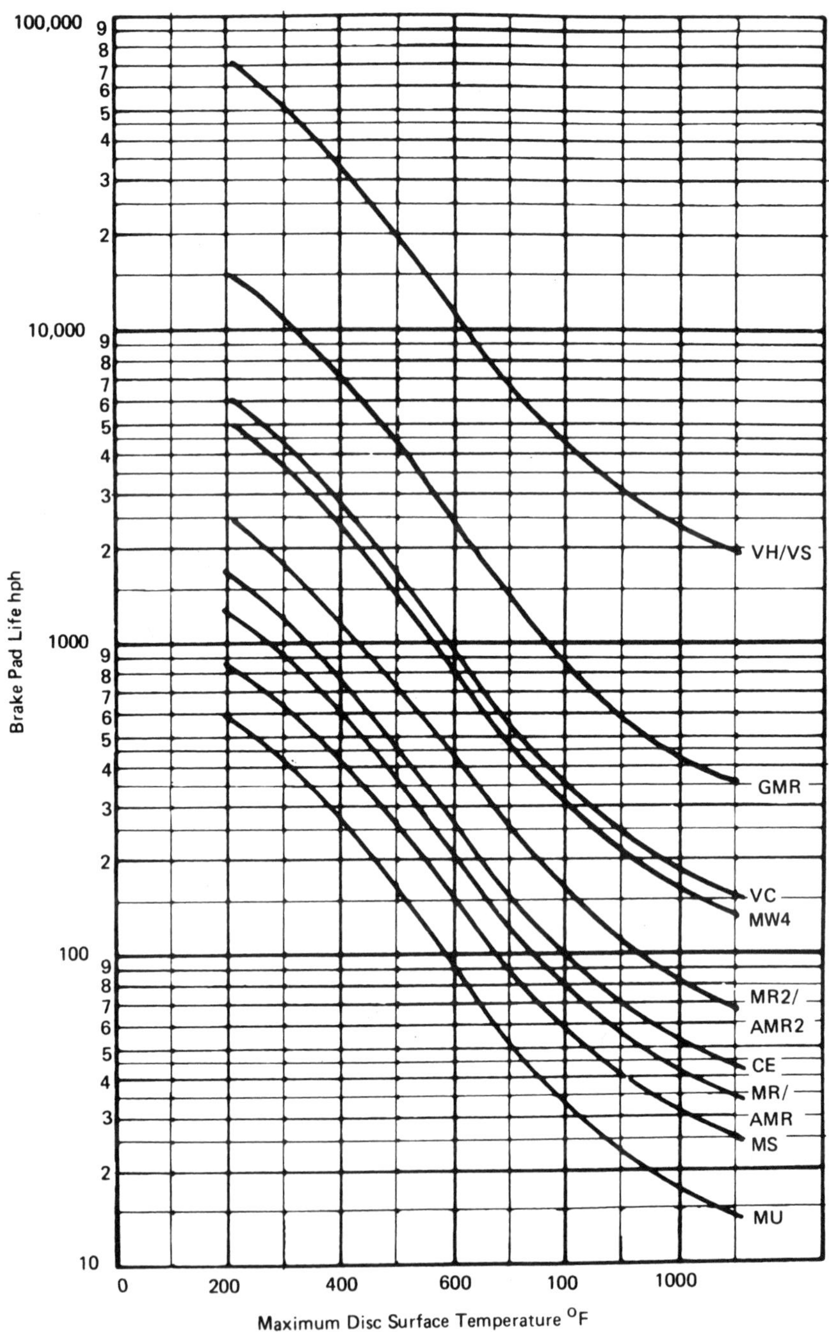

Figure 8 Approximate pad life as a function of the maximum disk pressure. (Courtesy of Twiflex Corp., Horseheads, NY.)

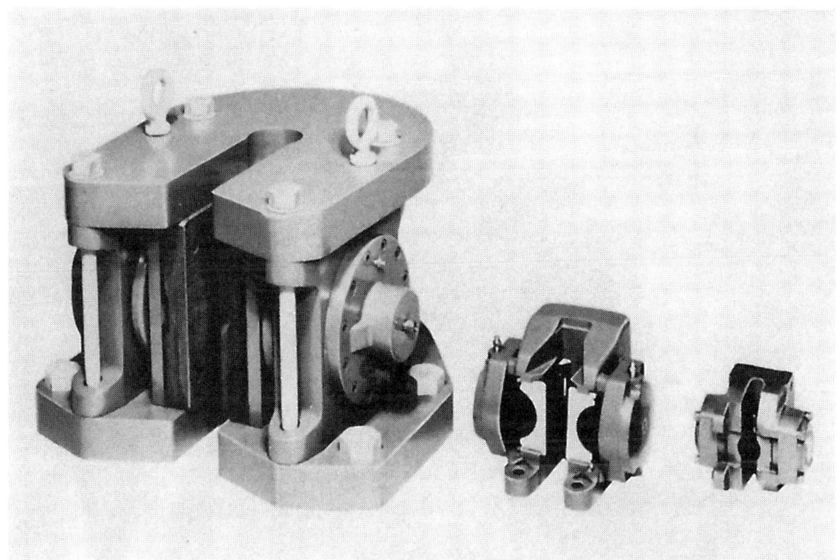

FIGURE 9 Large calipers for disk brakes–model vs. referenced in Figure 8. (Courtesy of Twiflex Corp., Horseheads, NY.)

This was unintentionally demonstrated by a winch manufacturer between 1970 and 1980, as will be described later. Because of these observations we shall turn our attention to finding the r_i that will produce the maximum torque before designing a face contact disk brake or clutch. Differentiation of equation (3-1) with respect to r_i and setting the derivative to zero yields

$$r_i = \frac{r_o}{\sqrt{3}} \tag{3-4}$$

as the theoretically optimum value of r_i, corresponding to torque and activating force given by

$$T = \frac{2}{3\sqrt{3}} \mu p_{\max} \pi r_o^3 \tag{3-5}$$

and

$$F = \frac{1}{\sqrt{3}}\left(1 - \frac{1}{\sqrt{3}}\right) 2\pi r_o^2 p_{\max} \tag{3-6}$$

for a single-face annular contact brake. Actual brake lining dimensions may differ somewhat from this inner radius because of concentric grooves in the

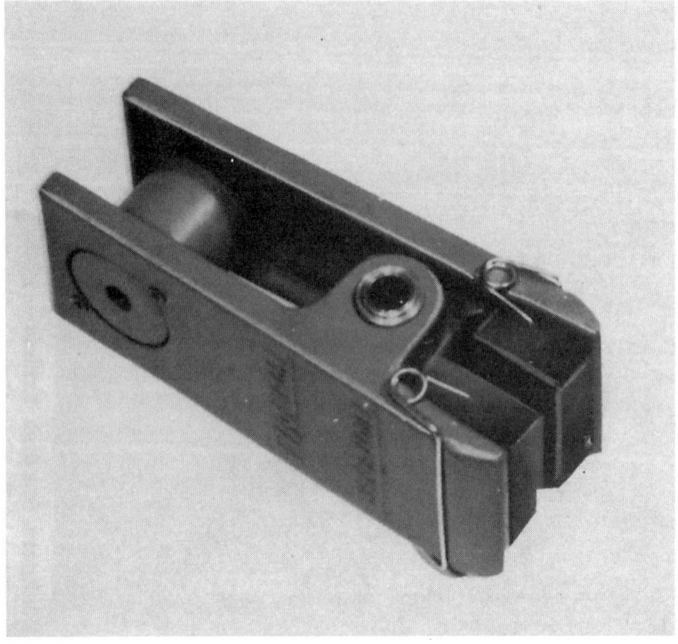

FIGURE 10 Small caliper for disk brakes–model MO referenced in Figure 8. (Courtesy of Twinflex, Corp., Horseheads, NY.)

lining and/or experimental data which may imply an effective pressure distribution different from that given in equation (1-3).

One advantage of multiple-plate brakes is that the torque increases in direct proportion to the number of plates added while the activation force theoretically remains unchanged. In mathematical terms,

$$T = \frac{2n}{3\sqrt{3}} \mu p_{\max} \pi r_o^3 \tag{3-7}$$

where n is the number of friction interfaces (8 in Figure 3, 4 in Figure 11). Adding springs to separate the plates when the brake is released will increase the activation force by the amount of the spring forces plus the friction forces generated by the motion of the plates along their lubricated splines.

B. Uniform Pressure

This assumption implies that either the disks or the lining or both are flexible enough to allow the deformation necessary for δ in equation (1-1) to vary with

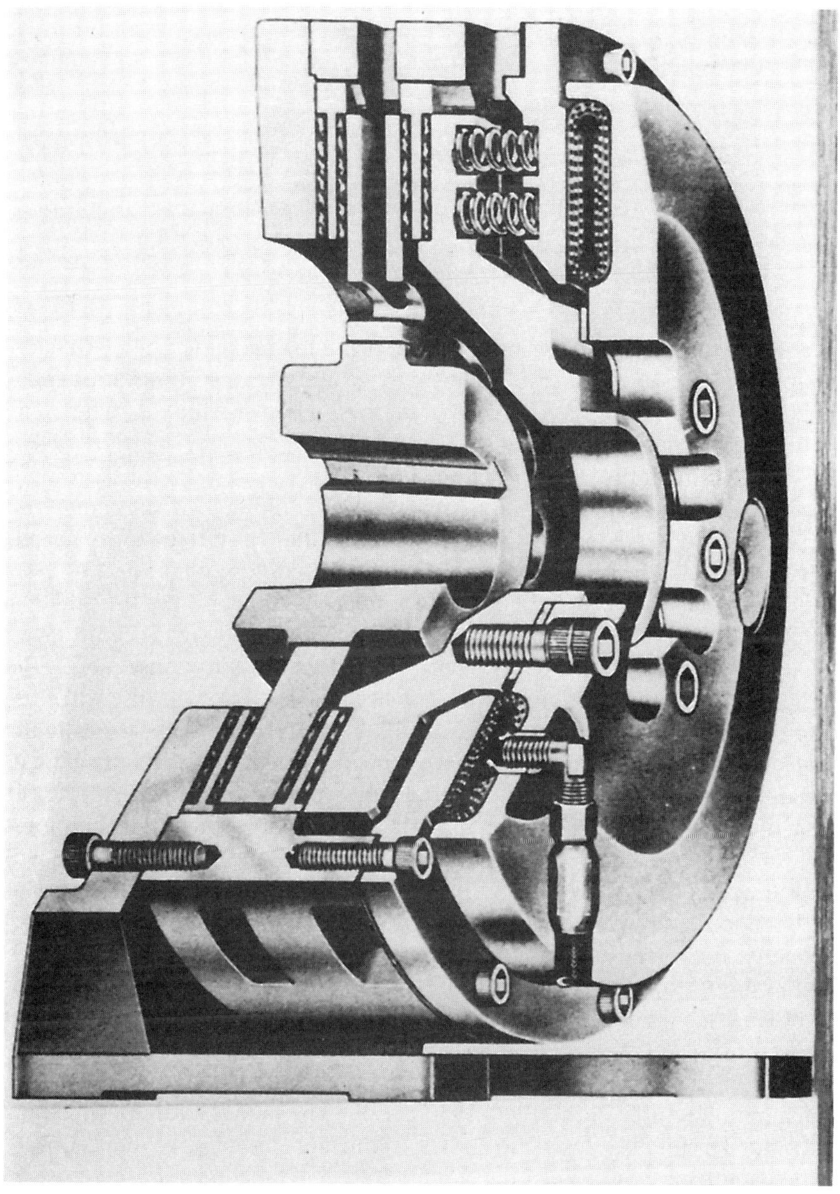

FIGURE 11 Dry multiple-disk brake, pneumatically activated. (Courtesy of Wichita Clutch Co., Dana Corp., Power Transmission Div., Toledo, OH.)

the radius such that the pressure can become constant. Whenever the pressure is uniform, equations (1-5) and (1-4) hold and may be easily integrated to give

$$T = \frac{2}{3}\pi\mu p(r_o^3 - r_i^3) \tag{3-8}$$

for the braking torque and

$$F = \pi p(r_o^2 - r_i^2) \tag{3-9}$$

as the activation force.

Since uniform pressure may require spring-loaded plates, plates of varying thickness, or some other mechanism to ensure no pressure variation, relations (3-7) and (3-8) may be restricted to single-plate brakes, where the additional mechanism may be added.

If an annular disc brake or clutch is replaced by one with full-faced rigid discs on both input and output shafts and with a lining, or facing, material that covers the entire face of one of the discs so that $r_i = 0$, the torque capability of the clutch, or brake, may be given initially by equation (3-8). Its torque capacity, however, will decrease with each application of the brake or clutch until it fails to transfer useful torque. This is because, according to equation (1-1), negligible wear will occur at and near $r_i = 0$. Consequently the lining will maintain its original thickness near the center of the disc while the lining beyond this region wears away. Eventually there will be negligible contact, and hence negligible pressure, outside of what has become a small raised circular region, or hump, centered at $r_i = 0$. The sharp peak expected at the center of the facing, or lining, material because of zero wear at that point will usually not be seen because the compressibility of the friction material will allow the peak to be mashed down by the mating plate. This compressibility of the lining material will extend the effective life of such a clutch or brake until the activating force is unable to compress the resulting central hump enough for the lining to contact the mating plate beyond this small central hump. Removing this small central region, however, will allow the brake or clutch to again transmit torque.

As noted earlier, this was unintentionally demonstrated by at least one winch manufacturer in the 1970–1980 period. The manufacturer's winch incorporated a clutch as described earlier in which one face was covered entirely by the facing material. When the clutch ultimately failed to transmit a useful torque, the manufacturer recommended replacing the facing material. Instead, the life of the facing could be, and was, more than doubled simply by removing the central region to produce an inner radius $r_i > 0$. If inner radius r_i, were made equal to that given by equation (3-4), then the torque capability would be restored to that given by equation (3-5).

Since greater torque enhancement can be obtained from multiple disk brakes that provide torque multiplication equal to the number of contacting friction surfaces, as indicated by equation (3-7), it follows that there is no motivation to try to devise some mechanism to assure uniform pressure between contacting annular plates.

IV. DESIGN EXAMPLES

Example 4.1

Estimate the torque and activation force for a floating, or sliding, caliper disk brake having circular pads 1 in. in diameter acting on a disk 11 in. in diameter, as shown in Figure 12. The expected friction coefficient is 0.32 and the maximum design pressure for the lining material is 300 psi. (A sliding caliper brake is held by a slide which allows its brake pads to be forced against opposite sides of the disk when its single piston is activated, as in Figure 1.)

From Figure 12 it is evident that $r_i = 4.5$ in., so from equation (1-8),

$$T = \pi\rho p_{\max} r_i \frac{d^2}{4} = \pi(0.32)(300)\frac{1}{4} = 339.292 \text{ in.} - \text{lb}$$

per caliper pad. Hence the total torque is 678.584 in.-lb.

According to Figures 12 and 13 and pad radius r_p it is evident that

$$r_i = r_o - 2r_p \qquad \text{and} \qquad r_c = r_o - r_p \qquad 0 \le p \le r_p$$

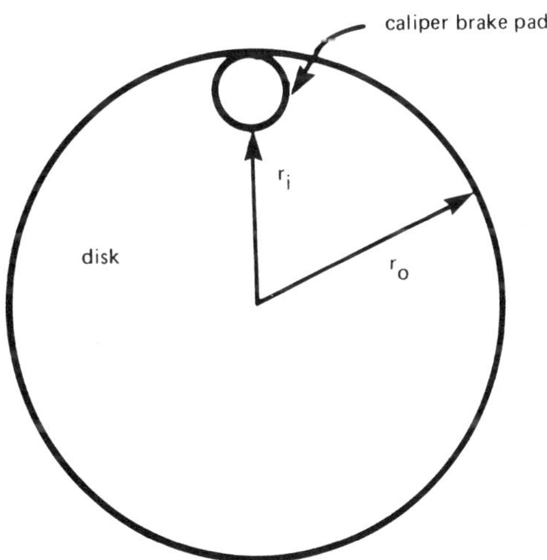

FIGURE 12 Circular lining pad of a caliper disk brake.

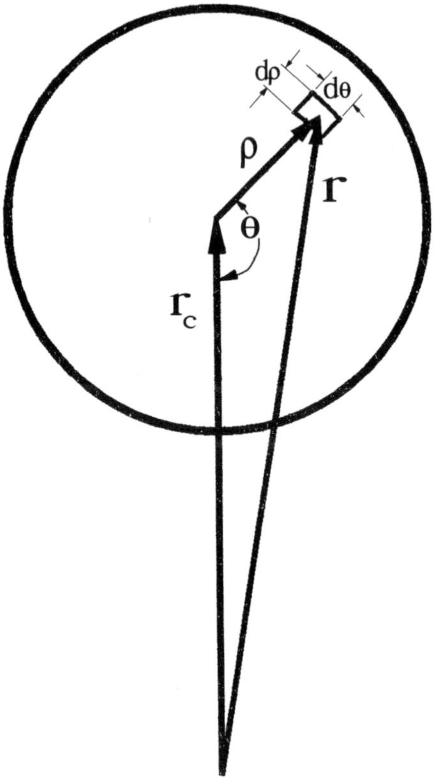

FIGURE 13 Geometric relations between ρ, θ, *r*, and r_c. Arc length at radius *r* over an angular incremet *d*θ is *r d*θ.

Also see p. 90. Use of these relations along with substitution of

$$p_{\max} = 300\ psi \quad \text{and} \quad r_p = 0.5 \text{ in.}$$

into equation (1-12) yields, after numerical integration,

$$F = 212.324 \text{ lb}$$

on each of the two opposing brake pads, corresponding to a hydraulic pressure of 270.339 psi.

Plot both the torque and the required activation force against the radius of the brake pad in order to answer the present question or any future questions of increasing the brake pad diameter. The resulting torque and the associated force on the brake pads as a function of the brake pad radius is shown in Figure 14.

Doubling the torque to 1357.168 in -lb by adding another caliper doubles the fluid flow volume but maintains the same pressure. Increasing the pad diameter to 1.50 in. provides a torque of 678.584 in -lb from each pad for the required total torque of 1357.168 in.-lb. The caliper frame must be strengthened to support a force of 447.841 lb on each pad, but the hydraulic system pressure may be reduced to 253.426 psi.

Existence of a maximum torque within the boundaries of the disk is consistent with the existence of a similar maximum found for annular disk brakes and clutches. In this particular case the maximum torque, of approximately $T = 1858.4$ in-lb, occurs in the vicinity of $r_p = 1.836$ in., as found with the aid of the Trace routine supplied by Mathcad. The corresponding force is close to 1639 lb. A plot of the force as a function of the brake pad radius, as in Figure 14, shows that it too reaches a slightly larger maximum of about

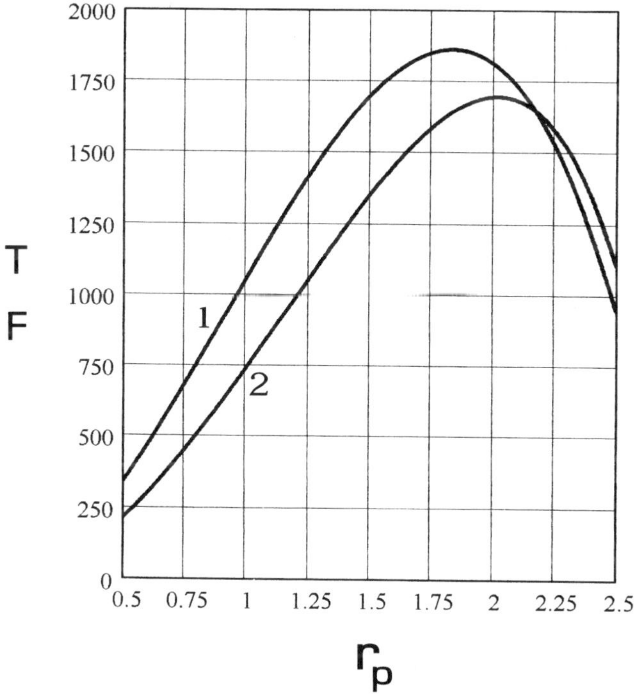

FIGURE 14 Torque in inch-pounds, curve 1, and force in pounds, curve 2, as functions of brake pad radius r_p in inches acting on a disk 11.0 inches in diameter for $p_{max} = 300$ psi.

1691.9 lb at a different value of r_p, at $r_p = 2.016$ in. Increased piston area will allow the line pressure to drop if the brake pad force is provided by a hydraulically driven piston whose diameter is equal to the pad diameter. The nature of this falloff in pressure with increased piston radius is illustrated in Figure 15.

Example 4.2

Estimate the torque and activation force for a caliper brake whose pad is a sector of an annular ring subtending the same angle at the center of the disk as subtended by the circular pad described in Example 4.1

According to the geometry of Figure 16, half of the subtended angle is given by

$$\frac{\theta}{2} = \sin^{-1}\frac{0.5}{5} = 0.1002 \text{ rad}$$

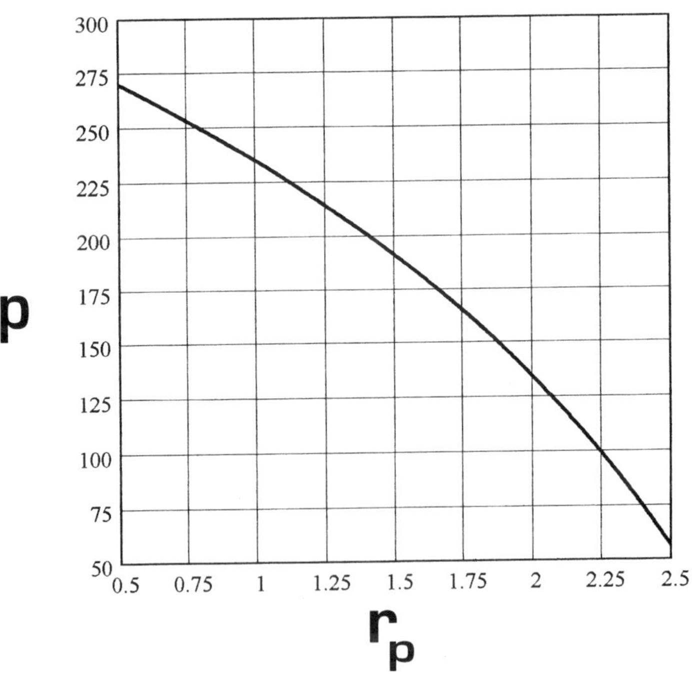

Figure 15 Hydraulic line pressure (psi) to a piston of radius r_p to provide the force shown in Figure 14.

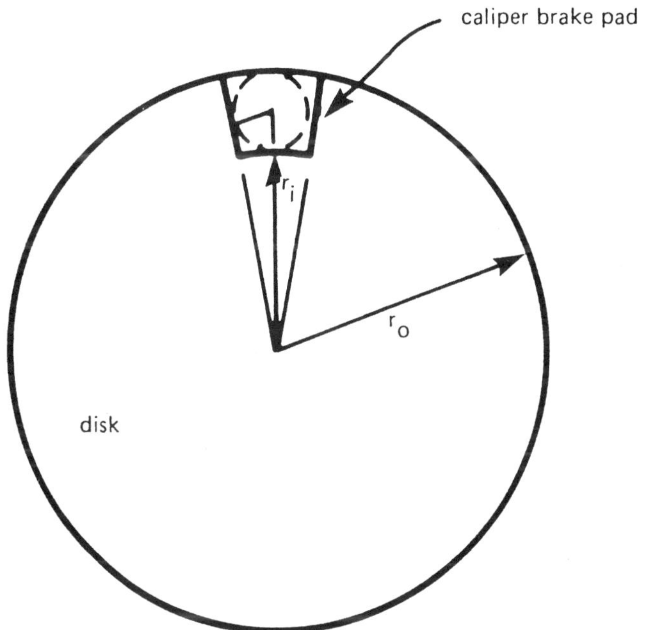

FIGURE 16 Caliper brake lining that is a sector of an annular ring.

So substitution into equation (1-7) yields a torque per pad of

$$T = \mu p_{\max} r_i \frac{\theta}{2}(r_o^2 - r_i^2) = 0.32\ (300)(4.50)(0.1002)(5.5^2 - 4.5^2)$$
$$= 432.862 \text{ in.} - \text{lb}$$

and substitution into equation (1-6) yields an activation force of

$$F = p_{\max} r_i \theta (r_o - r_i) = 300(4.5)(0.2004)$$
$$= 270.540 \text{ lb}$$

for a 28% increase in torque capacity and a 27% increase in the activation force.

Example 4.3

Compare the braking torque obtained from the caliper brake in Example 4.2 with that obtained from an annular, or face contact, disk brake for which r_i is determined from relation (3-4).

Substitution of $r_o = 5.50$ in. into equation (3-5) along with $p_{\max}$ and μ from Example 4.1 leads to

$$T = 19\ 313.536 \text{ in.-lb}$$

which is greater than the torque found in Example 4.2 by a factor of 44.6. A manufacturer of truck brakes has recently introduced a series of annular disk brakes to realize this advantage.

V. NOTATION

a	area (l^2)
d	diameter (l)
F	force (ml/t^2)
k	constant of proportionality
p	pressure (m/lt^2)
R, r	radius (l)
T	torque (ml^2/t^2)
δ	thickness (l)
θ	angle (1)
μ	friction coefficient (1)
ρ	radius (l)
φ	angle (1)

VI. FORMULA COLLECTION

Pressure distribution for uniform wear:

$$p = p_{\max} \frac{r}{r_i}$$

Activation force, caliper disk brake, annular sector pad:

$$F = p_{\max} r_i \theta (r_o - r_i)$$

Torque, caliper disk brake, annular sector pad:

$$T = \mu p_{\max} r_i \frac{\theta}{2} (r_D^2 - r_{-i}^2)$$

Activation force, caliper brake, circular pad:

$$F = p_{\max} r_i \int_0^{2\pi} \int_0^{r_p} \frac{\rho}{(\rho^2 + r_c^2 - 2\rho r_c \cos\theta)^{1/2}} d\rho \, d\theta$$

Torque, caliper brake, circular pad:

$$T = \mu p_{\max} r_i \frac{\pi}{4} d^2$$

Activation force, annular contact disk brake, uniform wear:

$$F = 2\pi p_{\max} r_i (r_o - r_i)$$

Torque, annular contact disk brake, uniform wear:

$$T = \mu p_{\max} \pi r_i (r_o^2 - r_i^2)$$

Activation force, annular contact disk brake, uniform pressure:

$$F = \pi p (r_o^2 - r_i^2)$$

Torque, annular contact disk brake, uniform pressure:

$$T = \frac{2}{3} \pi \mu p (r_o^3 - r_i^3)$$

6

Cone Brakes and Clutches

These brakes have the advantage of greater torque for a smaller axial force than either type of disk brake discussed in Chapter 5. The magnitude of the improvement is limited, however, by the observation that for small cone angles a disengagement force may be required, depending on the friction coefficient, because the inner and outer cones may tend to wedge together. This is because on engagement the inner cone is radially compressed and the outer cone is radially enlarged as the brake is engaged. For small cone angles the induced friction force dominates the normal force, which tends to expel the inner cone, so that an external force is required for separation. This characteristic, however, may be useful in those applications where a brake is to remain engaged in the presence of disengagement forces.

I. TORQUE AND ACTIVATION FORCE

The pertinent geometry of the cone brake is shown in Figure 1. If the inner and outer cones are concentric and rigid, the amount worn from the lining during engagement will be given by

$$\delta = kpr \tag{1-1}$$

where p denotes the pressure and r is the radius to the point where p acts. Proportionality constant k may be evaluated by observing that the form of relation (1-1) demands that the maximum pressure occur at the minimum radius. Hence

$$\delta = kp_{\max}r_i \tag{1-2}$$

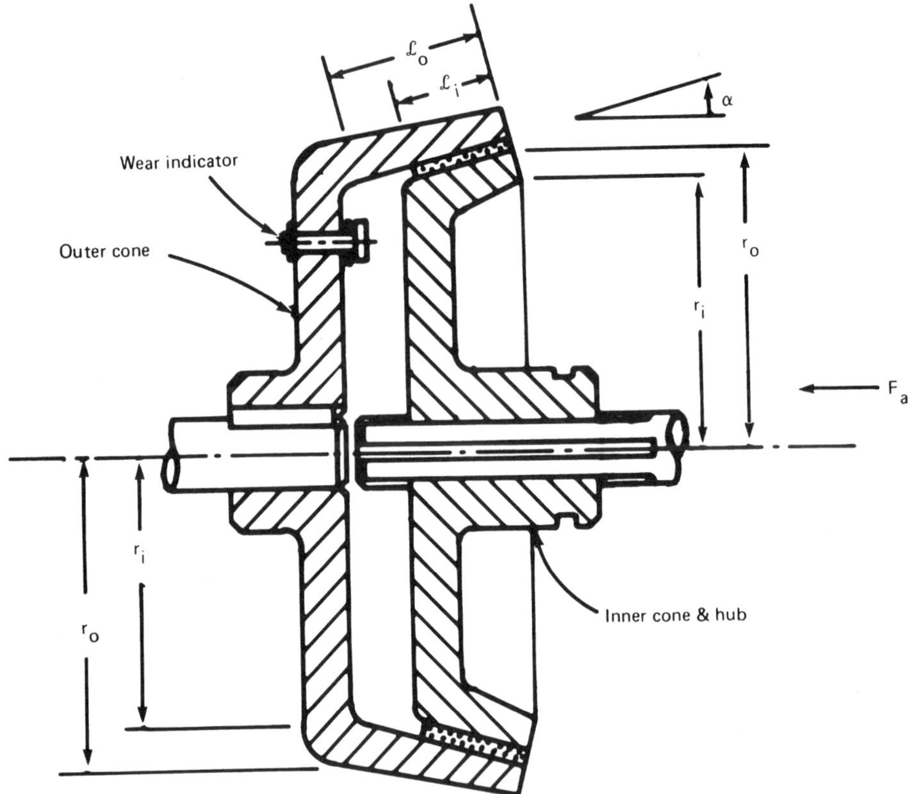

FIGURE 1 Cone brake and its geometry (partially worn lining).

Upon equating equations (1-1) and (1-2), we find that

$$p = p_{\max}\frac{r_i}{r} \tag{1-3}$$

Although the brake lining is more easily attached to the inner cone, with the torque acting at the inner surface of the outer cone, we shall derive formulas on the assumption that the torque acts on the outer surface of the inner cone because this will give a torque capacity that the brake can equal or exceed until the lining is destroyed. Thus

$$T = \mu\int_A pr\,da = \mu p_{\max} r_i \int_A da = \frac{2\mu\pi p_{\max} r_i}{\sin\alpha}\int_{r_i}^{r_o} r\,dr \tag{1-4}$$

where the element of area on the outside of the inner cone is given by

$$da = 2\pi r\, d\ell = 2\pi r \frac{dr}{\sin \alpha} \tag{1-5}$$

and where we have used $d\ell \sin \alpha = dr$ and the Pappus theorem for the area of a surface of revolution. Upon integration the expression for the torque becomes

$$T = \frac{\mu \pi p_{\max}}{\sin \alpha} r_i \left(r_0^2 - r_i^2\right) \tag{1-6}$$

Since this expression vanishes for $r_i = 0$ and for $r_i = r_o$ but not for intermediate values, we may set the derivative of T with respect to r_i equal to zero to find that the maximum torque may be obtained when

$$r_i = \frac{1}{\sqrt{3}} r_o \tag{1-7}$$

for which the torque is given by

$$T = \frac{2}{3\sqrt{3}} \mu \pi \frac{p_{\max}}{\sin \alpha} r_o^3 \tag{1-8}$$

To find the activation force, we return to Figure 1 to discover that it is given by

$$\begin{aligned} F_a &= \int_A (p \sin \alpha + \mu p \cos \alpha) da \\ &= (\sin \alpha + \mu \cos \alpha) p_{\max} r_i \int_A \frac{1}{r} 2\pi r \frac{dr}{\sin \alpha} \\ &= 2\pi p_{\max} \left(1 + \frac{\mu}{\tan \alpha}\right) r_i (r_o - r_i) \end{aligned} \tag{1-9}$$

When $\alpha = \pi/2$, equations (1-6) and (1-9) reduce to the correct expressions for the torque and activation force for an annular contact disk brake with a single friction surface.

Unlike plate clutch and brakes, it may take a retraction force to disengage a cone clutch or brake, just as it takes a force to remove a cork from a bottle. The magnitude of the retraction force, which we shall denote by F_r, may be derived from the force equilibrium condition in the axial direction for the forces shown in Figure 1. After replacing $\mu p\, da$ with $-\mu p\, da$, we find that the incremental retraction force dF_r is given by

$$dF_r = 2\pi r_i \frac{dr}{\sin \alpha} (\mu p \cos \alpha - p \sin \alpha) \tag{1-10}$$

where we again use the pressure p and element of area da as defined by equations (1-3) and (1-5), respectively. After performing the integration, we have

$$F_r = 2\pi p_{\max} r_i (r_o - r_i)\left(\frac{\mu}{\tan\alpha} - 1\right) \tag{1-11}$$

Clearly, a retraction force is necessary only when $(\mu/\tan\alpha - 1)$ is greater than zero. F_r vanishes if

$$\frac{\mu}{\tan\alpha} = 1$$

that is, if (1-12)

$$\mu = \tan\alpha$$

The ratio of torque to activation force for a cone clutch or brake may be obtained by dividing equation (1-6) by equation (1-9) to get

$$\frac{T}{F_a} = \frac{r_o + r_i}{2}\,\frac{\mu}{\sin\alpha + \mu\cos\alpha} \tag{1-13}$$

in which the ratio $(r_o + r_i)/2$ may be considered a magnification factor that operates upon the ratio

$$f(\mu, \alpha) = \frac{\mu}{\sin\alpha + \mu\cos\alpha} \tag{1-14}$$

To find an extreme value of $f(\mu,\alpha)$ with respect to the cone angle, differentiate it with respect to α to get

$$\frac{df}{d\alpha} = -\mu\frac{\cos\alpha - \mu\sin\alpha}{(\sin\alpha + \mu\cos\alpha)^2} = 0 \quad \text{whenever} \quad \cos\alpha = \mu\sin\alpha \tag{1-15}$$

Since the second derivative $d^2f/d\alpha^2$ is positive whenever equation (1-15) holds, $f(\mu,\alpha)$ is minimum along the curve

$$\mu = \frac{1}{\tan\alpha} \tag{1-16}$$

Because points on this curve represent the minimum torque that can be had from a cone brake or clutch, it is clear that a design for such a unit should not lie along this curve if it can be avoided.

Upon comparison of equation (3-3) with equation (1-8) we find that equation (1-8) reduced to equation (3-3) when $\alpha = \pi/2$. Consequently, we may find what configuration of a cone brake or clutch can equal or exceed the T/R ratio of a plate clutch or brake by solving

$$f(\mu, \alpha) = \mu \tag{1-17}$$

From equation (1-14) we find that equation (1-17) holds whenever $\sin\alpha + \mu\cos\alpha = 1$. Hence, designs for which μ is greater than

$$\mu = \frac{1 - \sin\alpha}{\cos\alpha} \tag{1-18}$$

usually should be avoided because a plate clutch having the same inner and outer radii will provide the same torque, but with smaller axial dimensions.

The last relation that is of interest in the design of a cone brake or clutch is the condition for which the retraction force is zero. From equation (1-11) it is clear that F_r vanishes when

$$\mu = \tan\alpha \tag{1-19}$$

Curves given by these last three relations are plotted in Figure 4. The dashed curve in this figure is the plot of relation (1-18), the dotted curve is the plot of equation (1-16), and the solid curve is the plot of equation (1-19). The surface described by equation (1-14) is shown in Figure 1, contour lines that depict elevations on that surface itself are shown in Figure 2. Upon

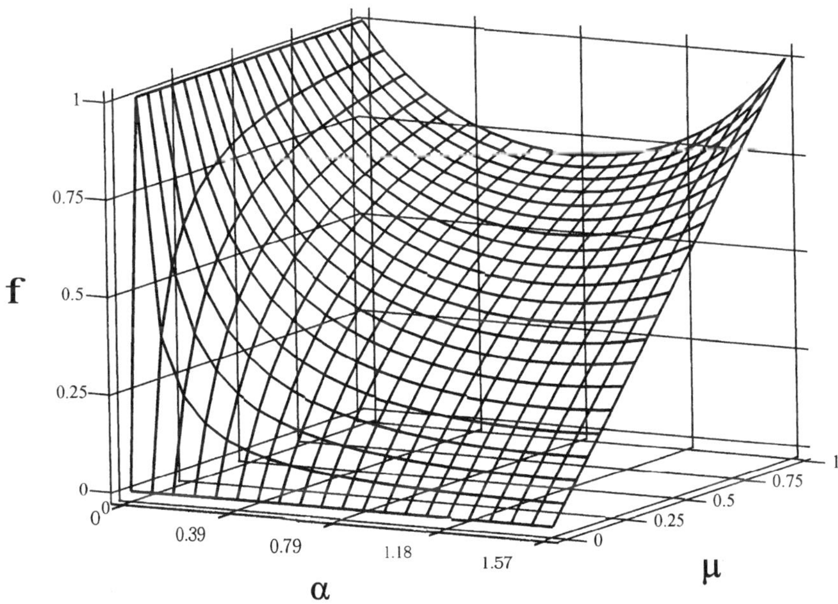

FIGURE 2 Surface defined by $f(\mu,\alpha)$ for $0 \le \mu \le 1$ and $0 \le \alpha \le \pi/2$.

comparison of the three figures, the minimum described by equation (1-16) and plotted in Figure 4 is qualitatively evident in Figures 3 and 4.

It is Figure 4 that is directly useful in the design of cone brakes and clutches, because we find from equation (1-19) that the regions to the left of the solid curve (regions 2 and 4) is where a retraction force is required; this is where $\mu \geq \tan \alpha$. Designs where μ and α are coordinates of points to the right of the solid curve that fall within regions 3 and 5 generally should be avoided because a greater torque-to-activation-force ratio (T/Fd) may be had with a plate clutch or brake. This leaves region 1, which lies below both the dotted curve and the dashed curve and to the right of the solid curve, as the only region where either a cone clutch or a cone brake is superior to either a single-plate clutch or to a single-plate brake, respectively, and where no retraction force is required.

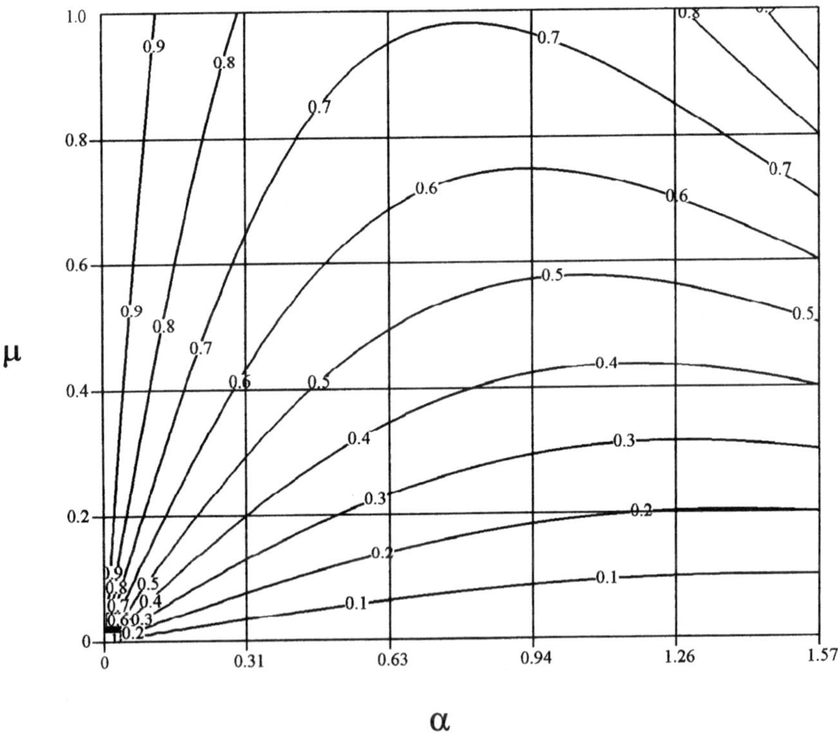

Figure 3 Contour plot of the surface $f(\mu,\alpha) = 2T/[(r_o + r_i)F_a]$.

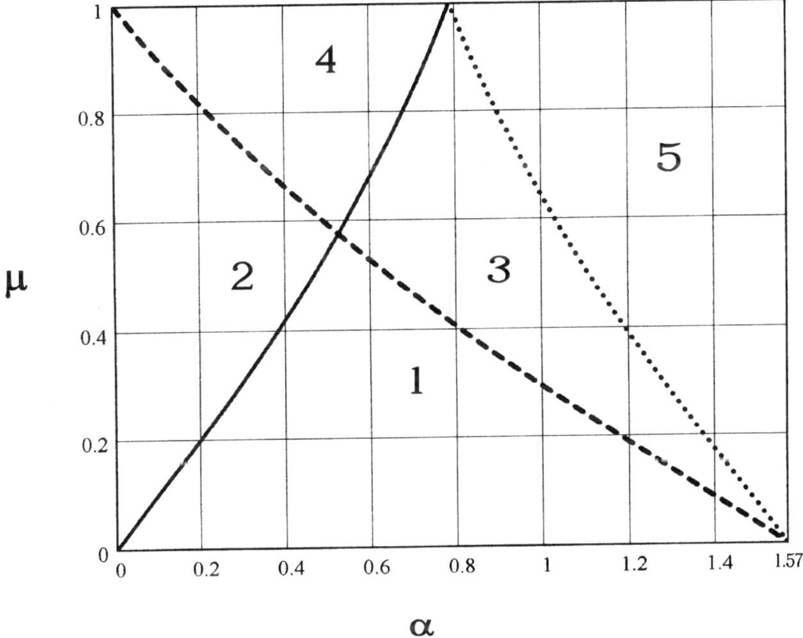

FIGURE 4 Design regions in the μ, α plane for cone clutches/brakes.

II. FOLDED CONE BRAKE

Prototype cone brakes have been designed and tested for a range of vehicle sizes, from tractors and trailers to subcompact automobiles [1]. Both the large and small sizes used a folded cone design, as illustrated in Figures 5 and 6, each with $\alpha = 27°$. Although the cone brake has fewer parts than drum brakes, this advantage must be balanced against the disadvantage of requiring an outboard wheel bearing.

Analysis of the folded cone brake with a sector shoe, shown in Figure 5, to obtain design formulas for the torque capability and the required activation force is quite similar to that used for simple cone brakes and clutches. Since the brakes illustrated in Figures 5 and 6 use a sector pad, we begin the analysis by observing from Figure 7(a) that an element of area on the conical surface may be written as

$$da = r\,d\theta \frac{dr}{\sin\alpha} \tag{2-1}$$

FIGURE 5 Truck cone brake and rotor (drum). (From reference 1. Reprinted with permission, © 1978 Society of Automotive Engineers, Inc.)

So the torque obtained due to a conical sector pad may be calculated from

$$
\begin{aligned}
T &= \mu p_{\max} r_i \int_A da = \frac{\mu p_{\max} r_i}{\sin\alpha} \int_0^{\theta} d\phi \int_{r_i}^{r_o} r\,dr \\
&= \frac{\mu p_{\max} r_i}{\sin\alpha}\,\theta\,\frac{r_o^2 - r_i^2}{2}
\end{aligned}
\tag{2-2}
$$

FIGURE 6 Cone brake on front-wheel-drive subcompact and the cone brake components. (From reference 1. Reprinted with permission, © Society of Automotive Engineers, Inc.)

Figure 6 Continued.

and the corresponding activating force on the sector pad may be calculated from

$$\begin{aligned} F_a &= p_{\max} r_i \frac{\sin\alpha + \mu\cos\alpha}{\sin\alpha} \int_0^\theta d\theta \int_{r_i}^{r_o} dr \\ &= p_{\max} r_i (1 + \mu\cot\alpha)\theta(r_o - r_i) \end{aligned} \tag{2-3}$$

Since the folded cone, shown by solid lines in Figure 7(b), is equivalent to two conical brakes, indicated by the dashed lines in that figure, it follows that the total torque and activating force may be found from

$$T = \frac{\mu p_{\max}}{\sin\alpha}\frac{\theta}{2}\left[r_{i_1}\left(r_{o_1}^2 - r_{i_1}^2\right) + r_{i_2}\left(r_{o_2}^2 - r_{i_2}^2\right)\right] \tag{2-4}$$

and

$$F_a = p_{\max}\theta r_i\left(1 + \frac{\mu}{\tan\alpha}\right)(r_{o_1} - r_{i_1} + r_{o_2} - r_{i_2}) \tag{2-5}$$

where θ is the angle subtended at the centerline by the lining sector.

III. DESIGN EXAMPLES

Example 3.1

Design a cone clutch to transmit a torque of 9050 N-mm or greater when fitted with a lining material having $\mu = 0.40$ and capable of supporting a maximum

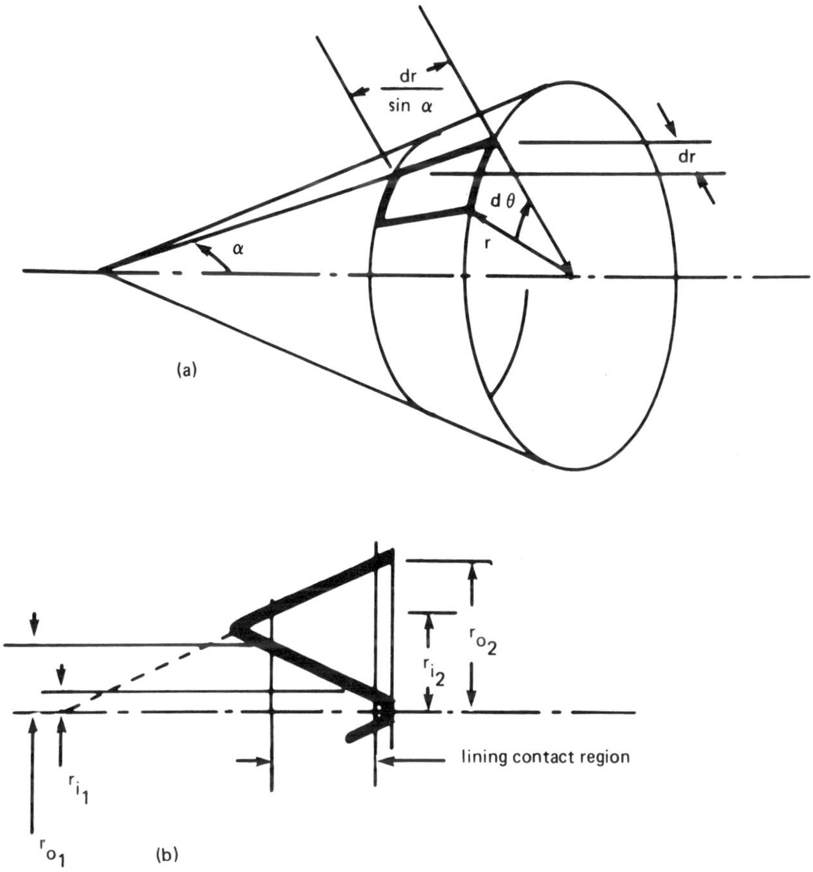

Figure 7 Cone geometry.

pressure of 4.22 MPa. The r_o value should be no larger then 35 mm and the clutch should release freely.

We shall begin by turning to Figure 4 and find that at $\mu = 0.40$, region 1 extends from $\alpha = 0.38485$ radians $= 22.051°$ to $\alpha = 0.79482$ radians $= 45.540°$ (as read with the aid of the Trace feature provided by Mathcad). From Figure 3 we note that the torque is greater at $\alpha = 22.051°$ than it is at $\alpha = 45.54°$, which suggests that a smaller α would be preferred. Hence, we shall initially consider two designs, one for $\alpha = 24°$ and one for $\alpha = 45°$. A slightly larger α was selected for the smaller of the two angles to ensure that no retraction force will be needed even with a manufacturing error of $-0.5°$.

Radius r_o was found by solving equation (1-8) for r_o. Activation force F_a was found from equation (1-9) after radius r_i was eliminated from it by using equation (1-7). Input data to these formulas was

$$T = 9050 \qquad \mu = 0.40 \qquad p_{\max} = 4.22 \qquad \beta(\alpha) = \alpha\frac{\pi}{180}$$

Here the variable β is introduced as the radian measure of angle α that is given in degrees, to avoid entering trigonometric arguments in the form (α·deg) that would otherwise be required by Mathcad. Thus,

$$r_o(\alpha) = \left[\frac{(T\,3r^{3/2}\sin(\beta(\alpha)))}{2\mu\pi p_{\max}}\right]^{1/3}$$

$$F_a\,(\alpha) = 2\pi p_{\max} r_o(\beta(\alpha))^2\left(\frac{1}{\sqrt{3}} - \frac{1}{3}\right)\left(1 + \frac{\mu}{\tan(\beta(\alpha))}\right)$$

$$d_o(\alpha) = 2r_o(\alpha)$$

$$d_o(24) = 24.344 \qquad d_o(45) = 29.272$$

$$F_a(24) = 124.872 \qquad F_a(45) = 140.022$$

Select the smaller diameter because of its smaller activation force.

Example 3.2

Examine the possibility of designing a cone brake that is to serve as a holding brake having a torque capacity of 40 ft-lb that can be released by a retraction force greater than 3 lb but no more than 10 lb if possible. The lining material characteristics are $\mu = 0.35$ and $p_{\max} = 220$ psi.

Begin by turning to Figure 4 and reading α at the intersection of the solid curve and grid line $\mu = 0.34941$ (error in μ of 0.00059). We find that the maximum α that will support a retraction force is 0.33615 rad $= 19.260°$. Select this value for our first trial and calculate the radius r_o from equation (1-8) and the retraction force from equation (1-10). The results are shown next in the Mathcad format, in which the base radius of the conical contact surface and the activation and the retraction forces are written as functions of the cone angle α and the coefficient of friction μ to facilitate considering a range of values for each of these variables. From their initial values we have

$$r_o(\alpha,\mu) = \left[\frac{(T\,3\sqrt{3}\sin(\beta(\alpha)))}{2\mu\pi p_{\max}}\right]^{1/3}$$

$$F_a(\alpha,\mu) = 2\pi p_{\max}\frac{r_o(\alpha,\mu)^2}{\sqrt{3}}\left(1 - \frac{1}{\sqrt{3}}\right)\left(1 + \frac{\mu}{\tan(\beta(\alpha))}\right)$$

$$F_r(\alpha,\mu) = 2\pi p_{\max}\frac{r_o(\alpha,\mu)^2}{\sqrt{3}}\left(1-\frac{1}{\sqrt{3}}\right)\left(\frac{\mu}{\tan(\beta(\alpha))}-1\right)$$

These relations yield

$$d_o(19, 0.35) = 2.377$$

$$F_o(19, 0.35) = 960.605$$

$$F_d(19, 0.35) = 7.848$$

Guided by the steep slope of the surface shown in Figure 2 in this region, a plot of the retraction force as a function of the cone angle for friction coefficients near 0.35 is shown in Figure 8. The extreme sensitivity of this cone brake to the cone angle and especially to the value of the friction coefficient requires that the friction coefficient of the material selected be independent of temperature over the temperature range expected during the operation of this brake. Moreover, the cone angle must be held within the range from 18.924° to 19.172° to meet the retraction force requirements.

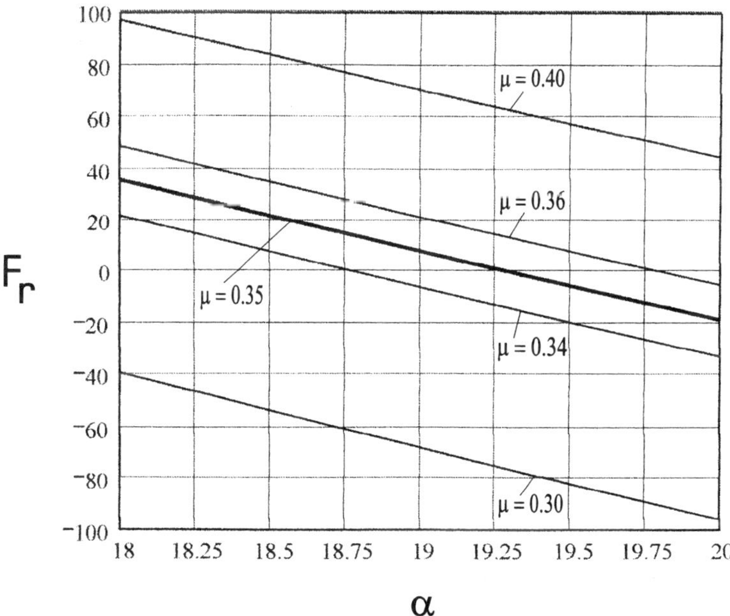

FIGURE 8 Retraction force F_r (lb) as a function of the cone angle (°) for the friction coefficients, μ, indicated.

Example 3.3

Calculate the change in torque and in the lining pressure due to wear for the clutch in Example 3.1 and the brake in Example 3.2 for lining thicknesses of 0.125 in. and lining wear of 0.05 in. Let δ in Figure 9 represent the thickness that has been worn away. Consider that lining wear may be as large as 0.5 mm for the clutch in Example 3.1 and as large as 0.02 in. for the brake in Example 3.2

Lining wear has an effect upon the torque limits for cone clutches and brakes because the reduced lining thickness due to wear affects the values of r_o and r_i by allowing the inner cone to move farther into the outer cone. Implicit in the previous analysis has been the notion that radii r_o and r_i, as illustrated in Figures 1 and 9, were the radii to the contacting surface between the inner and outer cones. Addition of a lining merely means that these radii pertain to the contact surface between one cone and the lining on the other.

In what follows we shall consider the case where the lining material is placed on the inside of the outer cone, as in Figure 9. Furthermore, let the inner cone dimensions be designed so that the inner cone will project beyond the outer cone when the lining is new and the clutch/brake is engaged. As the lining wears, the bases will approach one another and become even when the lining is so thin that it must be replaced. Thus, the entire lining surface will always be in contact with the inner cone when the clutch or brake is engaged.

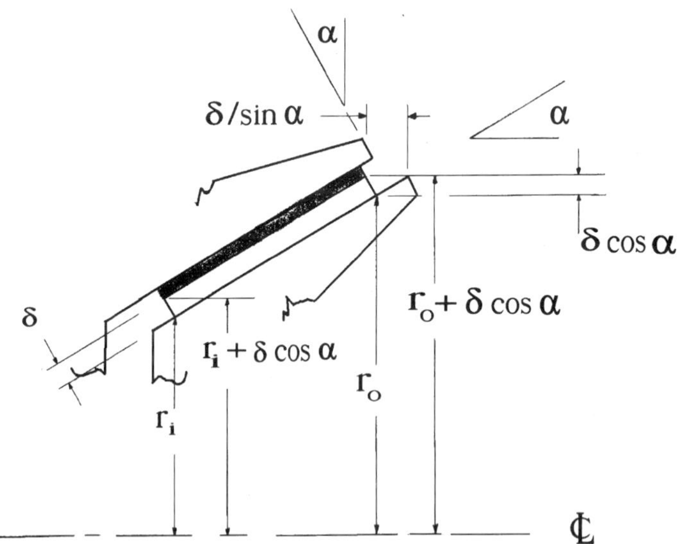

FIGURE 9 Geometry associated with lining wear in a cone clutch or brake.

When the lining has worn an amount δ, the inner cone will advance by the amount $\delta/(\sin\alpha)$, and radii r_i and r_o, measured on the conical surface that contacts the lining, will each increase by the amount $(\delta\cos\alpha)$. Consequently the smaller radius, which was initially given by $r_i = r_o/\sqrt{3}$, increases to

$$r_i = r_o/\sqrt{3} + \delta\cos\alpha \tag{3-1}$$

in terms of the lining wear δ and the cone half-angle α. The larger radius increases by the same amount, so

$$r_o \rightarrow r_o + \delta\cos\alpha \tag{3-2}$$

The maximum activation force that imposes pressure $p_{\max}$ on a new lining will impose a smaller maximum pressure on the worn lining because of its increased area. This smaller maximum pressure, denoted by p_{mw}, may be found by equating the activation force given by equation (1-9), here rewritten as

$$F_o = 2\pi p_{\max}\frac{r_o^2}{\sqrt{3}}\left(1+\frac{\mu}{\tan\alpha}\right)\left(1-\frac{1}{\sqrt{3}}\right) \tag{3-3}$$

with that obtained by replacing r_o and r_i in equation (1-9) with the values given by equations (3-1) and (1-2) to get

$$F_w = 2\pi p_{mw} r_o\left(1+\frac{\mu}{\tan\alpha}\right)\left(\frac{r_o}{\sqrt{3}}+\delta\cos\alpha\right)\left(1-\frac{1}{\sqrt{3}}\right) \tag{3-4}$$

in which $F_o(\alpha)$ represents the activation force as a function of cone angle α when the lining is new and $F_w(\alpha)$ represents an activation force of the same magnitude but one that now induces a maximum lining pressure of p_{mw}. Upon solving for p_{mw} we have

$$p_{mw} = \frac{p_{\max}}{1+\dfrac{\delta\sqrt{3}}{r_o}\cos\alpha} \tag{3-5}$$

So the torque delivered by a cone clutch or brake with a worn lining may be written as

$$T_w = \mu\pi\frac{p_{mw}}{\sin\alpha}r_1\left(r_2^2 - r_1^2\right) \tag{3-6}$$

where

$$r_1 = \frac{r_0}{\sqrt{3}} + \delta\cos\alpha \qquad r_2 = r_o + \delta\cos\alpha \tag{3-7}$$

The increase in length of the interior cone needed for it to contact the full length of the lining at it moves farther into the exterior cone as the lining wears is given by $\delta/\sin\alpha$, so the axial length of the interior cone measured from the plane containing r_i should be

$$l_o = r_o\left(1 - \frac{1}{\sqrt{3}}\right)\cot\alpha + \frac{\delta}{\sin\alpha} \tag{3-8}$$

according to the geometry displayed in Figure 9.

The notation used in Examples 3.1 and 3.2, such as $\sin(\beta(\alpha))$, is due to the Mathcad requirement that trigonometric arguments be in radian measure. This requirement may be satisfied by preceding trigonometric expressions with the relation $\beta(\alpha) = \alpha\pi/180$. The functional notation such as $p_{mw}(\alpha,\beta)$, is to allow new values for α and β to be entered directly rather than at a less convenient place elsewhere in the program.

Because torque varies as the radius cubed and the pressure change due to wear varies inversely with δ, the torque capability of cone clutches and brakes increases slightly with lining wear and the maximum lining pressure decreases slightly.

Turning first to Example 3.1, substitution into the preceding equations for the cone whose half-angle is 24° shows that the torque will increase to 10,097 N-mm after the lining thickness is reduced by 0.5 mm. The maximum lining pressure will be reduced to 3.96 MPa, and the interior cone's axial length, measured from the transverse plane containing r_i, should be 12.8 mm.

Torque increases to only 9717 N-mm for the cone having a 45° half-angle and the maximum lining pressure decreases to 4.05 MPa. That interior cone's axial length, measured as before, should be 6.9 mm.

Turning now to Example 3.2, substitution as before into equations (3-5) through (3-8) results in finding that the torque capability has increased to 489.6 in.-lb, or to 40.80 ft-lb, and the maximum lining pressure has decreased to 214.1 psi. The length of the interior cone should be increased from 1.44 in. to 1.50 in. to ensure that the interior cone contacts the full length of the lining after the lining thickness has decrease by 0.02 in.

Similar comments hold for a lining attached to the inner cone. The differences are that r_i and r_o would be measured to the surface of the lining at the outer cone and these radii would become $r_i - \delta$ and $r_o - \delta$ as the lining wears. Placing lining on the inner cone results in slightly less lining contact area as wear progresses, with a correspondingly slight increase heat per unit area to be dissipated for a given torque capacity.

IV. NOTATION

A	area (ℓ^2)
da	element of area (ℓ^2)
F_a	Activation force ($m\ell t^{-2}$)

F_r	retraction, or release, force $(m\ell t^{-2})$
f	friction function (1)
p	pressure $(m\ell^{-1}t^{-2})$
p_{max}	maximum pressure $(m\ell^{-1}t^{-2})$
r_i'	inner radius, inner cone (ℓ)
r_o'	outer radius, inner cone (ℓ)
r_i	inner radius, outer cone (ℓ)
r_o	outer radius, outer cone (ℓ)
T	torque $(m\ell^2 t^{-2})$
α	cone half-angle (1)
μ	friction coefficient between lining and cone (1)

V. FORMULA COLLECTION

Pressure distribution over lining:

$$p = p_{max}\frac{r_i}{r}$$

Torque in terms of r_o and r_i:

$$T = \frac{\mu\pi p_{max}}{\sin\alpha} r_i\left(r_o^2 - r_i^2\right)$$

Maximum torque:

$$T = \frac{2}{3\sqrt{3}}\mu\pi\frac{p_{max}}{\sin\alpha}r_o^3$$

Activation force in terms of r_o and r_i:

$$F_a = 2\pi p_{max}\left(1 + \frac{\mu}{\tan\alpha}\right)r_i(r_o - r_i)$$

Release force in terms of r_o and r_i:

$$F_r = 2\pi p_{max}\left(\frac{\mu}{\tan\alpha} - 1\right)r_i(r_o - r_i)$$

Pressure maximum on a worn lining:

$$p_{mw} = \frac{p_{max}}{1 + \dfrac{\sqrt{3}\delta}{r_o}\cos\alpha}$$

Radius associated with torque T and angle α:

$$r_o = \left(\frac{3\sqrt{3}T\sin\alpha}{2\mu\pi p_{max}}\right)^{1/3}$$

REFERENCES

1. Johnson, M. E. (1979). Testing the cone brake design, SAE Technical paper 790465. *Society of Automotive Engineers*. PA: Warrendale.
2. Spotts, M. F. (1978). *Design of Machine Elements*. 5th ed. Englewood Cliffs, NJ: Prentice-Hall.
3. Deutschmann, A. D., Michels, W. J., Wilson, C. E. (1975). *Machine Design*. New York: Macmillan.
4. Shigley, J. E., Mitchell, L. D. (1983). *Mechanical Engineering Design*. New York: McGraw-Hill.
5. Black, P. H., Adams, O. E. Jr. (1968). *Machine Design*. New York: McGraw-Hill.

7

Magnetic Particle, Hysteresis, and Eddy-Current Brakes and Clutches

All three of these brake or clutch types have no wearing parts because the torque is developed from electromagnetic reactions rather than mechanical friction. Electronic controls and a rectifier to provide direct current are required, however, for their operation. They are, nevertheless, not usually referred to as electric brakes because that term had been reserved earlier to denote friction brakes which are electromagnetically activated: those in which an electric current through a coil induces a magnetic field that engages a shoe and drum, as pictured in Chapter 4.

Because particular construction variations from manufacturer to manfacturer can have a strong effect on the performance characteristics of these brakes in terms of magnetic fringing and local variation of the electric fields, we limit our discussion of the theoretical background of these brakes to the underlying equations only. This is consistent with the design practices associated with these brakes. They are often designed in the laboratory by a combination of theory and trial and error because our present theory is not adequate to handle small geometric effects on the electric and magnetic fields between conductors that are very close to one another. Incidentally, these theoretical shortcomings are also evident in present-day design procedures for high-frequency antennas.

Since these formulas are not presented with sufficient detail for the reader to design magnetic particle, hysteresis, or eddy-current brakes, they will not be summarized at the end of the chapter.

I. THEORETICAL BACKGROUND

The basic equations that define the theory used in explaining the generation of eddy currents and of hysteresis loops are presented in the remainder of this section. A more complete discussion of the theory, beginning with Maxwell's equations, equations (1-1), along with the derivation of the subsequent relations may be found in Stratton [1] and in Lammeraner and Starl [2]. Units for the quantities involved will be given according to the MKS system (acronym for meters, kilograms, seconds).

Maxwell's equations (1-1) in vector form are generally taken as the starting point for the study of the interdependent electric and magnetic fields in free space sufficiently far from their generating electron flows. These two vector equations are

$$\nabla \times \mathbf{E} + \frac{\partial \mathbf{B}}{\partial t} = 0$$
$$\nabla \times \mathbf{H} - \frac{\partial \mathbf{D}}{\partial t} = \mathbf{J} \tag{1-1}$$

in which **i**, **j**, and **k** denote unit vectors in the positive x-, y-, and z-directions, respectively.

Here, **E** denotes the electric field intensity (volts/meter), **H** the magnetic field intensity (ampere-turns/meter), **B** the magnetic induction (webers). **J** the current density (amperes/meter^2, and t the time (seconds); the operator ∇ is defined by

$$\nabla \equiv \frac{\mathbf{i}\partial}{\partial x} + \frac{\mathbf{j}\partial}{\partial y} + \frac{\mathbf{k}\partial}{\partial z}$$

It can be shown [1] as well that the following relations hold in free space:

$$\nabla \cdot \mathbf{B} = 0 \qquad \text{and} \qquad \nabla \cdot \mathbf{D} = \rho$$
$$\mathbf{D} = \varepsilon_o \mathbf{E} \qquad \text{and} \qquad \mathbf{H} = \frac{\mathbf{B}}{\mu_o} \tag{1-2}$$

where ρ denotes the charge density (coulombs/meter^3) and constants ε_o and μ_o denote the electric and magnetic permeabilities of free space, respectively. In the MKS system, the units of ε_o are farads/meter and the units of μ_o are henries/meter.

Within an isotropic and homogeneous material, equations (1-1) are replaced by the following set of equations:

$$\nabla \times \mathbf{E} + \frac{\partial \mathbf{B}}{\partial t} = 0 \quad \nabla \times \mathbf{B} - \varepsilon_o \mu_o \frac{\partial \mathbf{E}}{\partial t} = \mu_o \left(\mathbf{J} + \frac{\partial \mathbf{P}}{\partial t} + \nabla \times \mathbf{M} \right)$$
$$\nabla \cdot \mathbf{B} = 0 \quad \nabla \cdot \mathbf{E} = \frac{1}{\varepsilon_o} (\rho - \nabla \cdot \mathbf{P}) \tag{1-3}$$

where polarization vector **P** and magnetization vector **M** are defined by

$$\mathbf{P} = \mathbf{D} - \varepsilon_o \mathbf{E} \qquad \text{and} \qquad \mathbf{B} = \mu_o (\mathbf{H} + \mathbf{M} + \mathbf{M}_o) \tag{1-4}$$

because both **P** and **M** vanish in free space. The last two of equations (1-2) are replaced by

$$\mathbf{D} = \varepsilon \mathbf{E} \qquad \text{and} \qquad \mathbf{H} = \frac{1 \mathbf{B}}{\mu} \tag{1-5}$$

in which ε and μ are called the inductive capacities of the medium.

After adding Ohm's law, which is that

$$\mathbf{I} = \frac{\mathbf{E}}{\Omega} \tag{1-6}$$

in a medium having resistance Ω(ohms), we have all of the relations that together explain the generation of an eddy current **I** and a hysteresis loop for **H** in a homogeneous, isotropic medium [2].

The electric current flowing across a surface in the material is given by

$$I = \int_S \mathbf{J} \cdot \bar{\boldsymbol{n}} \, ds \tag{1-7}$$

In our discussion of electric brakes that induce a magnetic field, which is the primary source of the braking torque, we shall be concerned only with equation (1-4) and the equation for the work done by cyclic changes in the magnetic induction within a material volume V, which is

$$W = -\int_V dv \oint \mathbf{B} \cdot d\mathbf{H} \tag{1-8}$$

Magnetic induction **B** in the material is induced by an external **H** field, which in turn is usually generated by a current **I** in a coil of wire according to

$$\mathbf{H} = NI \tag{1-9}$$

where N is the number of turns of wire in the coil.

Calculation of work W according to equation (1-8) involves substituting for $\mathbf{B}$ from equations (1-4) to get

$$W = -\int_V dv \left(\frac{1}{\mu} \oint \mathbf{B} \cdot d\mathbf{B} \right) \tag{1-10}$$

which is nonlinear because of the interdependence of $\mathbf{M}$, μ, and $\mathbf{B}$. Depending on the material, the relation between $\mathbf{B}$ and $\mathbf{H}$ may appear as in Figure 1(a) or (b). It is the nature of these curves that determines the torque-control current

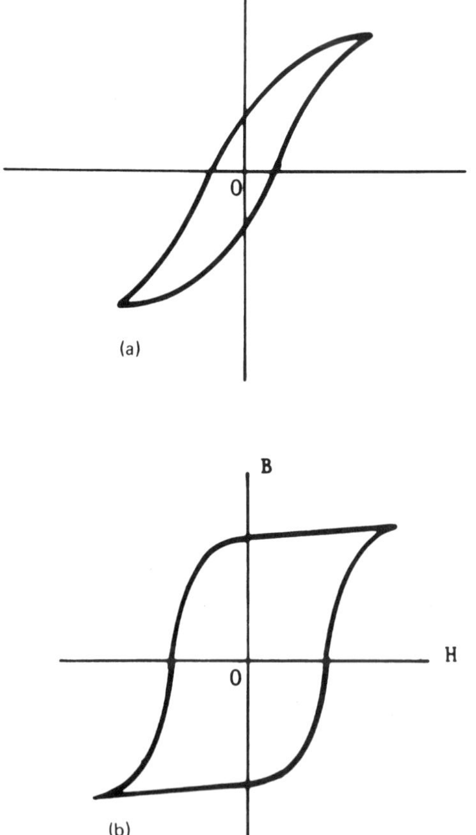

FIGURE 1 Representative hysteresis loops for (a) low-loss material and (b) high-loss material.

curve, represented by Figure 2, for a hysteresis brake. Techniques for generating the cyclic behavior of B and using it for braking are discussed in the sections devoted to individual brake designs.

Eddy currents are generated within a conducing material whenever the magnetic field changes, as implied by the relation for **J** in equations (1-3). For design purposes, the power P_e lost due to cyclic eddy-current variations in a flat plate may be estimated from

$$P_e = \frac{\pi \delta f \mathbf{B}_{\max}}{(Ck)} \tag{1-11}$$

where δ represents the plate thickness, f is the frequency of the cyclic variation, k is the specific resistance of the material, and C is a dimensional constant.

Although these relations indicate that hysteresis and eddy currents occur together in eddy-current and hysteresis brakes, one or the other may be made to dominate by selecting a material with the proper combination of μ and k.

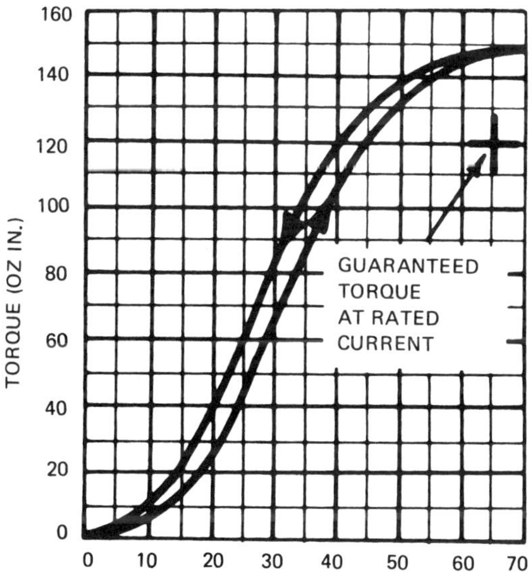

FIGURE 2 Typical torque control current curves for a hysteresis brake. Arrows indicate increasing or decreasing coil current. (Courtesy of Magnetrol, Inc., Buffalo, NY.)

II. MAGNETIC PARTICLE BRAKES AND CLUTCHES

These brakes are available in a range of sizes that include the 100-lb-ft model shown in Figure 3 and the 8-lb-ft model shown in Figure 4. Since these configurations are equally suited for clutches, they may be combined to form clutch-brake combinations, as in Figure 5. When used as a clutch, the unit has two moving parts; when used as a brake it has only one.

When used as a clutch, the configuration is as represented by the schematic in Figure 6(a). The input shaft is attached to a cylindrical drum, termed the *outer member*, or OM, which encases a smaller, inner cylinder, termed the *inner member*, or IM, which is attached to the output shaft. A dry, finely divided, proprietary magnetic material is contained in the region between the

FIGURE 3 Magnetic particle brake with a 100-lb-ft capacity. (Courtesy of Sperry Electro Components, Durham, NC.)

FIGURE 4 Hysteresis brake with a 8-lb-ft capacity. (Courtesy of Magnetic Power Systems, Inc., Fenton, MO.)

OM and the IM. The brake configuration differs from the clutch only in that the IM is rigidly attached to the brake frame.

An electromagnetic coil outside the OM and concentric with it is used to activate the brake or clutch. When the coil in energized by passing current through it a magnetic field is established which causes the particles to bridge the gap between the IM and the OM and form links between the two, as represented in Figure 6(b). These links are along the magnetic lines of force, which are made nearly perpendicular to the OM by the configuration of the OM and the coil housing, as shown in Figures 6 and 7.

Both the shear and tensile stresses in these links resist relative motion between the IM and the OM and so transmit torque for the brake/clutch. These shear and tensile stresses developed are dependent on the coil current

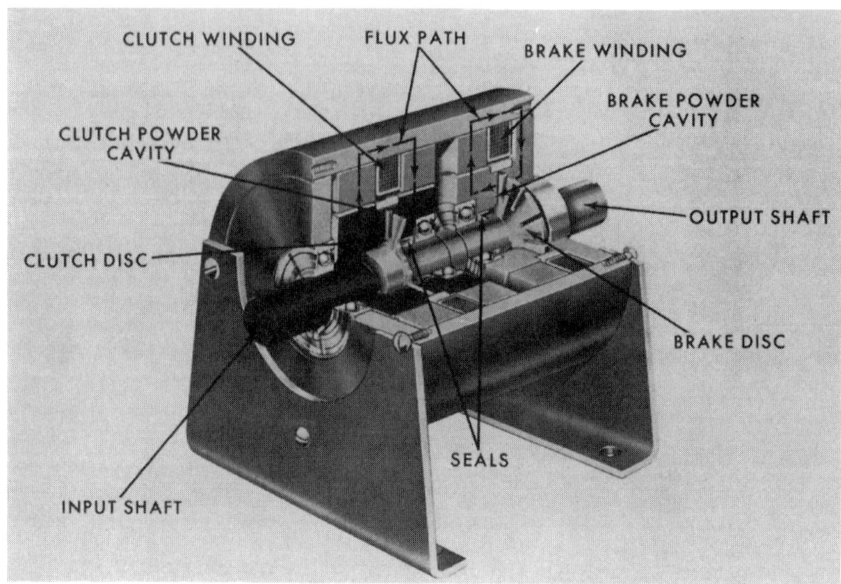

FIGURE 5 Magnetic particle clutch and brake combination. (Courtesy Simplatrol Dana Industrial, Webster, MA.)

and are independent of rotational speed. Typically, the torque varies with the coil current, as illustrated in Figure 8, while the torque remains constant regardless of the rotational speed of the OM, as shown in Figure 9.

III. HYSTERESIS BRAKES AND CLUTCHES

Construction of a hysteresis clutch, shown in Figure 10, differs from that of a hysteresis brake only in that the outer member, termed the OM, is prevented from rotating. This schematic implies that in the brake configuration the coil winding occupies a greater portion of the base of the cup-shaped OM, as indicated in the schematic in Figure 11.

In either construction the cup-shaped OM is fitted with a central post that fits within the smaller cup-shaped inner member, termed the IM. Magnetic field variation is accomplished by reticulating the OM wells and post, as indicated in Figure 12(a) to produce an alternating set of north and south magnetic poles when the OM is magnetized by current flowing through the coil in its base. At any instant the magnetic field from these poles induces a set of opposite poles in the walls of the IM. Rotation of the IM is, therefore,

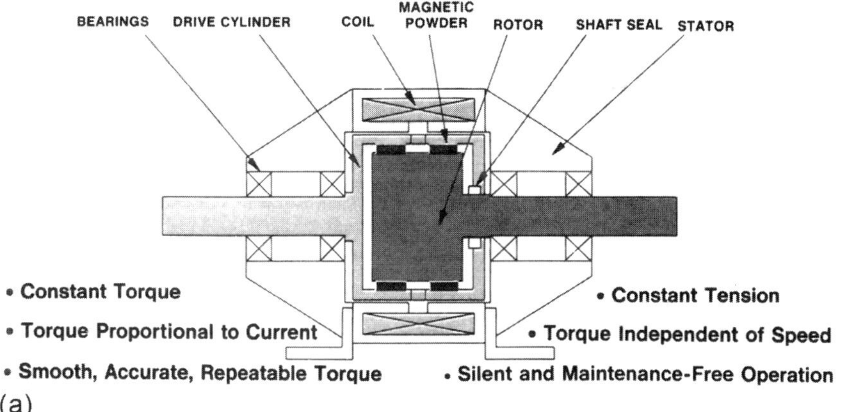

FIGURE 6 Schematic of a magnetic brake/clutch to display its operation. (a) Magnetic particle clutch. (b) Input shaft "R" and output shaft "N" are positioned within the electromagnetic coil. Magnetic particles lay loosely between input and output components. No current is applied to the coil. No torque is transmitted. (c) Here maximum current energizes the coil. The clutch now operates at 100% of clutch rating. Full transmission of torque occurs. Depending on coil current, any level between 0 and 100% torque transmission is possible. (Courtesy Magnetic Power Systems, Inc., Fenton, MO.)

opposed by the magnetic force between the induced poles in the IM and those in the OM because it disturbs this arrangement by forcing opposite poles apart and similar poles together. As the rotation continues due to external shaft torque, the magnetic field from the OM changes the magnetization of each point in the magnetized region of the IM so that the magnetic induction B at any point on the walls of the IM traverses the hysteresis loop as that point moves under the north to south to north pole of the OM's outer shell.

By forming the IM from a magnetically hard material (one that resists a change in magnetization as indicated by a small value of μ) which also has a large area enclosed by the hysteresis loop, the manufacturer can assure relatively large losses in the brake. The energy extracted from the input shaft in this manner heats the IM, which must be cooled to maintain the performance of the brake.

Figure 13 clearly shows that the braking torque is maximum for low rotational speed, including 0 rpm, and that as the speed increases a critical point is reached which corresponds to the maximum power that can be dissipated by the brake, based on its internal construction and the ambient temperature.

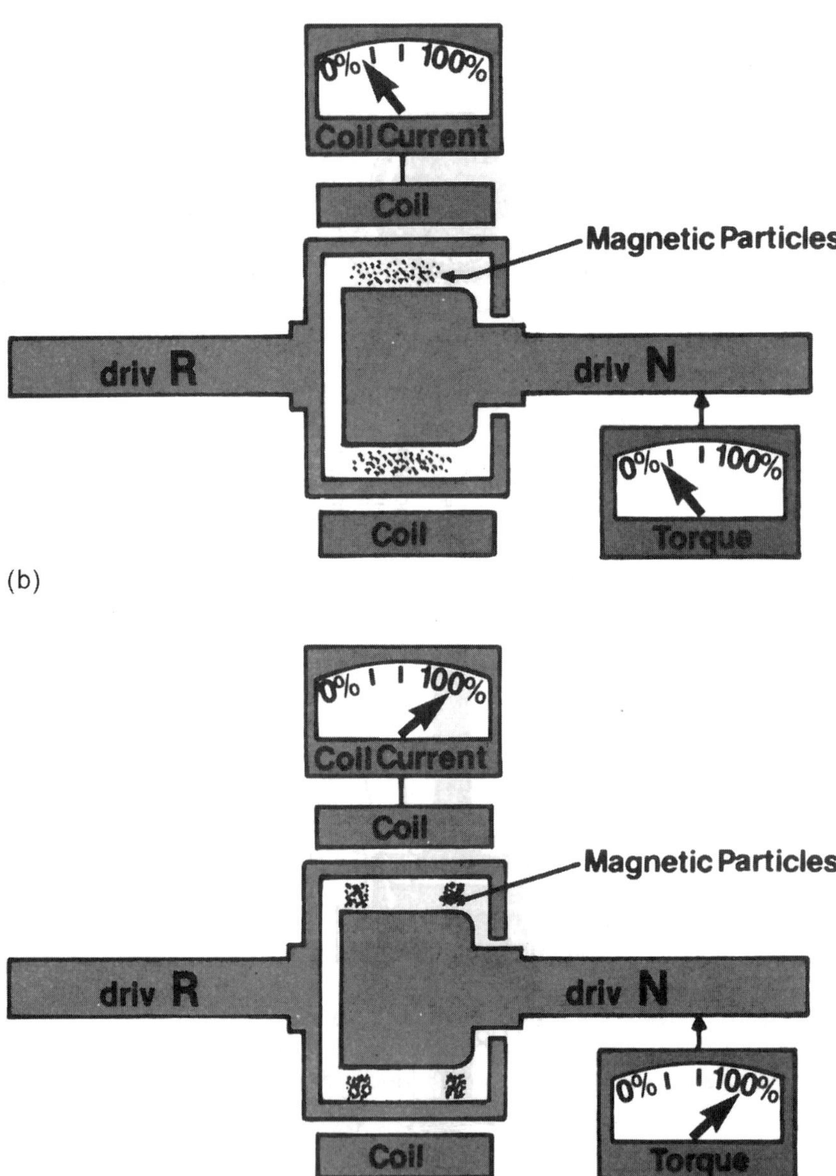

FIGURE 6 Continued.

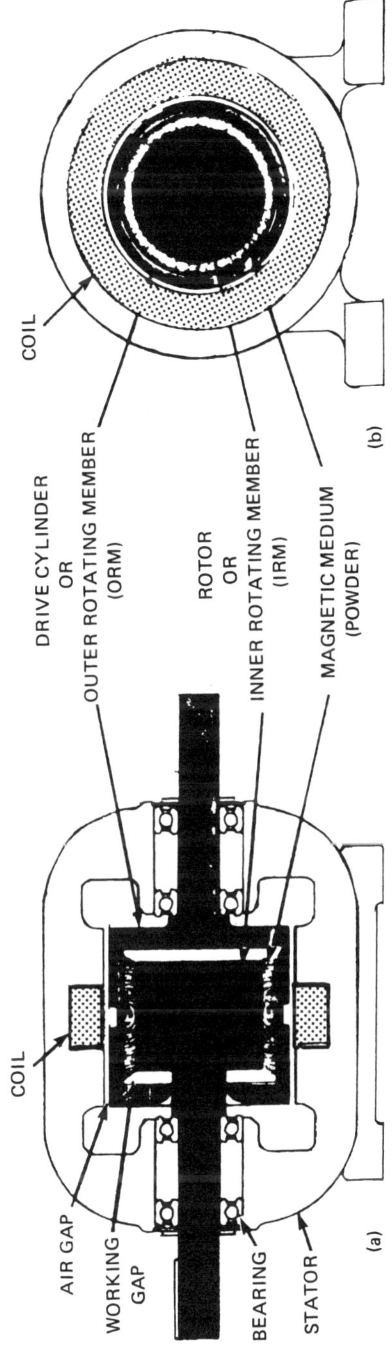

FIGURE 7 Magnetic lines of force linking the outer member (OM) and the inner member (IM). (Courtesy of Sperry Electro Components, Durham, NC.)

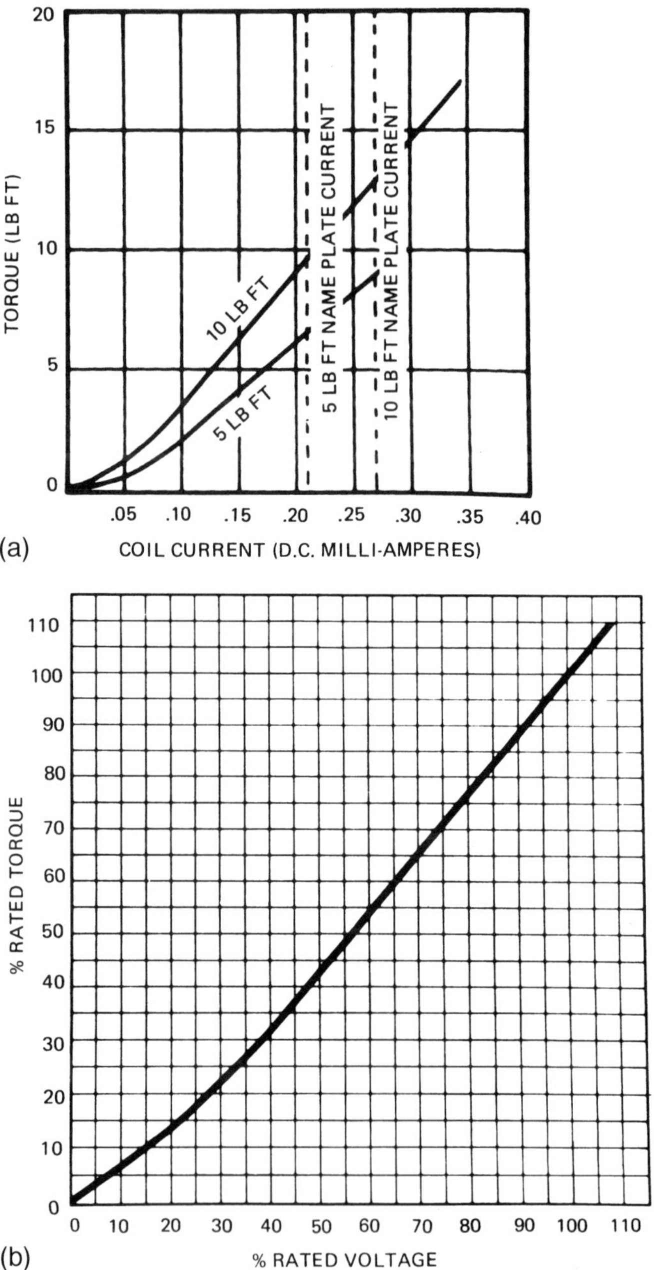

FIGURE 8 (a) Torque current curve for a particular brake; (b) torque voltage curve for a series of magnetic particle brakes. (Courtesy of Sperry Electro Components, Durham, NC, and Simplatrol Dana Industrial, Webster, MA.)

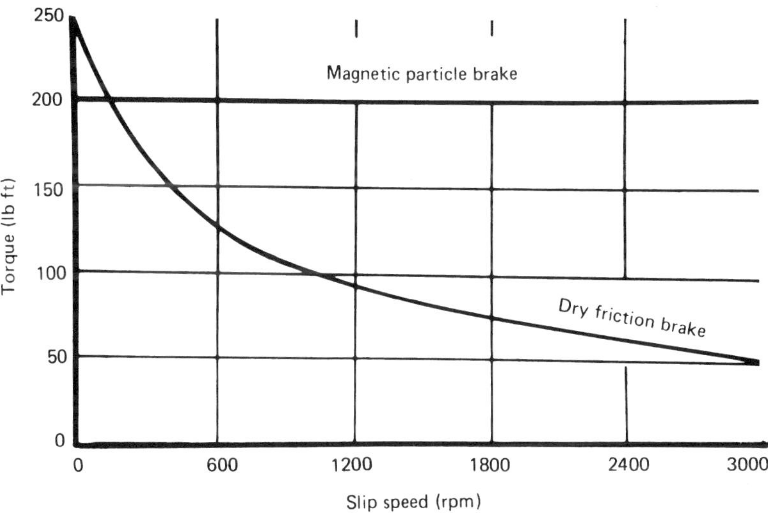

Figure 9 Torque-slip speed curves for dry friction and magnetic particle brakes (also clutches).

Beyond this point the torque decreases rapidly, as shown in the slip torque versus speed curve in Figure 13(a). Comparison with Figure 13(b) correctly implies that the shape of the decreasing-torque portion of the curve to the right of the critical point reflects both the change in the hysteresis loop with increasing temperature and the heat transfer characteristics of the cooling system (i.e., whether air or liquid and the temperature and velocity of the cooling medium). When these conditions are fixed the shape of the curve remains qualitatively invariant. Thus, as the brake torque increases from one size of brake to another, that portion of the curve to the left of the critical point decreases unless improved cooling is used to move the concave portion of the curve upward and to the right, thus moving the critical point to the right.

The magnitude of that portion of the curve which is independent of rotational speed to the left of the critical point in Figure 14a is, of course, also determined by the torque versus control current curve shown in Figure 14b. The difference between the torque obtained from increasing and decreasing control current is shown in Figure 2.

Use of the term *slip torque*, incidentally, is to emphasize that the torque acts between two mechanical parts which may be moving relative to one another because these brakes may be used as tension control devices as well as a means of stopping the rotation entirely.

Figure 10 Hysteresis clutch with cutout section showing the OM (which also forms the outer shell), the IM, and the electromagnetic coil. (Courtesy of Magnetrol Inc., Buffalo, NY.)

IV. EDDY-CURRENT BRAKES AND CLUTCHES

Construction of eddy-current brakes is physically similar to that of hysteresis brakes. The essential difference is that the IM is now made of a magnetically soft material (one having large μ, a small magnetization vector M, and therefore, easy magnetization) which also has a low specific resistance. Although there are small hysteresis losses in eddy-current clutches and brakes, just as there are small eddy-current losses in hysteresis clutches and brakes, the primary source of power loss in these brakes is in the generation of eddy

Hysteresis Brakes

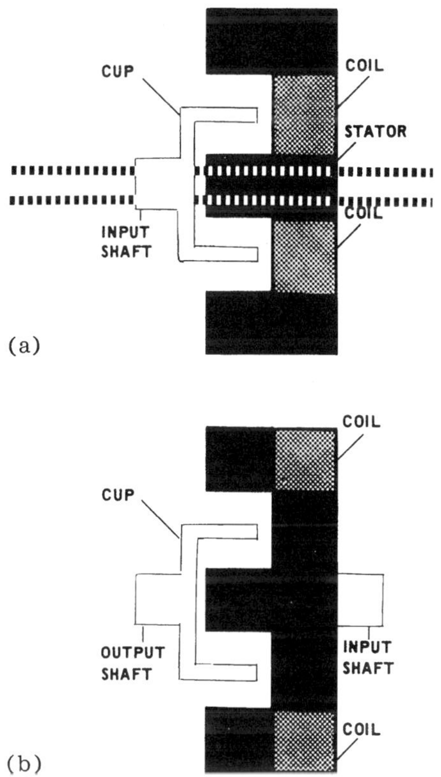

FIGURE 11 Schematic of (a) a hysteresis brake and (b) a hysteresis clutch. The E-shaped cross section represents the cross section of the OM and its inner post (the outer shell in Figure 10). (Courtesy of Magnetrol, Inc., Buffalo, NY.)

currents in the IM. These eddy currents, which are often represented as small current loops, as illustrated in Figure 15, are generated in a direction to oppose the change in the magnetic field whenever there is a change in the magnetic field crossing the IM. Pole geometry for an eddy-current brake/ clutch is shown in Figure 12 where the outer ring a is the cup, or OM, and the inner cylinder a is the central post (Figure 11), which completes the magnetic circuit, and the intermediate ring b is the IM, which rotates in the magnetic field between the cup and the inner post. The rate of change of the magnetic field due to relative rotation between the IM and the OM is

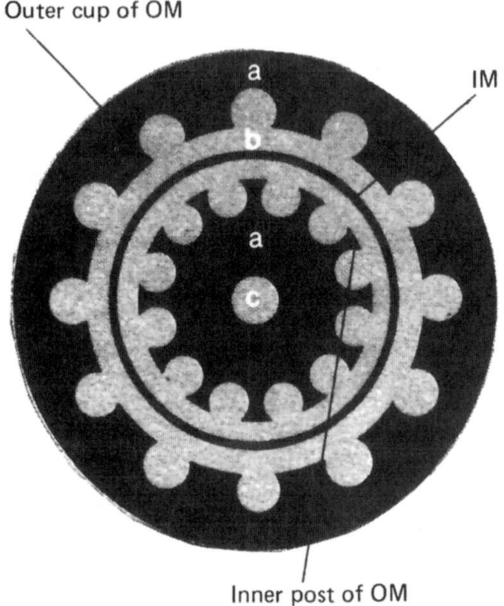

(a)

(b)

Figure 12 Schematic of a cross section of a hysteresis brake in a plane perpendicular to the shaft axis-showing reticulation of the OM cup walls and inner post. (Courtesy of Magnetrol, Inc., Buffalo, NY.)

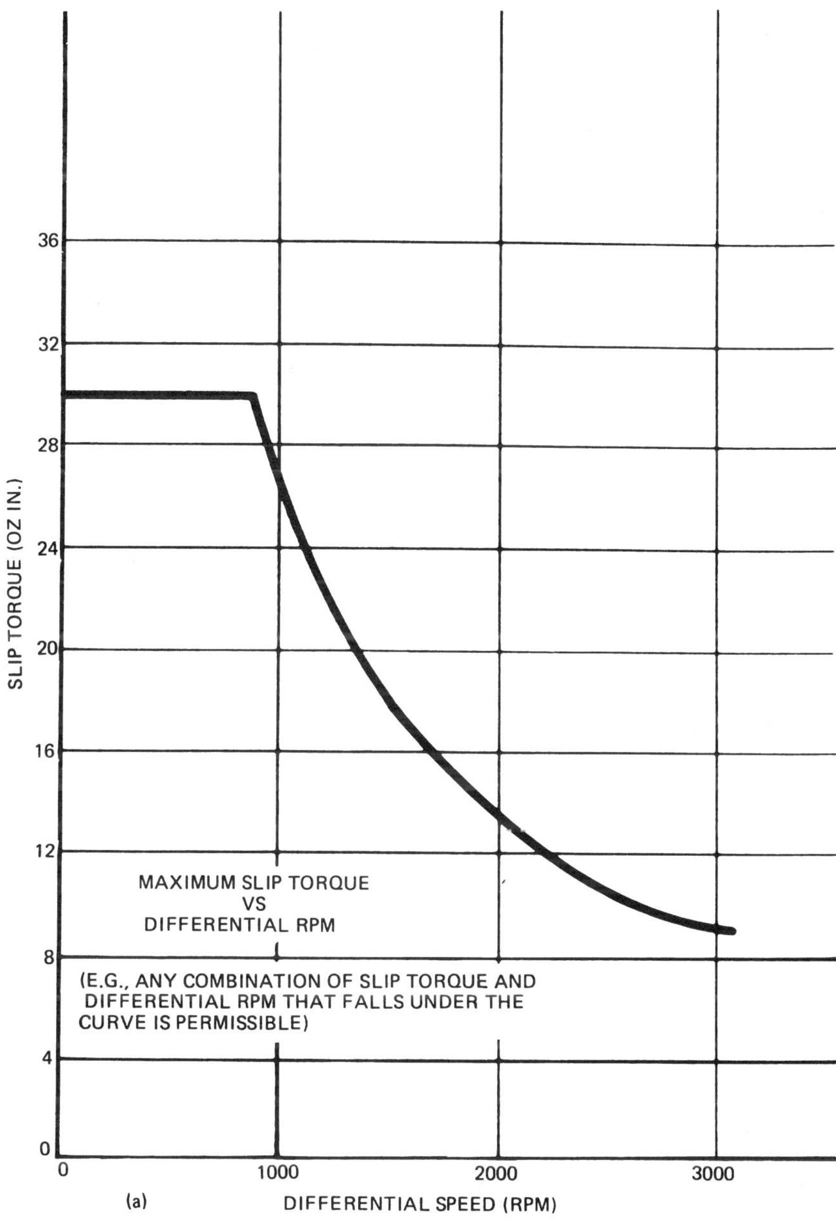

FIGURE 13 Torque (also termed slip torque) differential speed (or slip speed) for hysteresis brakes of different capacity. The dashed line shows the effect of increased cooling. (Courtesy of General Electro-Mechanical Corp., Buffalo, NY.)

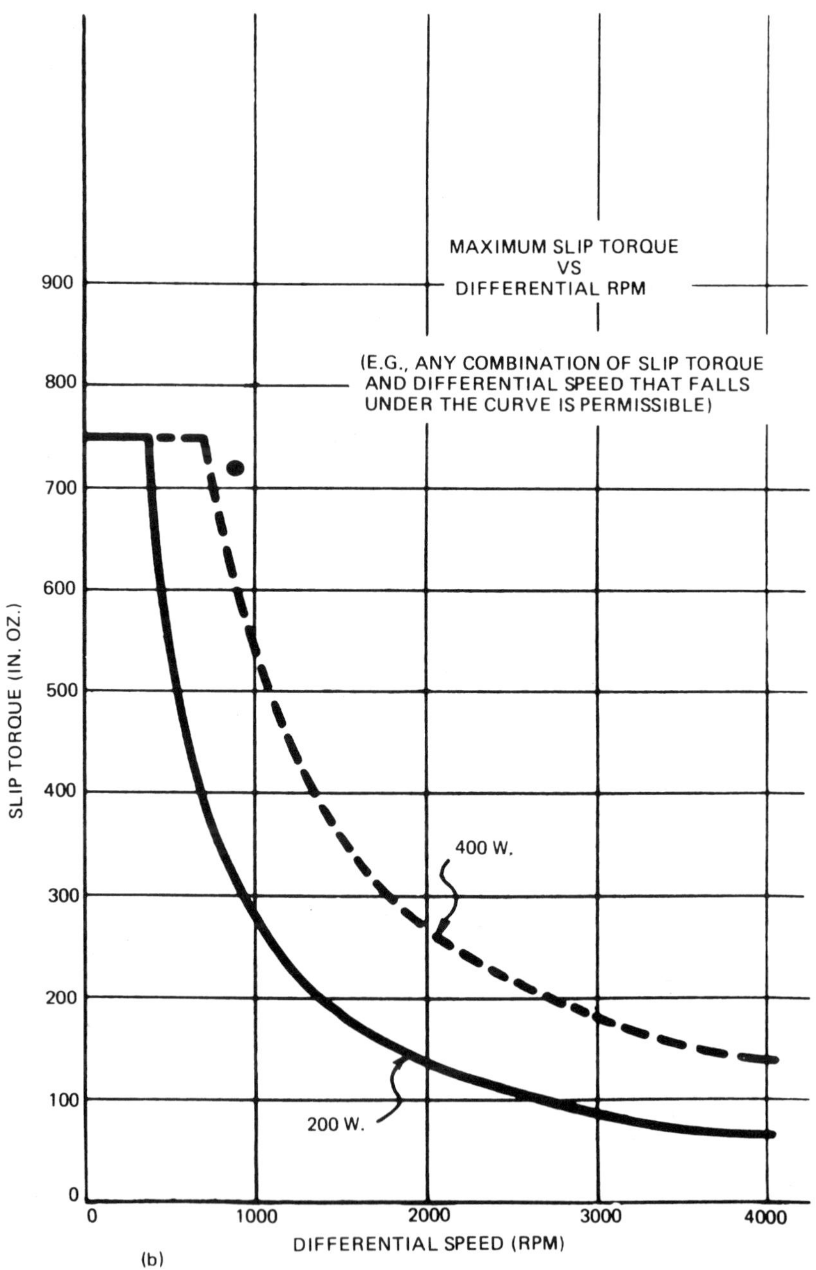

Figure 13 Continued.

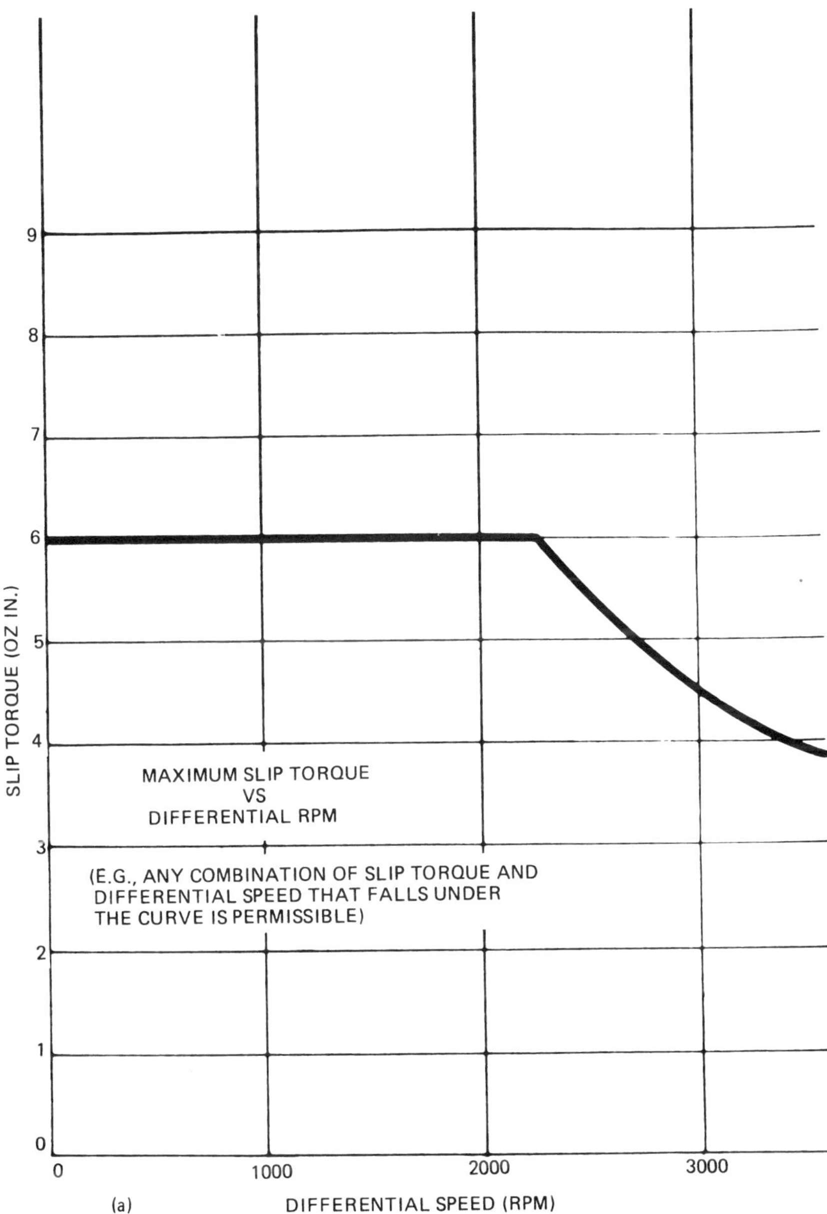

FIGURE 14 Torque versus differential speed (a) and torque versus control current (b) for a particular hysteresis brake. Torque differential speed curve shown corresponds to approximately 30 mA of control current through a 1900-Ω coil. (Courtesy of General Electro Mechanical Corp., Buffalo, NY.)

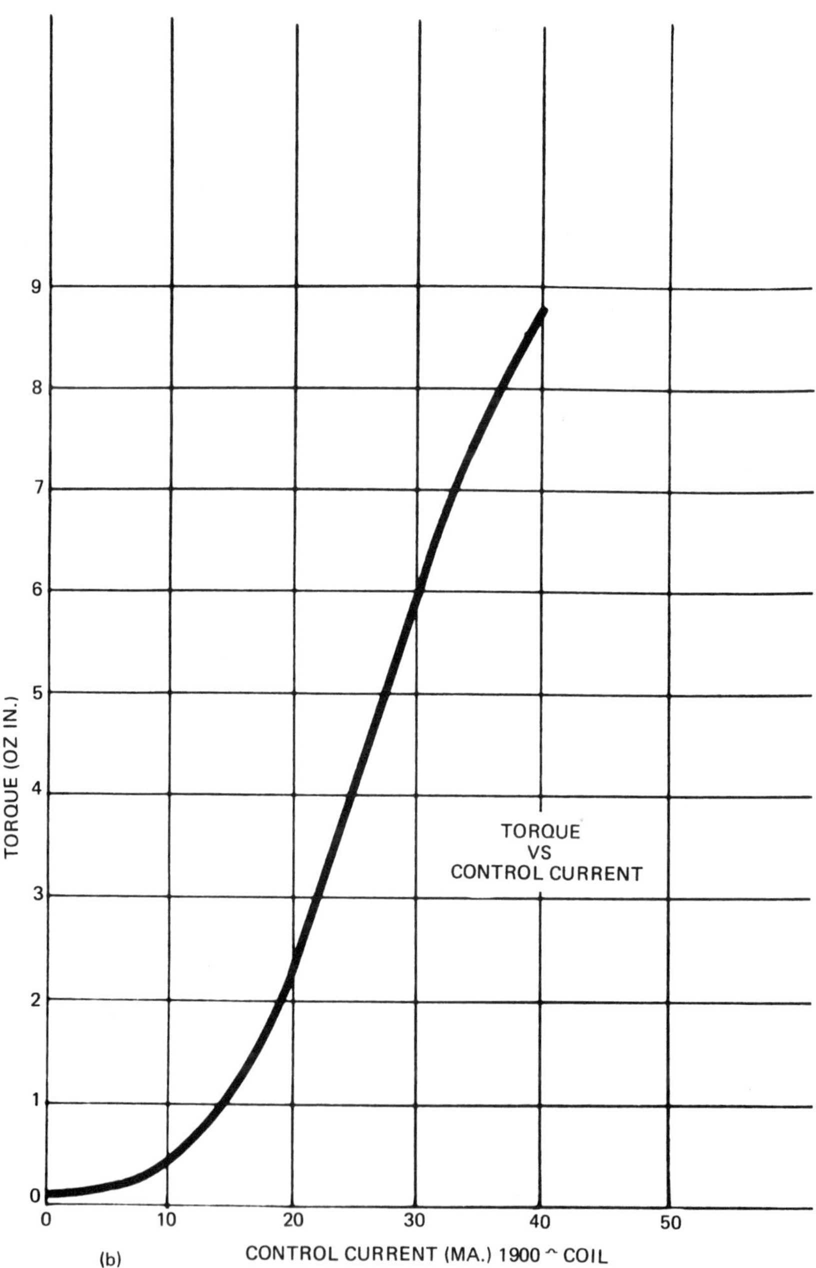

Figure 14 Continued.

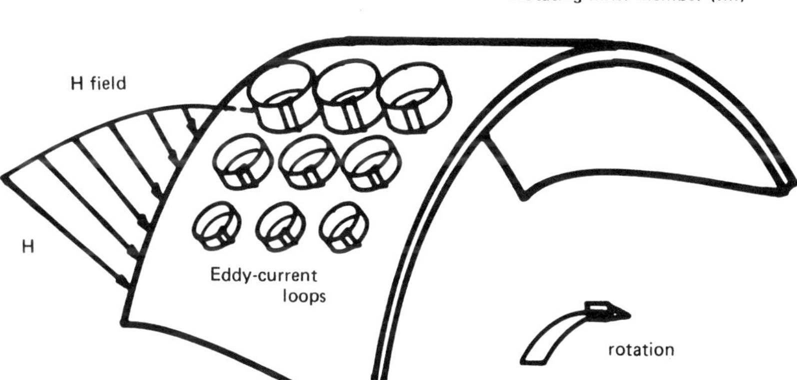

FIGURE 15 Eddy-current loops induced in the IM by the changing **H** field in an eddy-current brake.

determined by the number of poles in the OM and the rotational speed of the IM. From the frequency term f in equation (1-11) we see that the power dissipated is, therefore, proportional to the number of poles and the rotational speed. Although the braking torque is zero at 0 rpm, it does not increase linearly with the rotational speed for speeds at the upper end of the operating range because of effects not explicitly shown in equation (1-11), as demonstrated by the torque versus rotational speed curves shown in Figure 16. Notice that the torque maxima in these curves are directly related to the percent excitation, so that they provide current versus torque data as well.

Figure 17 illustrates a model of air-cooled eddy-current brakes produced in sizes having heat dissipation capacities from 5 to 100 hp and braking torque capacity from 60 to about 1800 lb-ft. Larger eddy-current brakes with dissipation capacities up to 4000 hp are liquid cooled, while smaller brakes, with capacities of several ounce-inches, require no cooling other than local convection air currents.

These brakes are used in applications where tension is to be maintained either by preventing a shaft from overspeeding due to external torque or by controlling tension between two sets of roller by having one set rotate opposite the direction of applied torque, thus stretching the material between these two sets of rollers. Small torque models are used for controlling tension in filiment manufacture and in magnetic tape drives, while the larger models find applications in laying cables, winding sheet metal rolls, and in conveyor controls.

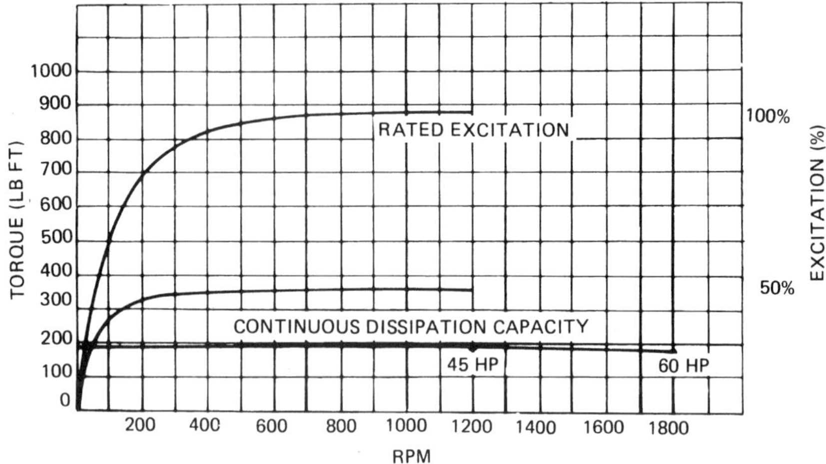

FIGURE 16 Combined torque rotational speed curve and torque excitation curves for eddy-current brakes. (Courtesy of Eaton Power Transmission Systems, Industrial Drives Operations, Kenosha, WI.)

Simplatrol Dana produces a small-capacity (under 8 oz-in.) unit designed to have an adjustable torque range and to use the construction similarities between eddy-current and hysteresis brakes/clutches. In it the IM and OM are replaced by a permanent-magnet disk and either an eddy-current or hysteresis disk. Torque capacity may be adjusted by means of the flux gate placed between them, as shown in the brake version in Figure 18. Manual rotation of the flux gate relative to the magnetic disk determines the strength of the magnetic field that acts on either the hysteresis or eddy-current disk attached to the front, unthreaded, shaft on the assembly shown. The rear disk, the magnetic plate, and the flux gate rotate together in the case of a clutch, or remain stationary in the case of a brake. Clutch and brake units differ only in that the rear shaft of the brake is threaded, as shown in the figure.

The torque versus speed curves for an eddy-current brake in Figure 16 may also be used to deduce the characteristics of an eddy-current clutch; namely, that an eddy-current clutch can provide a controlled soft start between a driver and a driven unit by controlling the excitation current as a function of the speed difference. Likewise, an eddy-current clutch may also be considered when a driven machine may experience speed changes of several hundred rpm that should not impose a large torque change on driving machine. Since the torque available to accelerate a driven machine back up to speed will be small, an eddy-current clutch will be suitable only if prolonged

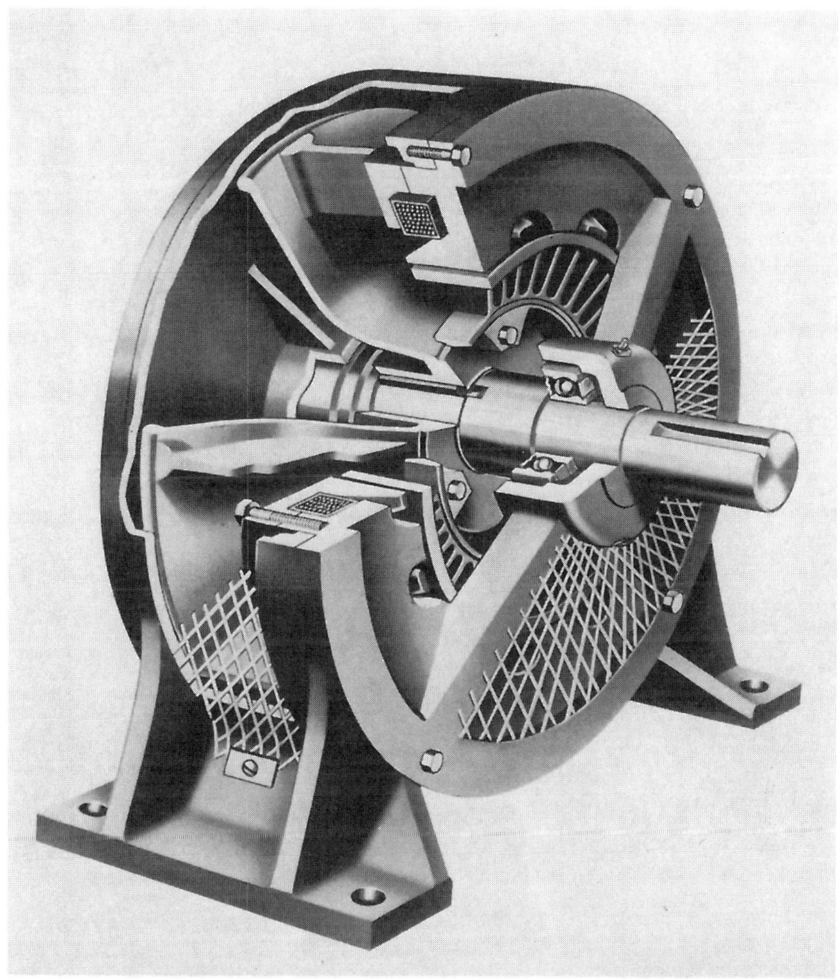

FIGURE 17 Air-cooled eddy-current brakes with torque capacities from 5 to 1740 lb-ft and power dissipation from 0.75 to 100 hp. (Courtesy of Eaton Power Transmission Systems, Industrial Drives Operations, Kenosha, WI.)

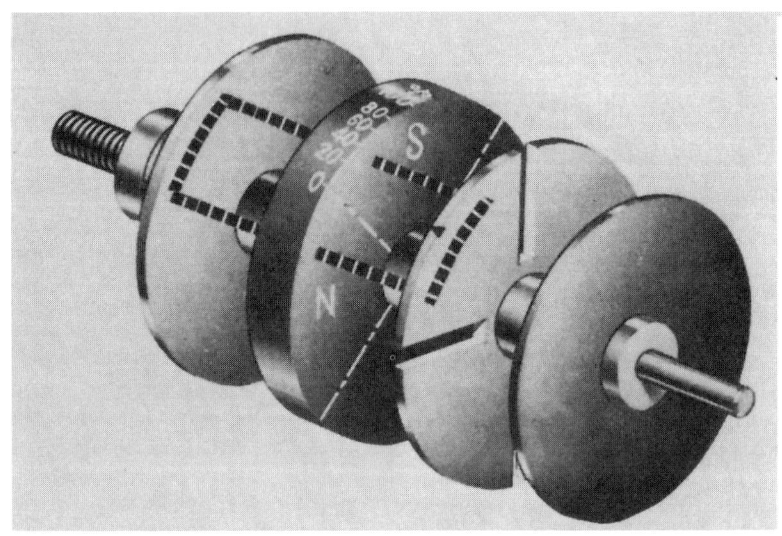

FIGURE 18 Combination hysteresis/eddy-current brake/clutch. (Courtesy of Simpatrol Dana Industrial, Webster, MA.)

periods of speed deviation are acceptable. Eddy-current clutches and brakes may, for example, be used in tape recorders to provide both a soft start to the tape drive and a gentle, programmed, control of the tape speed and to prevent over-speeding of the supply reel.

V. NOTATION

$\mathbf{B}$	magnetic induction
$\mathbf{D}$	electric displacement
$\mathbf{E}$	electric field intensity
f	frequency
$\mathbf{H}$	magnetic field intensity
I	current
$\mathbf{J}$	current density
k	specific resistance of a material
$\mathbf{M}$	magnetization vector
N	number of turns
$\boldsymbol{n}$	unit vector normal to surface S
$\mathbf{P}$	polarization vector
P_e	power loss due to eddy currents
S	surface
t	time
V	volume
W	work
x,y,z	spatial coordinates
δ	plate thickness
ε	electric permeability
μ	magnetic permeability
ρ	charge density

REFERENCES

1. Stratton, J. M. (1941). *Electromagnetic Theory*. New York: McGraw-Hill.
2. Lammeraner, J., Stafl, M. (1996). English translation. In: Toombs, G. A., ed. *Eddy Currents*. London: Iliffe Books, Ltd.

8

Acceleration Time and Heat Dissipation Calculations

Brake and clutch design or selection from a manufacturer's catalog both require that we design or select a brake or clutch which has the torque capability necessary to stop or start either a machine or a mechanical component in a specified amount of time and also has the ability to dissipate the heat generated.

Torque capability depends, as we have found, on the particular brake or clutch design. The heat to be dissipated does not; it depends only on the machinery being stopped and is, therefore, independent of the brake or clutch used.

In this chapter we are concerned with the related problems of estimating stop or startup times and the amount of heat generated. Both problems may be analyzed in terms of the energy supplied by the driving unit, the energy transmitted to the driven unit, and the energy dissipated as heat by either the brake or clutch. Although the energy considerations are independent of the particular brake/clutch design involved, the resulting formulas may be used to compare various brake/clutch design suitability for any mechanical system.

Calculation of heat dissipation by a mechanical system involving a clutch or brake may be divided into two parts: the mechanical energy converted to heat in the clutch or brake, and the rate of transfer of this heat to the surroundings. In the remainder of this chapter we shall be concerned only with the first of these two problems. Those readers who may be concerned with the second problem as well are referred to existing books devoted to the calculation of heat transfer by conduction, convection, and radiation, along with the specific heats for common cooling fluids, including air, the methods for determining the coefficients involved, and the numerical techniques required for solving practical heat transfer problems.

I. ENERGY DISSIPATED IN BRAKING

The heat dissipated in any mechanical system is equal to the energy withdrawn from the system as it is either stopped or slowed by a brake or as it is accelerated by a clutch, plus any work done on the system during the time a brake or a clutch is being applied. This equality is the foundation of the formulas to be developed and demonstrated.

Following industry practice in the United States we shall measure heat in terms of its mechanical equivalent pound feet (foot-pounds) in old english (OE) units or in joules (newton-meters) in SI units, rather than in terms of calories or Btu. This may be converted to the temperature rise in the brake components by converting to kilocalories or Btu using the joule equivalent, which is that 1.0 kilocalorie = 4186 N-m and that 1.0 Btu = 778.26 foot-pounds and using the relation that

$$\left(\frac{\partial Q}{\partial \Theta}\right)_P = C_P$$

or

$$\Theta_2 - \Theta_1 = \frac{1}{C_P}\int_{Q_1}^{Q_2} dQ$$

where Θ represents the temperature, Θ_1 and Θ_2 are the temperatures before and after the amount of heat Q is added to the system, and C_p denotes the specific heat at constant pressure for the material involved.

The mechanical equivalent of the heat, Q_m to be dissipated is given by

$$Q_m = \mathrm{KE}_2 - \mathrm{KE}_1 + W_a \tag{1-1}$$

where KE_1 and KE_2 represent the kinetic energy of the system at the beginning and at the end of the interval during which either a brake or a clutch is applied and W_a is the work added to the system during that interval. Heat Q_m is also equal to the integral of the work done on the brakes during the braking interval, so

$$Q_m = \int_{t_1}^{t_2} \frac{dW_a}{dt}\,dt \tag{1-2}$$

This last relation, in somewhat modified form, may be used to estimate the relation between the torque to be exerted by a brake or clutch, the time the brake or clutch must act, and the heat dissipated during the time the brake or clutch acts.

Before we can equate the energy in a moving mechanical system to the work done by a brake or a clutch in changing the rotational speed of a mechanical system, we must have expressions for total energy in the system and for the work done by a brake or clutch. These matters are considered in the next two sections in that order.

II. MECHANICAL ENERGY OF REPRESENTATIVE SYSTEMS

To apply equation (1-1) we need to obtain expressions for the kinetic energy for three typical mechanical systems: geared systems; translating and rotating systems, exemplified by vehicles and conveyor belts; and systems involving a change in potential energy, as exemplified by cranes and hoists. All formulas will initially be given in terms of the physical quantities involved and will subsequently be rewritten in terms of commonly used OE and SI units in the Formula Collection at the end of the chapter.

A. Geared Systems

Whenever a geared system similar to that illustrated in Figure 1(a) involving a single gear train is to be stopped, or slowed, by a brake acting on shaft 1 rotating at speed ω_1, the kinetic energy to be dissipated in reducing the rotational speed from ω_{1a} to ω_{1b} may be expressed in terms of the gear ratios n_{21} and the moments of inertia of each rotating member as

$$KE = \frac{1}{2}(I_1 + I_2 n_{21}^2)(\omega_{1_a}^2 - \omega_{1_b}^2) \tag{2-1}$$

where I_1 is the total moment of inertia of all masses rotating with shaft 1, that is, the sum of the moments of inertia of the brake drum or disk, shaft 1 itself, and gear 1. Similarly, I_2 represents the total moment of inertia of gear 2, shaft 2, and whatever mass rotates with shaft 2. The speed ratio n_{21} is defined by

$$n_{21} = \frac{\omega_2}{\omega_1} \tag{2-2}$$

where ω_1 and ω_2 denote the rational speeds of shafts 1 and 2, respectively, at any instant. In a more complicated case, as illustrated in Figure 1(b), the kinetic energy to be dissipated in slowing or stopping the rotation is given by

$$\text{KE} = \frac{1}{2}(\omega_{1_a}^2 - \omega_{1_b}^2)(I_1 + I_2 n_{21}^2 + I_3 n_{31}^2 + I_4 n_{41}^2) \tag{2-3}$$

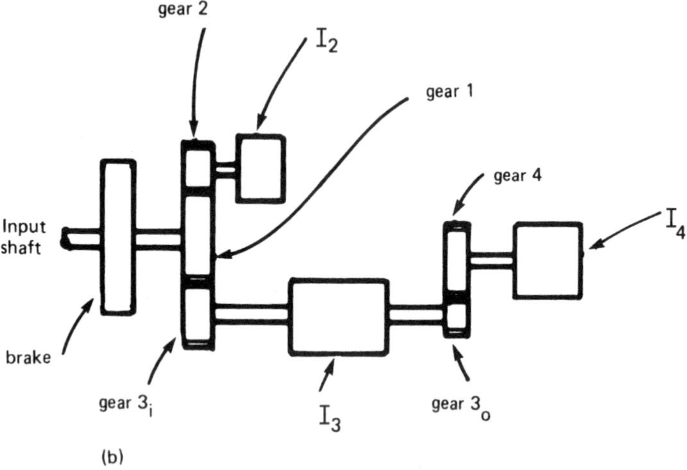

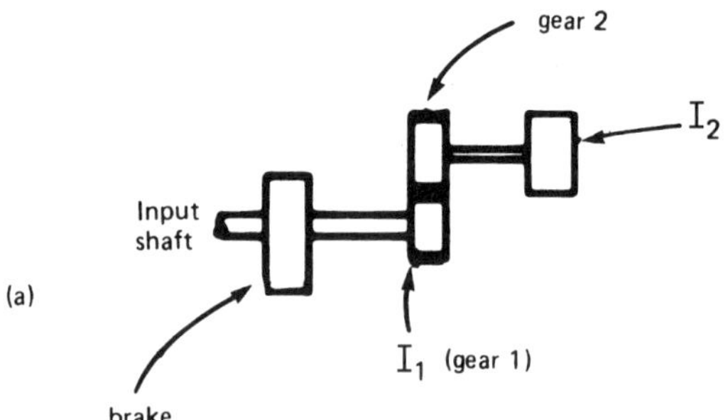

FIGURE 1 Brake and gear train schematic. Moments of inertia I_i include moments of inertia of all masses rotating with shaft i (i.e., gears and shaft itself).

where n_{41} may be written in terms of n_{43} and n_{31} as

$$n_{41} = n_{43}n_{31} \tag{2-3}$$

In summary, the kinetic energy to be dissipated from a geared system may be written as

$$\text{KE} = \frac{1}{2}(\omega_{1_a}^2 - \omega_{1_b}^2)(I_1 + \sum_{i=2}^{k} I_i n_{i1}^2) \tag{2-4}$$

for moments of inertia I_i rotating at speeds ratios n_{i1} relative to shaft 1, where the brake is located.

For simplicity the moment of inertia of most rotating mechanical components is often given in terms of the radius of gyration r_g, which is defined by

$$I = mr_g^2 \tag{2-5}$$

where $m = W/g$ in terms of the weight of the component and the acceleration due to gravity, usually taken as 32.2 ft/sec^2 or 9.81 m/sec^2.

Returning to equation (2-1), we note that if n_{21} is less than 1, i.e., if ω_2 is less than ω_1, the contribution of I_2 to the kinetic energy is reduced by the square of n_{21}. Guided by this observation, we may conclude that it is generally advantageous to place the brake on the fastest of all of the shafts involved so that the torque requirement for the brake is reduced.

B. Combined Translation and Rotation

When translation is present, as in the case of a moving vehicle, the kinetic energy due to linear motion must also be included to obtain the total kinetic energy that must be dissipated by the brakes. In the case of a vehicle, if we take the rotation of one of the road wheels as our reference, the translational velocity is given by

$$v = r\dot{\phi} = r\omega \tag{2-6}$$

where $\dot{\phi} = d\phi/dt = \omega$ is in rad/sec so that v is in terms of the units of r per second. If the motor is not disconnected as the brakes are applied, its effect must also be included, either as a retarder, which adds to the braking effect, or as a driver, which opposes the brakes. In some vehicles and machines the motor may act as retarder for some operating conditions and as a driver in others. In either event, the contribution of the motor is usually included in the W_a term, so the energy to be dissipated in slowing from v_a to v_b may be written as

$$E = \frac{1}{2}N_w I_w \omega^2 + \frac{1}{2}mv^2 + W_a = \frac{1}{2}\left[N_w m_w \left(\frac{r_g}{r_w}\right)^2 + m\right](v_a^2 - v_b^2) + W_a \tag{2-7}$$

where m represents the total mass of the vehicle and its cargo. This relation holds if each of the N_w wheels has a mass m_w, a radius of gyration r_g, and an outside radius r_w. W_a is positive if it represents the work done by the motor during braking and negative if it represents the work dissipated either by the motor itself or by a retarder.

Although equations (2-6) and (2-7) have been discussed in terms of vehicle motion, they apply equally well to conveyors having N_w similar rollers of mass m_w, radius of gyration r_g, and radius r.

Often, the kinctic energy due to wheel rotation is negligible compared to the translational kinetic energy of the cargo, so that the rotational terms in equation (1-10) are usually omitted from the brake selection formulas found in a manufacturer's catalog.

C. Braking with Changes in Potential Energy: Cranes and Hoists

Since motion is assumed to be in the vertical direction, the energy change due to braking or clutching when a load is either raised or lowered is the sum of the changes in kinetic and potential energy and the work W_h done on the system by motors and retarders. Thus energy E may be written as

$$E = \frac{1}{2}\sum_{i=1}^{k} m_i(v_{ia}^2 - v_{ib}^2) + \frac{1}{2}\sum_{i=1}^{m} I_i(\omega_{ia}^2 - \omega_{ib}^2) + \sum_{i=1}^{n} W_i(h_{ia} - h_{ib}) + W_a \tag{2-8}$$

which is an extended version of equation (2-7) by including k masses m_i, m rotating components, each having moment of inertia I_i, n weights $\hat{W}_i$ and their elevation changes, and including nonzero values of velocity v_i, and angular rotation ω_i.

III. BRAKING AND CLUTCHING TIME AND TORQUE

Work done by a brake in slowing or stopping a mechanical system is converted to heat at the mechanical interface in friction brakes or in the inner and outer members in eddy-current, hysteresis, or magnetic particle brakes. Regardless of the particular brake design, the work done is equal to

$$W = \int_{t_1}^{t_2} \omega T\, dt = \int_{\phi_1}^{\phi_2} T\, d\phi \tag{3-1}$$

where T denotes the braking torque, $\omega = d\phi/dt$, ϕ represents the angular rotation of the active braking element (drum, disk, outer member), and t denotes time.

Preliminary design or selection of a brake is often predicted on constant torque, constant load, and therefore, constant deceleration. For this condition,

$$\omega = \omega_0 - \alpha t \tag{3-2}$$

so substitution into equation (3-1) with $t_1 = 0$ and $t_2 = t$ yields

$$\begin{aligned} W &= \int_0^t T(\omega_0 - \alpha s)ds = T(\omega_0 - \frac{\alpha t}{2})t \\ &= \Delta E = \Delta \mathrm{KE} + \Delta \mathrm{PE} + \Delta W_a \end{aligned} \tag{3-3}$$

when time is measured from that instant when the brake was first applied. If the brake is to stop or slow the rotation of a component, this work must equal the energy that must be dissipated in bringing that component to the new rotational speed. Upon substitution for αt from equation (3-2) into equation (3-3), we find that

$$W = Tt\frac{\omega_0 + \omega_1}{2} \tag{3-3a}$$

Hence equation (3-3) may be written as

$$Tt = \frac{1}{\omega_0 + \omega_1}\left[\sum_{i=1}^{k} m_i(v_{i_a}^2 - v_{i_b}^2) + \sum_{i=1}^{m} I_i(\omega_{i_a}^2 - \omega_{i_b}^2) + 2\sum_{i=1}^{n} W_i(h_{i_a} - h_{i_b}) + 2W_{h_a}\right] \tag{3-4}$$

If a single rotating moment of inertia I is involved, KE $= (1/2)\, I\, (\omega_0^2 - \omega_1^2)$ and

$$T = I\frac{\omega_0^2 - \omega_1^2}{(\omega_0 + \omega_1)t} = I\frac{\omega_0^2 - \omega_1^2}{2\omega_{av}t} = \frac{I}{t}(\omega_0 - \omega_1) \tag{3-5}$$

Finally, if all rotation is to be stopped, $\omega_1 = 0$ and equation (3-5) becomes

$$T = \frac{I\omega_0}{t} \tag{3-6}$$

Moments of inertia for other than geometrically simple objects–such as a solid, homogeneous cylinder–are generally given in terms of the mass m of the rotating object when SI units are implied (i.e., kilograms) and in terms of the weight W when OE units are implied (i.e., pounds). According to this practice, I will be presented in terms of mass m and radius of gyration r_g as

$$I = mr_g^2 = \frac{W}{g}r_g^2 \tag{3-7}$$

Returning now to equation (3-6), it frequently appears in design guides in different terms. Its modified form may be found by replacing ω in rad/sec by n, the initial rotational speed in rpm, according to

$$\omega = \frac{2\pi n}{60} \tag{3-8}$$

and by replacing I by Wr^2/g according to equation (3-7). The result is

$$\begin{aligned} T &= 2\pi\frac{mr_g^2 n}{60t} \cong \frac{mr_g^2 n}{10t} \qquad \text{(SI)} \\ T &= 2\pi\frac{Wr_g^2 n}{60gt} \cong \frac{Wr_g^2 n}{307t} \cong \frac{Wr_g^2 n}{308t} \qquad \text{(OE)} \end{aligned} \tag{3-9}$$

Our previous discussion has been concerned with brake design without specific knowledge of the friction and heat dissipation characteristics of the brake as a function of the slip speed, which is the rotational speed difference between the engaging faces of the brake or clutch. When that information is known from catalog data, as represented by Figure 2, we can use it, together with the governing equation of motion, to obtain a more realistic estimate of the activation time and the heat dissipated for a viscously damped system, as shown schematically in Figure 3(a), where the viscous damping is due to the process itself, or in Figure 3(b), where the viscous damping is supplied by a retarder used to add to the energy dissipated during stopping. Except for the brake itself, Coulomb, or dry friction, damping is generally suppressed in the remainder of the system and elastic effects are generally negligible.

From this figure we find the governing equation to be

$$I\frac{d\omega}{dt} = -T(\omega) - c\omega \tag{3-10}$$

where $T(\omega)$ is negative because it acts to slow the motion (i.e., to cause $d\omega/dt$ to be negative) and where ω denotes the instantaneous angular velocity of the system as it is being stopped or retarded and I denotes the moment of inertia of all masses in the system when written in terms of the angular velocity of the shaft on which the brake acts. Integration of equation (3-10) yields

$$t_1 - t_2 = I\int_{\omega_2}^{\omega_1} \frac{d\omega}{T(\omega) + c\omega} \tag{3-11}$$

which relates the deceleration time t to:

1. The net torque $T(\omega)$, which includes the torque transferred across the brake (positive), as given by curves similar to those shown in Figure 2, as well as any torque (negative) due to motors or other drivers that may continue to supply torque while the brake is applied
2. The damping $c\omega$ supplied by a retarded (described in Chap. 11), damping in the system itself, or both.

In equation (3-11), I represents both the rotational and translational inertia, where the translational velocity is expressed in terms of ω and the appropriate radius according to $v = r\omega$.

Equation (3-11) may be used to obtain an estimate of the relation between the torque and the braking time whenever $T(\omega)$ is known from data such as that shown in Figure 2. This will be demonstrated in one of the following examples. To show that this equation produces relation (3-6) when the torque is constant, it may be integrated to give

$$t_2 - t_1 = \frac{I}{c}\ln\frac{T + c\omega_1}{T + c\omega_2} \tag{3-12}$$

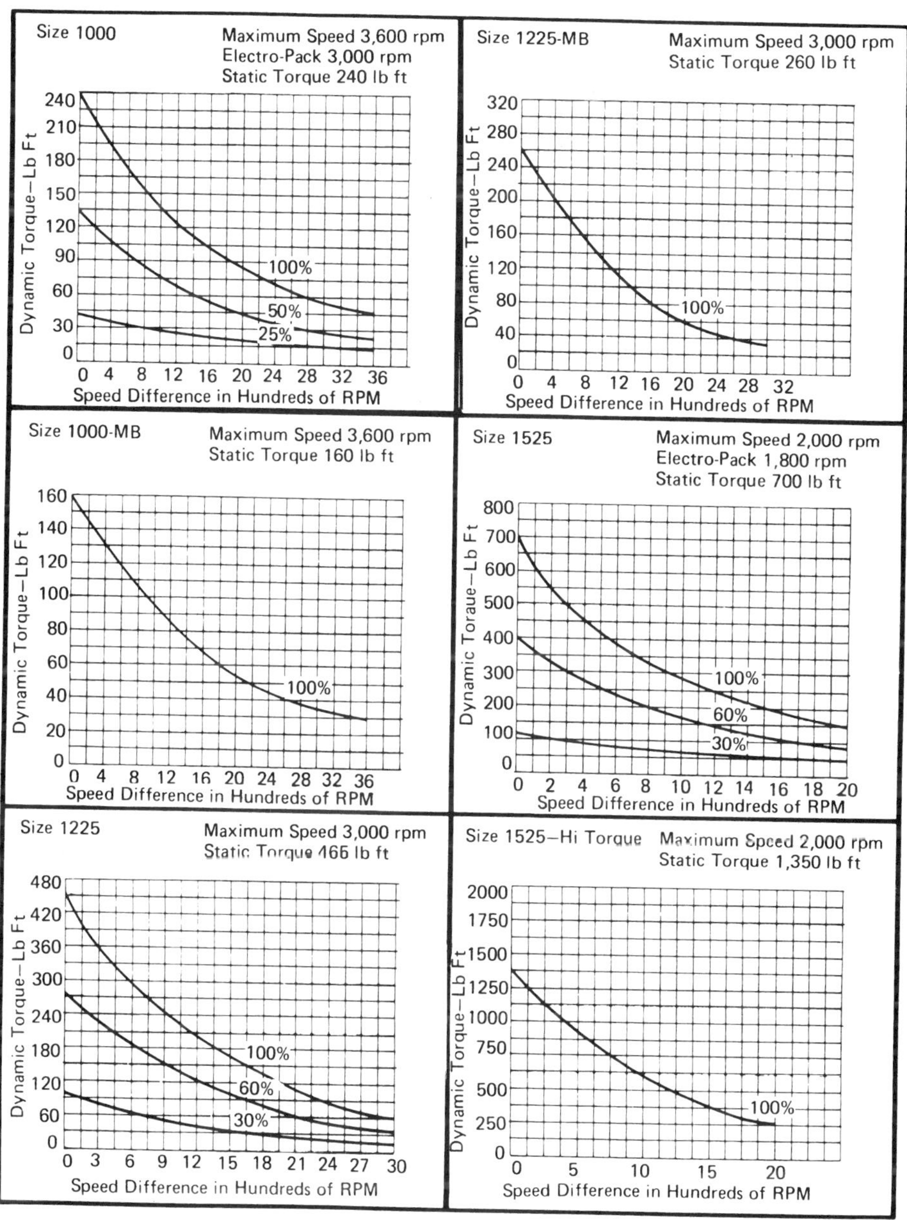

FIGURE 2 Dynamic torque as a function of the speed difference, or slip speed, between input and output shafts. (Courtesy of Warner Electric Brake & Clutch Co., South Beloit, IL.)

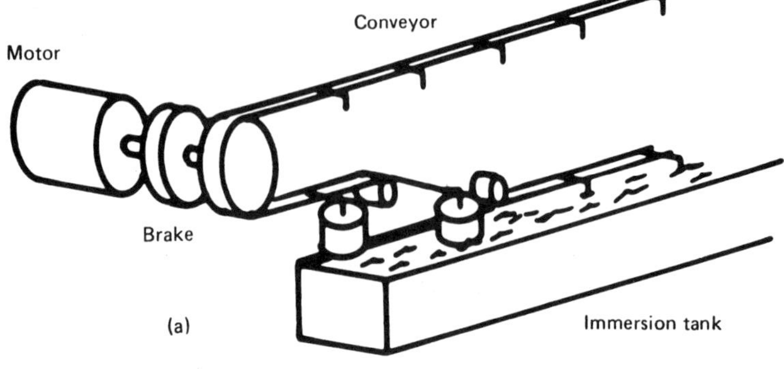

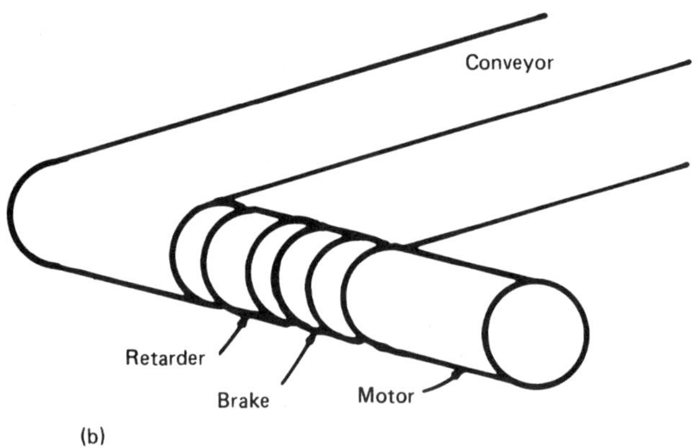

FIGURE 3 Schematic conveyor systems where viscous damping is due to (a) the process itself or (b) a retarder to aid in stopping.

which may also be written to give the required torque as

$$T = c\frac{\omega_1 - \omega_2 e^{(c/I)(t_2-t_1)}}{e^{(c/I)(t_2-t_1)} - 1} \tag{3-13}$$

If time is measured from the instant the brake is applied so that $t_1 = 0$ and if the system is brought to rest so that $\omega_2 = 0$, equation (3-13) simplifies to

$$T = \frac{c\omega_1}{e^{(c/I)t_2} - 1} \tag{3-14}$$

Finally, after expansion of the exponential in equation (3-14) according to

$$e^x - 1 = x + \frac{x^2}{2!} + \frac{x^3}{3!} + \frac{x^4}{4!} + \dots$$

and setting $x = ct_2/J$, we see that, if c is small enough for c^2 to be negligible compared to c, we then have

$$T = \frac{c\omega_1}{(c/I)t_2 + \dots} \cong \frac{I\omega_1}{t_2} \tag{3-15}$$

in agreement with equation (3-6), since ω_1 and t_2 in this equation play the role of ω_0 and t in equation (3-6).

IV. CLUTCH TORQUE AND ACCELERATION TIME

Many of the formulas developed in Sections 1 and 2 apply equally well to clutch applications. Only their use differs, in that now they are used to determine the work that must be done by the clutch on the load to accelerate it to the required speed.

The equations that may be used for either a clutch or a brake are (3-4) through (3-9). In the case of a clutch, equation (3-10) is replaced by

$$I\frac{d\omega}{dt} + c\omega = T(\omega) \tag{4-1}$$

which then requires that equation (3-11) be replaced by

$$t_2 - t_1 = I\int_{\omega_1}^{\omega_2} \frac{d\omega}{T(\omega) - c\omega} \tag{4-2}$$

as the relation between the torque, the damping, and the inertia of the system, both linear and rotational. When applied to a clutch, however, the time interval $t_2 - t_1$ in equation (4-2) applies to the time interval required for the clutch to bring the load up to speed. After the load is at operating speed, $d\omega/dt$ in equation (4-2) goes to zero, so the torque $T(\omega) = c\omega$ holds as long as the operating speed and load are constant (Figure 4).

Whenever T is constant, differential equation (4-1) may be integrated to give

$$t_2 - t_1 = -\frac{I}{c}\ln\frac{T - c\omega_2}{T - c\omega_1} \tag{4-3}$$

which differs from equation (3-12) only in the algebraic sign of c. Equation (4-3) may be solved for T to get

$$T = c\frac{\omega_1 e^{(-c/I)(t_2 - t_1)} - \omega_2}{e^{(-c/I)(t_2 - t_1)} - 1} \tag{4-4}$$

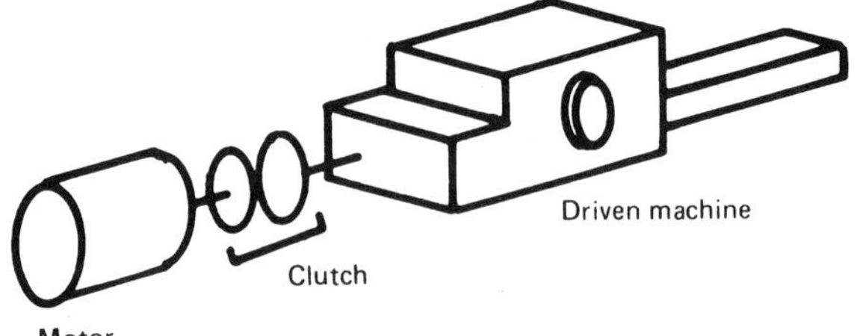

FIGURE 4 Schematic of a typical motor, clutch, machine configuration.

As a check on equation (4-4), note that if the clutch were applied at time $t_2 = 0$ when $\omega_1 = 0$, then equation (4-4) may be written as

$$T = c\frac{-\omega_2}{e^{(-c/I)t_2} - 1} \tag{4-5}$$

If we again use the series expansion for e^x given in the previous section, but with x now replaced by $-ct_2/I$ we find

$$T \cong \frac{I\omega_2}{t_2} \tag{4-6}$$

as in the case of a brake.

V. EXAMPLE 1: GRINDING WHEEL

Find the minimum torque capacity for a brake to be added to a twin-wheel motor grinder turning at 1725 rpm such that when either guard is raised the motor and two grinding wheels will stop within 0.1 sec. The moment of inertia of the motor rotor is 0.0137 slug-ft and each grinding wheel weights 10 lb and has a radius of gyration of 4.00 in.

Since all the rotating masses are on a single shaft, equation (3-6) applies, where I represents the sum of the moments of inertia for the grinding wheels and the rotor. From equation (3-7) we find that the moment of inertia for each grinding wheel is

$$I_w = \frac{w}{g}r^2 = \frac{10}{32.2}\left(\frac{4}{12}\right)^2 = 0.0345 \text{ slug-ft}^2 \tag{5-1}$$

so the total moment of inertia is

$$I = 2(0.0345) + 0.0137 = 0.0827 \text{ slug-ft}^2 \tag{5-2}$$

With the rotational speed in rad/sec given by

$$\omega = \frac{2\pi\,\text{rpm}}{60} = \frac{\pi(1725)}{30} = 180.6416\ \text{rad/sec}$$

substitution for I from equation (5-2) into equation (3-6) yields

$$T = \frac{0.0827(180.6416)}{0.1} = 149.3906 \sim 150\ \text{lb-ft} \tag{5-3}$$

as the required torque.

VI. EXAMPLE 2: CONVEYOR BRAKE

Recommend the torque requirement for a brake for the conveyor belt shown schematically in Figure 5. It is rated for a total load of 180 lb (the combined weight of all items conveyed by the conveyor). The conveyor belt weight is 50 lb, the end rollers weigh 22 lb each, and the 20 intermediate rollers weigh 4.0 lb each. The diameter of each end roller is 8.750 in. and the radius of gyration of each end roller is 4.0 in. The intermediate rollers are 2.00 in. in diameter and each has a radius of gyration of 0.8 in. The reduction ratio of the gear train is 5.488, the maximum conveyor velocity is 90 ft/min, and the brake is mounted between the driving gear motor and the gear train. The motor is disconnected from the drive line when the brake is engaged and the conveyor is to be stopped in the minimum time which will not cause the packages on the conveyor to slide along the belt. All the products to be conveyed have such a low center of gravity that tipping is not a problem. The friction coefficient is 0.30.

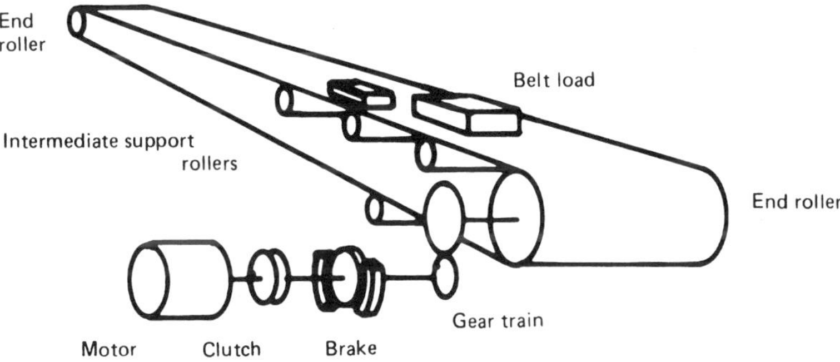

FIGURE 5 Conveyor belt schematic.

Kinetic energy due to rotation of the end and intermediate rollers, translation of the belt load, and translation of the belt itself will be considered; kinetic energy contributed by the gears and shafts in the gear train will be ignored because their combined moments of inertia is less than that of one of the intermediate rollers.

From equation (3-4) we find that the governing equation for a conveyor with k_J rotating masses and k_m translating masses is given by

$$T = \frac{\omega_0}{t}\left[\sum_{i=1}^{k_J} I_i n_i^2 + \left(\frac{d}{2}n_1\right)^2 \sum_{i=1}^{k_m} m_i\right] \tag{6-1}$$

where n_i is the ratio of rotational speed of roller i to the rotational speed of the shaft on which the brake is mounted and d is the diameter of the drive roller, whose speed ratio is represented by n_1

From equation (3-7) we find the moment of inertia of an end roller to be

$$I_e = mr_g^2 = \frac{22}{32.2}\left(\frac{4}{12}\right)^2 = 0.0759 \text{ slug-ft}^2 \tag{6-2}$$

and the moment of inertia of an intermediate roller to be

$$I_i = \frac{4.1}{32.2}\left(\frac{0.8}{12}\right)^2 = 0.0006 \text{ slug-ft}^2 \tag{6-3}$$

The rotational speed of the end rollers may be found from equation (2-6) to be

$$\omega = \frac{90}{60}\left(\frac{12}{4.375}\right) = 4.1143 \text{ rad/sec} \tag{6-4}$$

from which it follows that the speed of the input shaft to the gear train is

$$\omega_b = \omega n_1 = 22.5792 \text{ rad/sec} \tag{6-5}$$

for $n_1 = 5.488$. Since the intermediate rollers that support the belt along its length have radii of 1.00 in., their angular velocity is 18 rad/sec for an effective speed reduction factor of 1.254 relative to the input shaft to the gear train.

Since the belt moves with the same velocity as the product being conveyed, we can group them together so that $k_m = 1 + 1 = 2$. The two end rollers and the 20 intermediate rollers give $k_J = 20 + 2 = 22$. With all masses and moments of inertia known, we may substitute into equation (6-1) once we select a stopping time t. To find the minimum stopping time without slip between the product and the conveyor belt, recall that the stopping force of the product is μmg, so the maximum deceleration becomes μg. If this force is constant, the stopping time may be found from $t = v/a = 90/[60(0.3)32.2] = 0.1553$ sec. Substitution into equation (6-4) yields

$$T = \frac{22.579}{0.1553}\left[\frac{2(0.0759)}{5.488^2} + \frac{20(0.000556)}{1.254^2} + \frac{230}{32.2}\,\frac{4.375^2}{(12^2)(5.488^2)}\right]$$

$$= 6.344 \text{ lb-ft} \qquad (6\text{-}6)$$

$$= 76.130 \text{ lb-in.}$$

where $n_i = 1/5.488$ for the end rollers and $n_i = 1/1.254$ for the intermediate rollers.

If the brake had been mounted on either of the end roller shafts, equation (6-6) would have been replaced by

$$T = \frac{4.114}{0.1553}\left[2(0.0759) + 20(0.000556)(4.375)^2 + \frac{230}{32.2}\left(\frac{4.375}{12}\right)^2\right]$$

$$= 34.811 \text{ lb-ft} \qquad (6\text{-}7)$$

and the braking torque requirement would have been $n = 5.488$ times larger than that found by equation (6-6). This comparison is an example of the general rule that the brake should usually be placed in the faster shaft.

VII. EXAMPLE 3: ROTARY KILN

The curves in Figure 6 clearly imply that efficient use of a clutch by reducing the power loss due to heat generation, along with wear, requires that the speeds of its input and output shafts should be nearly equal. Accordingly, depending upon the power source (electric or hydraulic motor, turbine, or internal combustion engine), a clutch may be used to change gear ratios, to change from one power source to another when the speeds are nearly equal, or to disconnect the power source before braking.

This example will consider a load that is essentially rotational in order to concentrate on clutch and brake selection when dynamic torque and brake heating curves are available. Both clutch and brake analyses will display some of the calculation involved when the speeds of the input and output shafts are not almost equal.

A rotary kiln is to be driven by a 15-hp three-phase motor operating at 870 rpm and rated to deliver a torque of 240 lb-ft with a K factor (overload factor for starting) of 2.64. The motor, clutch, gear train with a 28.4 speed-reduction ratio, and rotary kiln are arranged as shown in Figure 7. The overall damping coefficient is approximately 0.10. The starting moment of inertia of

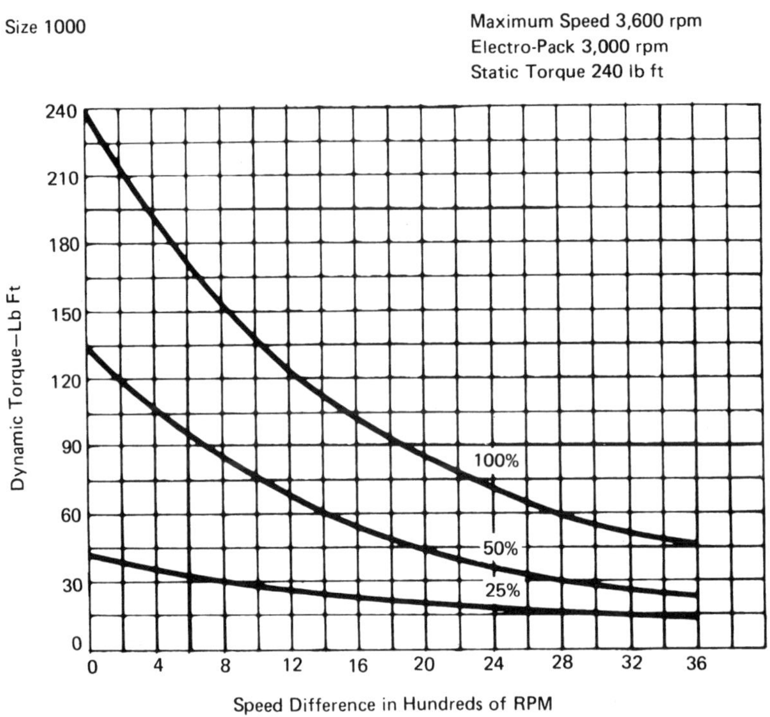

FIGURE 6 Dynamic torque as a function of the speed difference between input and output shafts. (Courtesy of Warner Electric Brake & Clutch Co., South Beloit, IL.)

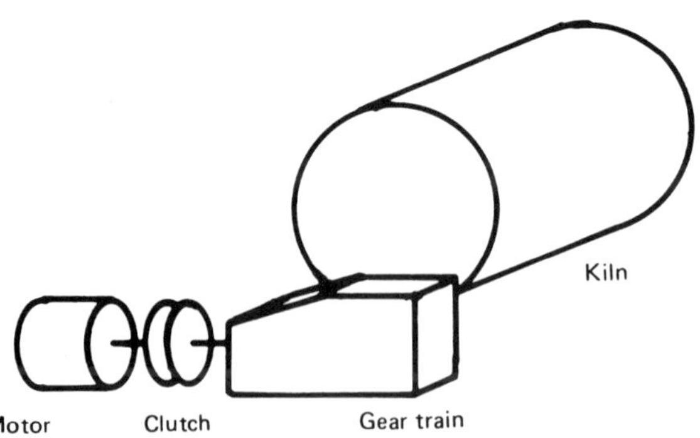

FIGURE 7 Schematic of motor, clutch, gear train, and kiln.

the kiln is equivalent to a weight of 31,832 lb and a radius of gyration of 2.8 ft, the clutch characteristics are given in Figure 6. A brake with similar characteristics will be used to stop kiln rotation.

The moments of inertia of the gears in the gear train will be neglected for simplicity. They will be considered for a different gear train in a subsequent example.

Conversion from horsepower (hp) and revolutions per minute (n) to torque (T) in ft-lb according to

$$T = \frac{(16,500\ \text{hp})K}{\pi n}$$

yields

$$T = \frac{(16,500)15(2.64)}{870\ \pi} = 239.0617\ \text{lb-ft}$$

as the required starting torque

Upon calculating the moment of inertia of the kiln according to equation (3-7), we find

$$I = \frac{31,832}{32.2}(2.8)^2 = 7750.4\ \text{slug-ft}^2$$

From equation (2-1), the equivalent moment of inertia at the clutch is given by

$$In_{21}^2 = \frac{7750.4}{28.4^2} = 9.6092\quad \text{slug-ft}^2$$

According to equation (3-6), the approximate time for the motor to bring the kiln up to speed is

$$t = \frac{I\omega_o}{240} = \frac{9.6092}{240}\left(\frac{870\pi}{60}\right) = 1.82\ \text{sec}$$

For a more precise calculation of the time to get up to speed, we may turn to equation (4-2), which requires the input data shown in Table 1, as read and calculated from the 100% speed difference curve in Figure 6.

Upon turning to a TK Solver routine* for the numerical integration of a integral whose integrand is given as a series of data points, we find that

---

*Enter L in the Type column after entering a name (i.e., time) in the Function Sheet and enter the data in Table 1 in the List Function Sheet. On the Rule Sheet type "value = integral ('time, x_1, x_2" where x_1 and x_2 are the lower and upper limits of integration, respectively.

TABLE 1 Input Data and Intermediate Values for Integrands in Equations (3-11) and (4-2)

Δn (rpm)	ω (rad/sec)	$c\omega$ (lb-ft)	$T(\omega)$ (lb-ft)	$T(\omega) - c\omega$ (lb-ft)	$T(\omega) + c\omega$ (lb-ft)	$\frac{1000}{T(\omega) - c\omega}$	$\frac{1000}{T(\omega) + c\omega}$
870	0	0	145	145.0000	145.0000	6.8966	6.8966
800	7.3304	0.73304	153	152.2670	153.7333	6.5674	6.5048
700	17.8024	1.78024	160	158.2198	161.7802	6.3203	6.1812
600	28.2743	2.82743	170	167.1726	172.8274	5.9818	5.7861
500	38.7463	3.87463	180	176.1254	183.8763	5.6778	5.4385
400	49.2183	4.92183	190	185.0782	194.9218	5.4031	5.1303
300	59.6903	5.96903	202	196.0310	207.9690	5.1020	4.8084
200	70.1622	7.01622	213	205.9838	220.0162	4.8548	4.5451
100	80.6342	8.06342	225	216.9366	233.0643	4.6096	4.2907
0	91.1062	9.11062	240	230.8894	249.11062	4.3311	4.0143

Note: Entries in the two right-hand-most columns have been multiplied by 1000 to avoid including 10^{-3} after each entry.

evaluation of equation (3-11) for a brake and equation (4-2) for a clutch gives start-up times $\tau = (t_2 - t_1)$ of:

$$\tau = 4.6396 \rightarrow 4.6 \text{ seconds for start-up}$$
$$\tau = 4.8397 \rightarrow 4.8 \text{ seconds for stopping}$$

The difference between these values and the time of 1.8 seconds given by equation (3-6) is, of course, largely due to the omission of damping in equation (3-6). Heat transferred to the surroundings for the surface temperatures shown in Figure 8 may be read directly from these curves by interpolating for surface temperatures between 250°F and 300°F.

Heat dissipation in the absence of curves similar to Figure 8 may be estimated from the work dissipated according to the relation

$$W(\omega) = T_0\,\omega_i\,\tau - I \cdot \int_0^{\omega_i} \frac{T(\omega)\,\omega}{T(\omega) - c\,\omega}\,d\omega \tag{7-1}$$

where $T_0\omega_i\tau$ represents the work done on the clutch and its load by the input shaft rotating at angular velocity ω_i, where τ denotes the time required for the load speed to reach ω_i, and where the integral represents the work done by the clutch in accelerating the load. Heat generated per cycle by the clutch determines the cooling method to be used so that the

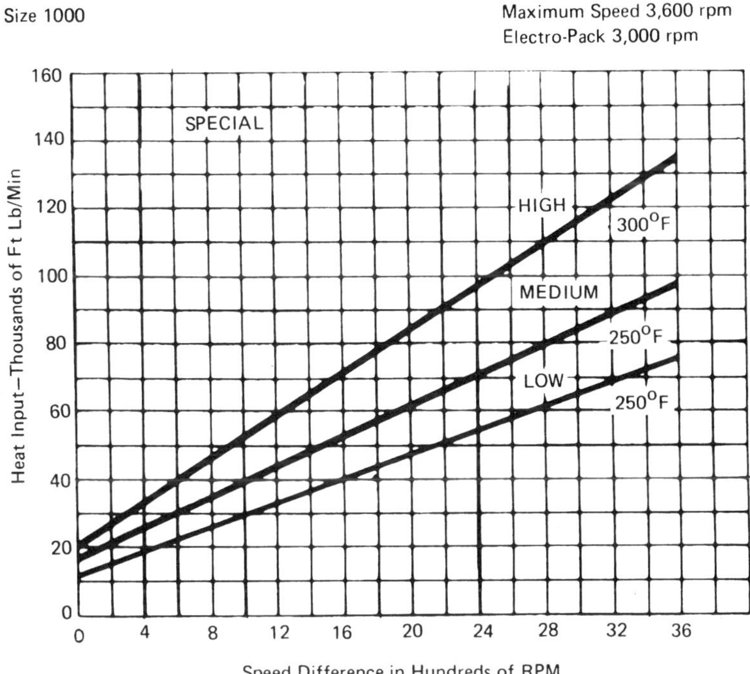

FIGURE 8 Heat input that can be transferred by radiation and convection for the surface temperatures shown in the rotational speed range of the rotating element. (*Speed difference* refers to the speed of the rotating element relative to the stationary element.) (Courtesy of Warner Electric Brake & Clutch Co., South Beloit, IL.)

heat can be transferred from the clutch per cycle for the expected ambient temperature.

Use of 31,832 lb for the average gross weight of the kiln instead of 31,800 lb may be justified by noting that it takes no more keystrokes to enter nonzero values. Carrying four digits to the right of the decimal point simply gives a more precise basis for the final round-off of the result to practical values than may be had when carrying fewer digits.

VIII. EXAMPLE 4: CRANE

Select a brake for a crane rated for a maximum load of 2800 kg as limited by the load rating for the 19 × 7 nonrotating wire rope used. The rope diameter is

21 mm, its weight is 2.069 kg/m, and the maximum drop of the cable is 30 m. The grooved cable drum is 0.50 m in diameter at the base of the grooves, weighs 696 kg, will accept 20 turns of wire rope, and has a radius of gyration of 0.23 m. The drum is driven by a gear train, illustrated in Figure 9 for which the gear data are as follows:

Gear number	Pitch diameter (m)	Radius of gyration (m)	Mass (k)
1	2.00	0.81	1278
2	0.40	0.17	276
3	0.80	0.32	721
4	0.25	0.09	231

Four turns remain on the drum when the load is 30 m below the top of the crane. The rope length from the drum to the top of the crane, Figure 10 is 21 m. Motor speed is 485 rpm, and a descending maximum load is to be stopped within 2.00 seconds after the brake is applied. The motor is disengaged by means of a clutch immediately before the brake is applied.

We may begin by calculating the angular velocity of each of the gears, their gear ratios relative to the input shaft on which the brake is mounted,

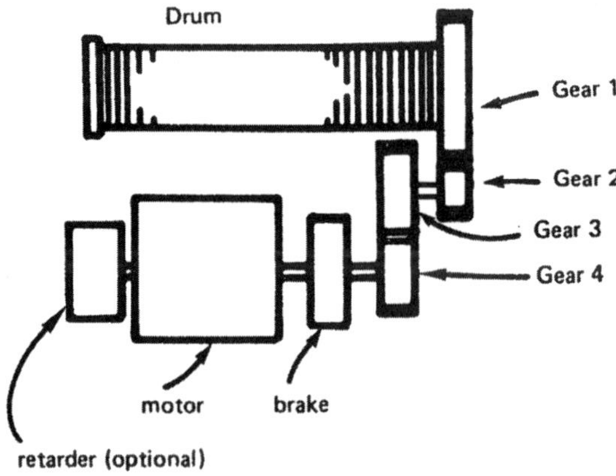

FIGURE 9 Schematic of motor, gear train, and drum for a typical crane. A retarder, if used, may be added at either end of the motor shaft.

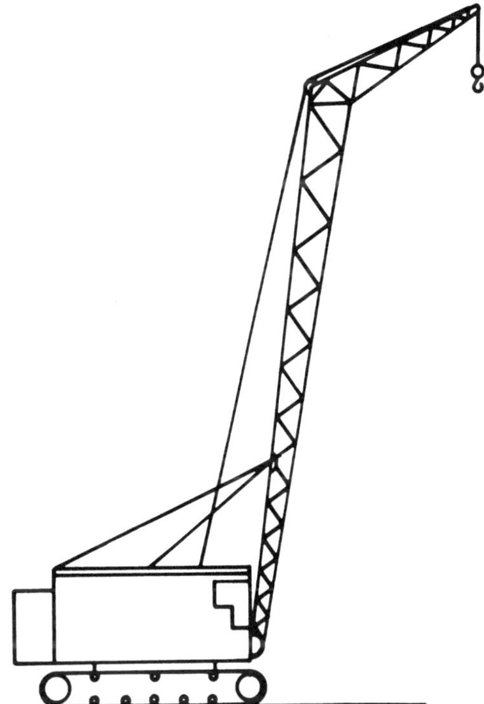

FIGURE 10 Sketch of a hoist showing the wire rope extension beyond the drum.

shaft 4, and the polar moment of inertia of each gear. The results are as follows:

Gear number	Speed ratio	Speed (rad/sec)	Polar moment of inertia (kg-m^2)
1	1:16.0	3.1743	838.496
2	1:3.2	15.8716	7.976
3	1:3.2	15.8716	73.830
4	1:1.0	50.7891	1.871

where all speed ratios have been calculated relative to the motor speed.

The dimensions from the drum to the top of the crane and the maximum drop include that portion of the cable over the pulley, or sheave, at the top of the crane, for a total cable length of 63 meters. Hence, the input data for the following formulas are:

$m_L = 2800\,\text{kg}$	$d_r = 0.021\,\text{m}$	$\gamma_r = 2.069\,\text{kg/m}$	$g = 9.8067\,\text{m/sec}^2$
$d_d = 0.50\,\text{m}$	$m_d = 696\,\text{kg}$	$r_{dg} = 0.23$ m	$n = 485$ rpm
$N_t = 4$ turns	$l_o = 63$ m	$t = 2.0\,\text{sec}$	$y_o = 30\,\text{m}$

The moment of inertia of the drum is given by

$$I_d = m_d r_{dg}^2 = 36.818\ kg\text{-}m^2 \tag{8-1}$$

The angular velocity of the drum and the mass of the rope are given by

$$\omega_m = \pi \frac{485}{30} \qquad m_r = [4\pi(d_d + d_r) + l_o]\gamma_i \tag{8-2}$$

respectively, to give $\omega_m = 50.789$ rad/sec and $m_r = 143.893$ kg. Rope velocity is given by

$$v_r = \frac{d_d + d_r}{2} \frac{\omega_m}{16} = 0.827\ \text{m/sec} \tag{8-3}$$

Next, estimate the distance the load will descend during its deceleration due to braking by integrating $a = d^2x/dt^2$ twice, subject to the initial conditions that $x(0) = 0$ and $dx(0)/dt = 0$ under the assumption that the deceleration is constant. From the resulting formulas,

$$v = at \qquad \text{and} \qquad s = ½\ at^2$$

it follows that if the load is to stop 2.0 seconds after the brake is applied, the values of acceleration a and distance s must be

$$a = 0.413\ \text{m/sec}^2 \qquad \text{and} \qquad s = 0.827\ \text{m}$$

Now that distance s is known, we can calculate the potential energy change for the rope as it extends from $y_1 = y_o - s$ to $y_2 = y_o$ by integrating over this length to get

$$\text{PE} = \gamma_r \int_{y_1}^{y_2} y\,dy = \frac{\gamma_r g}{2}\left[y_o^2 - (y_o - s)^2\right] = 496.405\ \text{N-m} \tag{8-4}$$

The potential energy for the load is given by

$$\text{PE}_L = m_L g s = 22,705.916\ \text{N-m} \tag{8-5}$$

The kinetic energies may be found from

$$n_{14} = \frac{1}{16} \qquad n_{24} = \frac{1}{3.2} \qquad n_{34} = \frac{1}{3.2}$$

$$m_1 = 1278 \qquad m_2 = 276 \qquad m_3 = 721 \qquad m_4 = 231$$

$$r_{1g} = 0.81 \qquad r_{2g} = 0.17 \qquad r_{3g} = 0.32 \qquad r_{4g} = 0.09$$

$$I_{g1} = m_1 r_{1g}^2 \qquad I_{g2} = m_2 r_{2g}^2 \qquad I_{g3} = m_3 r_{3g}^2 \qquad I_{g4} = m_4 r_{4g}^2 \qquad I_d = m_d r_{dg}^2$$

$$\text{ke}_d = \frac{1}{2} I_d n_{14}^2 \qquad \text{ke}_{g1} = \frac{1}{2} I_{g1} n_{14}^2 \qquad \text{ke}_{g2} = \frac{1}{2} I_{g2} n_{24}^2 \qquad \text{ke}_{g3} = \frac{1}{2} I_{g3} n_{34}^2$$

$$\text{ke}_{g4} = \frac{1}{2} I_{g4}$$

where m_1 through m_4 are the masses of gears 1 through 4, respectively, and r_{1g} through r_{4g} are their respective radii of gyration. Thus,

$$\text{KE}_d = \text{ke}_d \omega_m^2 = 185.5 \text{ N-m} \qquad \text{KE}_L = \frac{1}{2} m_L v_c^2 = 957.3 \text{ N-m}$$

$$\text{KE}_r = \frac{1}{2} m_r v_c^2 = 49.2 \text{ N-m} \qquad \text{KE}_{g1} = \text{ke}_{g1} \omega_m^2 = 4224.5 \text{ N-m}$$

$$\text{KE}_{g2} = \text{ke}_{g2} \omega_m^2 = 1004.7 \text{ N-m} \qquad \text{KE}_{g3} = \text{ke}_{g3} \omega_m^2 = 9299.2 \text{ N-m}$$

$$\text{KE}_{g4} = \text{ke}_{g4} \omega_m^2 = 2413.3 \text{ N-m}$$

Addition of these gives

$$\text{KE} = (\text{ke}_{g1} + \text{ke}_{g2} + \text{ke}_{g3} + \text{ke}_{g4} + \text{ke}_d)\omega_m^2 + \text{KE}_L + \text{KE}_r$$
$$= 18{,}133.6 \text{ N-m}$$

So upon adding this to the total potential energy of

$$\text{PE} = 23,202.3 \text{ N-m}$$

the torque required may be found from

$$T_o = \frac{\text{PE} + \text{KE}}{(\omega_m/2)t} = 813.9 \quad \text{N-m} \tag{8-6}$$

in which the average motor speed during braking was taken to be $\omega_m/2$.

The braking requirement of 813.9 N-m may be met by using a variety of brakes, such as band, external linear, annular caliper, and annular disk brakes. To choose among these, recall equation (1-10) from Chapter 1, equation (2-1) from Chapter 4, and equations (1-7) and (3-5) from Chapter 5 corresponding to the foregoing order, and let the internal radius, r_o, for both the annular caliper and annular disk be given by equation. Accordingly, evaluate the formulas

$$T = p_{\max} w r_o^2 (1 - e^{-\mu\phi}) \tag{8-7}$$

for a band brake,

$$T = 2\mu p_{\max} w r_o^2 \sin\left(\frac{\phi_o}{2}\right) \tag{8-8}$$

for either two opposing internal or external linearly acting brake shoes,

$$T = 2\mu p_{\max} \frac{r_o^3}{3\sqrt{3}} \phi_o \tag{8-9}$$

for two opposing disc brake pads, each subtending angle φ_o, and

$$T = 4\pi\mu p_{\max} \frac{r_o^3}{3\sqrt{3}} \tag{8-10}$$

for two complete annular pads in which $\varphi_o = 2\pi$ in equation (8-9).

We shall also consider an external pivoted drum brake with a leading and trailing shoe that may be evaluated by invoking the program used in Chapter 3. In all of these calculations assume a friction coefficient of 0.3, and set the width for the band, the linearly acting drum brake, and the externally pivoted brake to 5 cm. Limit the maximum lining pressure for the band brake and for the externally pivoted brake to 2.0 MPa, and limit the pressure for the other linings to 3.0 MPa, which may be either formed or solid. Lining pressure for the externally pivoted brake was taken to be 2.0 MPa, merely for comparison with the band brake.

Figure 11(a) shows the torque capacity in newton-meters as a function of angle φ subtended by each shoe for a drum diameter of 300 mm, and Figure 11(b) shows the torque capacity in newton-meters for band, linearly acting, caliper, and annular brakes as a function of the drum or disc diameter in millimeters.

Although the linearly acting drum brake is clearly more effective than the other brakes shown in Figure 11(b), it and all of the other three brakes in that figure require more hardware than does the band brake. Therefore, select the band brake, because it can provide the necessary torque capability with mechanical simplicity.

External dual-shoe drum brakes are the next simplest. Increasing the maximum lining pressure to 3.0 MPa for an externally pivoted dual-shoe brake allows the drum diameter to be reduced to 170 mm and the radial distance to the shoe pivot to be reduced to 100 mm, from the 150 mm associated with Figure 11(a), to get a torque vs. angle curve similar in shape and magnitude to that in Figure 11(a). Thus, either an externally pivoted

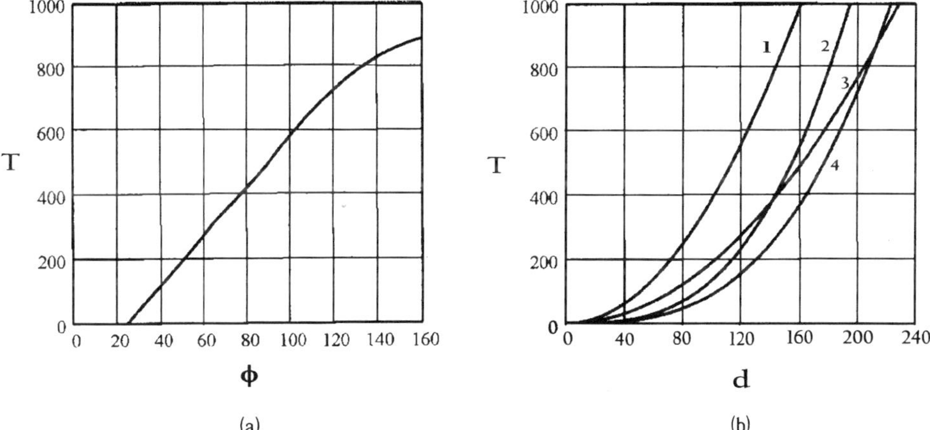

FIGURE 11 (a) Torque (N-m) as a function of shoe subtended angle for a drum diameter of 200 mm. (b) Torque (N-m) as a function of disk or drum diameter *d* (mm) (1) a linearly acting drum brake, (2) an annular disk brake, (3) a band brake, (4) a caliper brake.

dual-shoe brake or an external linearly acting dual-shoe brake might be recommended if space considerations are more important than mechanical simplicity.

IX. EXAMPLE 5: MAGNETIC PARTICLE OR HYSTERESIS BRAKE DYNAMOMETER

The dynamometer application is represented schematically in Figure 12, wherein either a magnetic particle or hysteresis clutch is used. Torque is independent of rotational speed throughout the range of a magnetic particle clutch and is independent of rotational speed to within about 0.003% per rpm for a hysteresis clutch for rotational speeds from 0 to a speed that is dependent on the cooling provided, as illustrated in Figure 13. Since the torque acts continuously, brake heating is expressed in terms of the dissipated power in units of watts, given by

$$P_d = \begin{cases} T_\omega = \dfrac{\pi T n}{30} & \text{(SI units)} \\ \dfrac{\pi T n}{22.126.5} & \text{(OE units)} \end{cases} \qquad (9\text{-}1)$$

where P_d is in watts, often termed slip watts, ω is in rad/sec, and n is in rev/min. Input torque T is in kg-m in the SI system and in lb-ft in the old English system

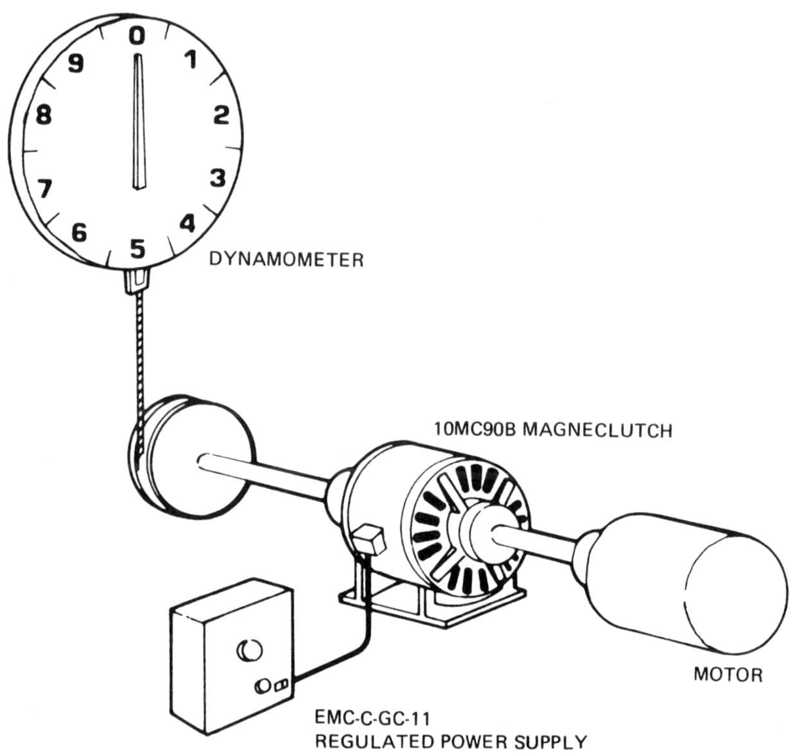

FIGURE 12 Magnetic particle brake dynamometer. (Courtesy of Sperry Electro Components, Durham, NC.)

of units. A typical slip watts–rpm curve showing the heat dissipation capability of a magnetic particle clutch is presented in Figure 14.

Equation (9-1) can also be applied to a clutch if n is redefined to be the difference in rpm between the speed of the input and output shafts. It also gives the power transmitted if T is redefined as the output torque and n is redefined as the speed of the output shaft.

Calculation of the power dissipated by either magnetic particle or hysteresis brakes is very simple. For example, consider a dynamometer as shown in Figure 12, where the motor runs at 890 rpm and the force reads 429.182 N for a 0.500-m lever arm. The torque is $431.342 \times 0.500 = 215.671$ N-m and the power dissipated, according to the first of equations (9-1), is

$$P_d = \frac{215.671(890)\pi}{30} = 20.101 \text{ kW}$$

which is, of course, equal to the power delivered by the motor at 890 rpm.

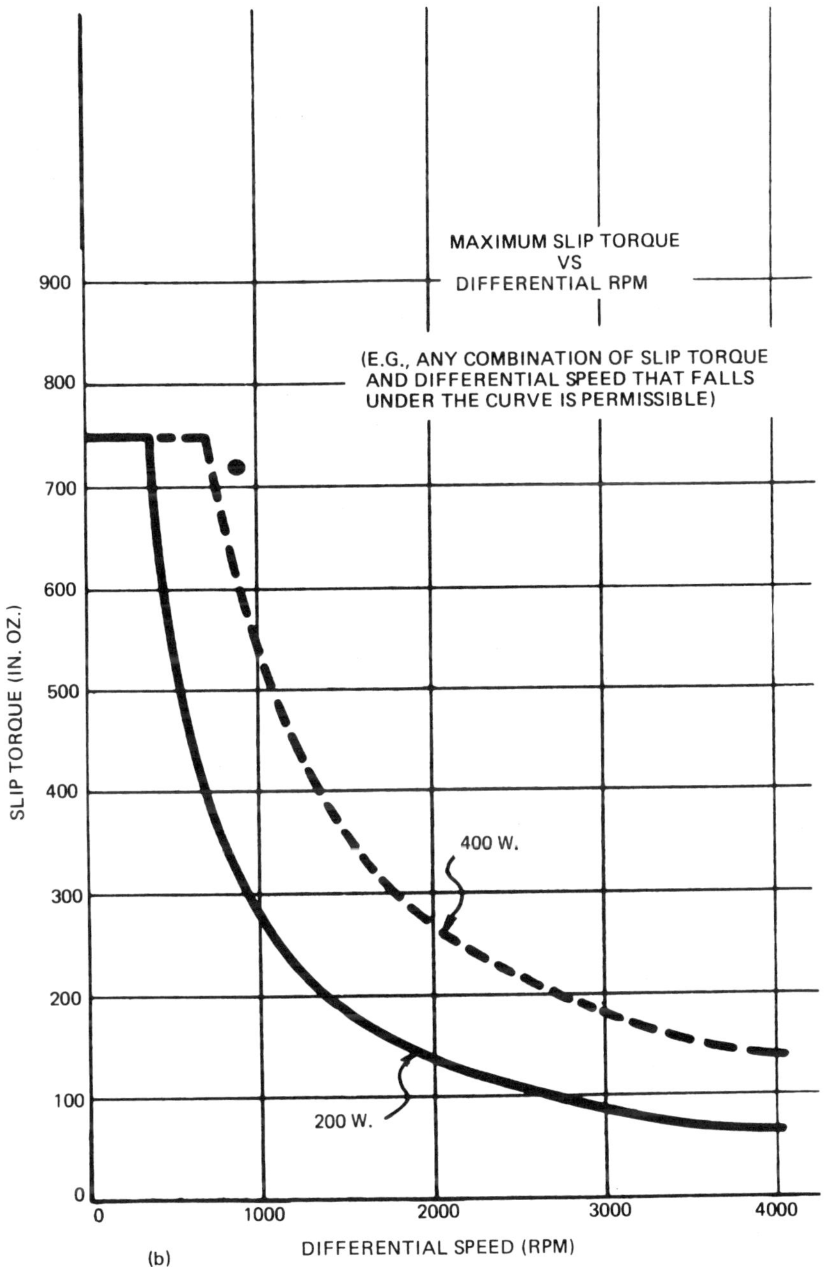

Figure 13 Representative torque-slip speed curve for hysteresis brake showing the effect of improved cooling. (Courtesy of General Electro-Mechanical Corp., Buffalo, NY.)

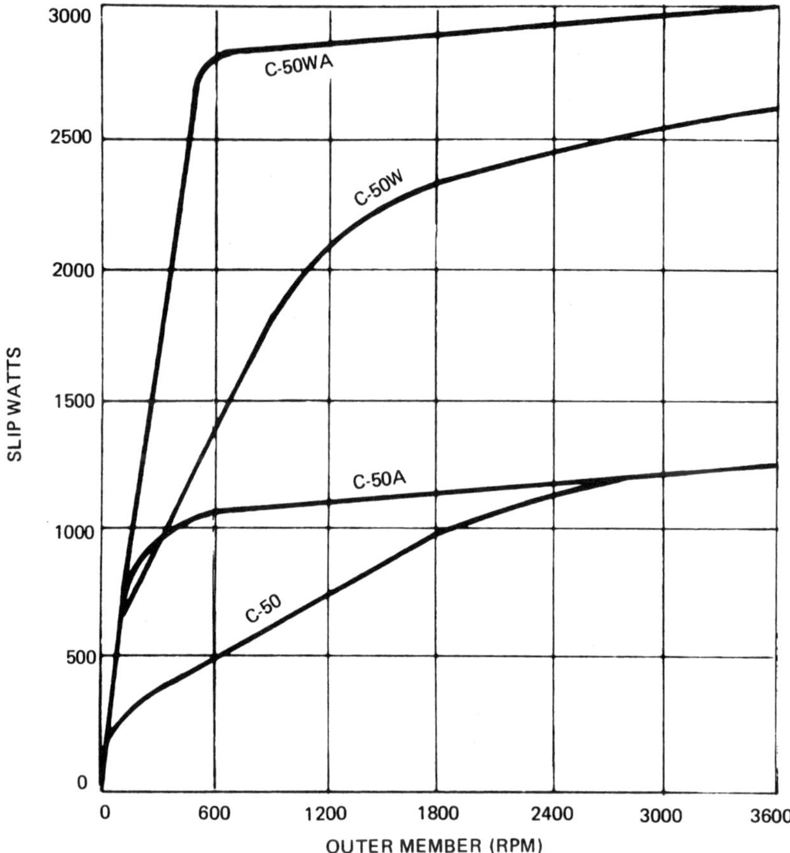

Figure 14 Typical slip watts-rpm curve for a magnetic particle clutch for various means of cooling. A force air; W, circulated water; otherwise, radiation and convection to ambient air. (Courtesy of Magnetic Power Systems, Inc., Fenton, MO.)

X. EXAMPLE 6: TENSION CONTROL

Tension control is often used in manufacturing processes that involve drawing, coating, slitting, printing, and winding of sheet material and in the formation of wires and filaments. Selection of magnetic particle or hysteresis brakes for such an application is usually based on the torque required and the brake's steady-state power dissipation capacity because the braking is generally continuous in these operations.

Suppose we are to select brakes to be used for the two draw rolls shown in Figure 15(a). The drive motor provides 1.5 kW at 950 rpm to drive rollers

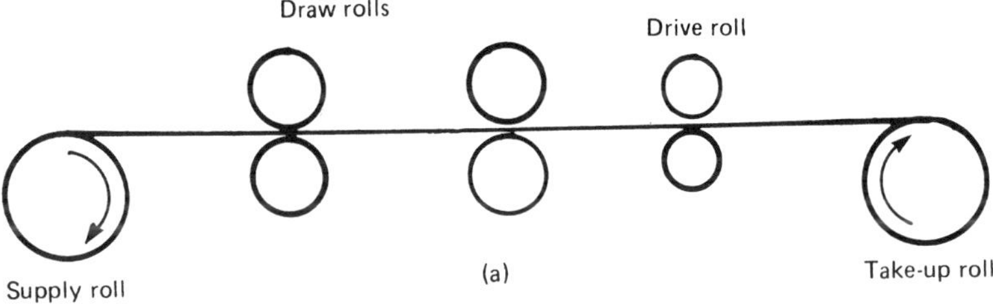

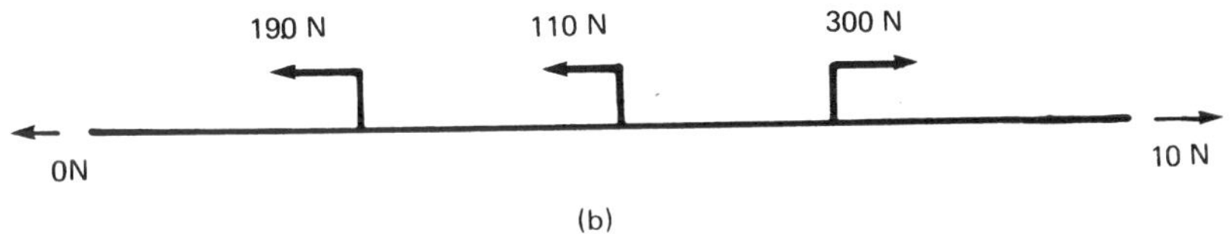

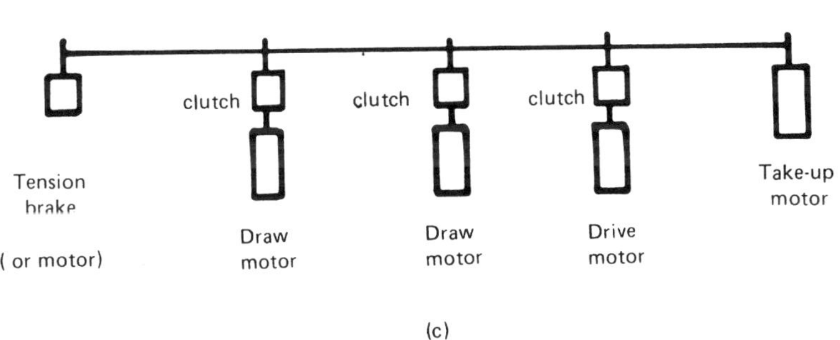

FIGURE 15 (a) Schematic of a tension-control drawing process; (b) the corresponding diagram of the forces acting on the web; (c) draw and tension motors used in braking.

100 mm in diameter. Draw rollers are 130 mm in diameter and web tension provided by the take-up roll motor is 10 N. Lab test results are available to aid in estimating the elongation of the web due to drawing.

From the force diagram shown in Figure 15(b) we observe that the drive rollers rotate with the speed of the web and that although both sets of draw rollers rotate in the same direction as the drive rollers, the torque on these rollers opposes the motion of the web. The clutch at the drive rolls may be selected on the basis of torque alone because it will experience only slight heating due to coil losses as long as the web moves at the design velocity.

Web velocity at the drive rolls may be calculated from

$$v = \pi dn = \pi(0.1)(950) = 298.451 \text{ m/min}$$

Based on lab results we estimate that web velocity at draw roller 1 will be 297.141 m/min, corresponding to a rotational speed of

$$n_1 = \frac{v}{\pi d} = \frac{297.141}{\pi(0.130)} = 727.561 \text{ rpm}$$

The torque requirement at draw rolls 1 is given by

$$T = rF = 0.065(110) = 7.150 \text{ N-m}$$

Cooling requirements at the brakes may be greatly reduced if the differential speed at the brakes is reduced by installing them between the draw rolls and a motor that is controlled to resist rotational speeds greater than a specified value, as illustrated in Figure 18. If these motors are to operate at 950 rpm, the power dissipated at the draw rolls may be estimated from equations (9-1), with the rotational speed replaced by the differential speed, as

$$P_d = \frac{\pi T(n_r - n_1)}{30} \tag{10-1}$$

At draw rolls 1, therefore,

$$P_{d1} = \frac{\pi(7.150)(950.000 - 727.561)}{30} = 166.550 \quad \text{slip watts}$$

At draw rolls 2, the web velocity is estimated to be 296.920 m/min, so $n = 296.920/0.130\pi = 727.020$ rpm, which implies that the power dissipated by the brake at draw rolls 2 may be

$$P_{d2} = \frac{\pi(12.350)(950.000 - 727.020)}{30} = 288.378 \quad \text{slip watts}$$

XI. EXAMPLE 7: TORQUE AND SPEED CONTROL

Control of both output torque and output speed for a constant input speed may be accomplished with a magnetic particle, eddy-current, or hysteresis clutch, simply by controlling the coil current. This capability allows us to drive a machine using a motor whose torque-speed curve would otherwise be incompatible with that of the prime mover if they were directly connected.

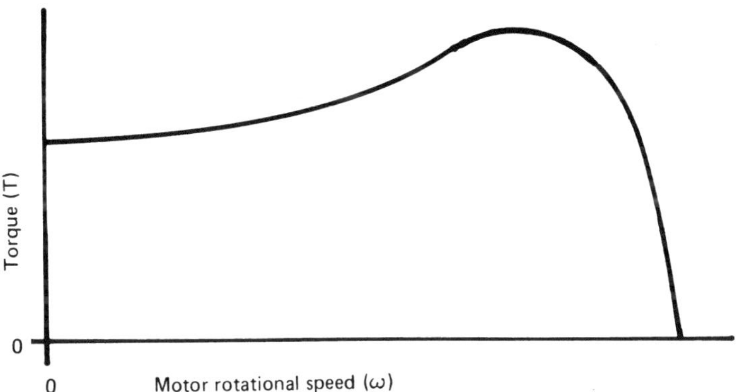

Figure 16 Torque-speed curve for the prime mover, an electric motor.

Supposed, for example, that the prime mover is an electric motor with the torque-speed curve shown in Figure 16 and that the desired torque-speed curve for the load is that shown in Figure 17.

In this example an eddy-current clutch will be selected because the design considerations in its use are somewhat more complicated than those associated with either a magnetic particle or a hysterests clutch.

To transfer power from the motor to the load, the eddy-current clutch must have a torque curve at 100% excitation whose maximum torque equals or exceeds the maximum torque required by the load, as illustrated in Figure 18. Selection from eddy-current clutches with curves represented by curve c, d, or e in Figure 18(a) depends on the degree of control required and the precision required for the maximum torque between points 1 and 2 in Figure 18(b).

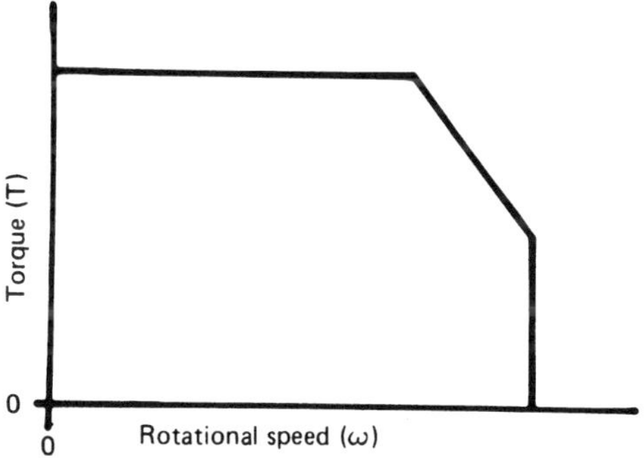

Figure 17 Torque-speed curve for the load.

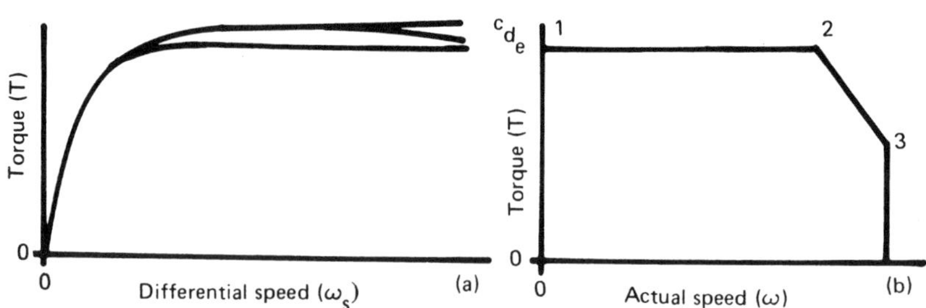

Figure 18 Typical eddy-current clutch curves c, d, and e in (a), which may be used to drive the load in (b). In (a) the slip speed is represented by ω_S, and in (b) the load speed is represented by ω.

In what follows we shall assume that curve e in Figure 18(a) has been selected so that the controller monitoring the speed and torque between points 1 and 2, where the slope is slightly positive, may uniquely relate speed to torque.

Minimum motor speeds at the required torques for this clutch may be found from Figure 19 by reading the minimum slip speeds at these torques from the clutch torque-slip speed curve as shown. The dashed lines represent the family of curves obtained by coil excitation less than 100%, as labeled. Thus torque and load combination at point 3 in Figure 19 requires a slip speed of ω_{s3}, while the combination at 2 requires a slip speed of ω_{s2}. Note that since we selected curve (c) in Figure 18(a), other, larger, slip speeds may also be used to achieve this torque by reducing the coil excitation current. (The implicit

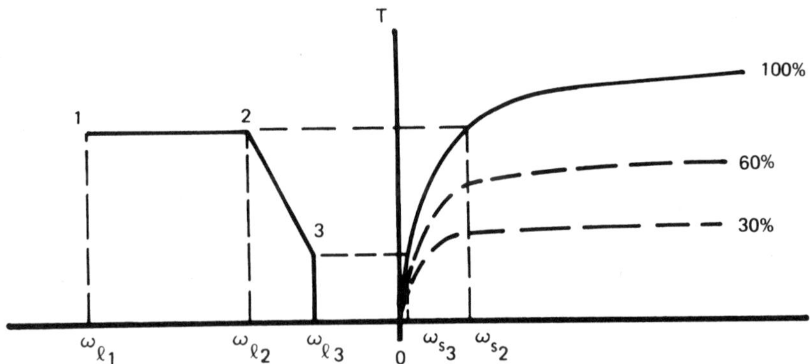

Figure 19 Load torque-speed and clutch torque-slip speed curves used to find minimum slip speed for load levels 1, 2, and 3 based on 100% coil excitation.

assumption that the torque-speed curves do not change character as the excitation current is reduced is not always true.)

Upon superimposing the slip speed obtained from Figure 19 to the operating speed of the load, we may find the minimum operating speed of the motor that will enable the clutch to deliver the specified torques, as has been done in Figure 20. This figure also clearly shows that by using an eddy-current clutch, we are able to operate at a higher torque at low load speeds than would have been possible with the motor alone.

In this example the load torque-speed curve was such that each torque-speed curve of the clutch crossed it only once. Where a single coil current may correspond to more than one torque-speed combination, as shown in Figure 21, it may be advisable for some applications to increase the motor speed to provide the curves shown in Figure 22 in order to reestablish a unique torque-speed relation.

This example, mentioned at the outset, was constructed to show the considerations involved in the use of an eddy-current clutch. Obviously, the controls would generally have been simpler if a magnetic particle or a hysteresis clutch had been used because the torque would have been constant

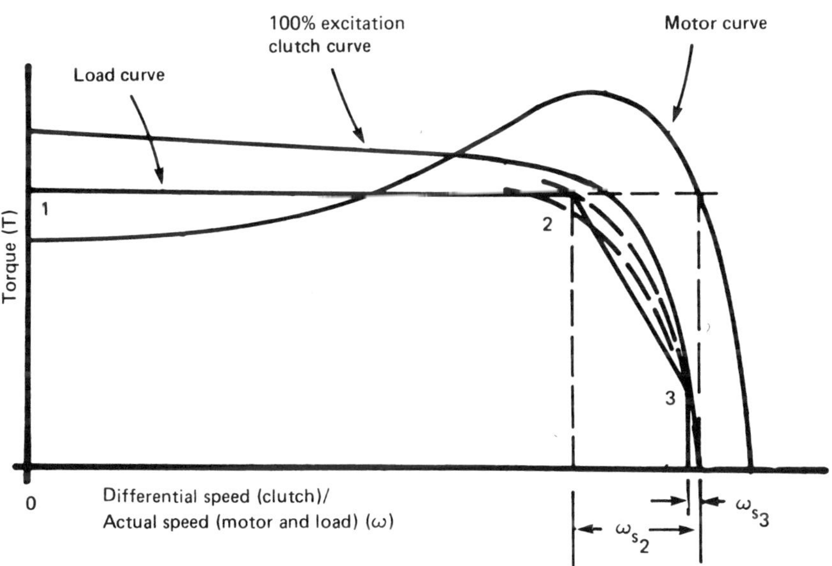

FIGURE 20 Graphical relation between the motor operating speed, the minimum eddy-current clutch slip speed, and the load curve. Less than 100% coil excitation curves are dashed.

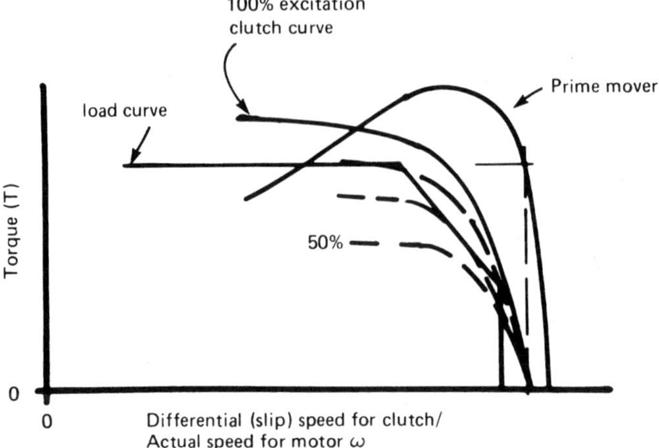

FIGURE 21 Coil current and slip speed combinations that permit more than one torque-slip speed combination for some coil excitation values.

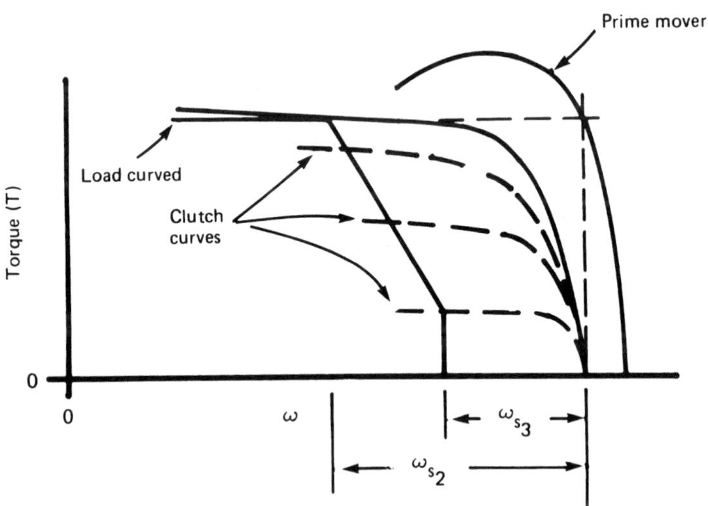

FIGURE 22 Increased slip speeds to obtain unique coil current values for each point on the torque-speed curve for the load when using an eddy-current clutch.

over the range for a given coil current. The slightly more complicated controls for eddy-current clutches are justified in those applications where the torque is to vanish whenever the driver and driven units approach equal speeds.

XII. EXAMPLE 8: SOFT START

The term *soft start* denotes starting without an initial shock, as may occur when a friction clutch is engaged too quickly. Soft starts may be had by using a torque converter, a fluid coupling, a magnetic particle clutch, a hysteresis clutch, or an eddy-current clutch. In the case of either a torque converter or a fluid coupling the torque transferred for a given input torque may be controlled by controlling the amount of fluid pumped into the converter or coupling. The same effect may be had from a magnetic particle, hysteresis, or eddy-current clutch by controlling the field current.

Generally, torque converters and fluid couplings are used in portable equipment, such as oil field drilling rigs, and in vehicles, such as trucks, buses, and automobiles, while magnetic particle, hysteresis, and eddy-current clutches are usually used in factories and mills where electrical power is available and where data from remote sensors may be processed to control brakes and clutches on machinery such as printing presses, tape transports, conveyor belts, and extrusion equipment.

Soft starts are perhaps most easily accomplished by using a clutch in which the torque is constant over a differential speed range that equals or exceeds the operating speed of the driven machine. Magnetic particle and hysteresis clutches fulfill this requirement and do not require that the motor speed exceed the driven speed by a minimum amount, as in the case of an eddy-current clutch.

Since the driving torque is constant over the operating range of the driven machine, we may in principle prescribe any coil-current versus time relation we wish to in order to prescribe the torque, and hence the acceleration as a function of time.

With these comments in mind, recommend a coil current profile for a magnetic particle clutch so that the acceleration of the take-up roll on a tape winder will increase slowly at the beginning of the acceleration period and will decrease slowly at the end of the acceleration period such that the first derivative of the acceleration, known as the jerk, will be zero at the beginning and end of the acceleration period. Assume that the damping in the system is negligible.

To provide a soft start we may consider providing a torque to the driven load that varies as the load torque versus time curve shown in the upper right-hand panel in Figure 23, in which the torque increases smoothly from zero to a maximum and then decreases to the steady-state torque when the machine is up to speed.

Before writing a program to find the required variation of coil current with time to transmit this torque profile, it may be instructive to demonstrate the procedure graphically. Upon entering the load torque versus time curve at time t_1, say, we project upward to the curve to read to corresponding torque. By projecting this torque to the clutch torque versus coil current curve we may read downward from the intersection to find the required current, say, i_1. If we now plot coil current and time axes as shown in the lower left-hand panel in Figure 23, we may locate the corresponding time on this second time axis by projecting downward from the time axis for the load torque curve to a 45° line and then project horizontally to the left from the 45° line as shown. The intersection of this projection with the vertical projection from the coil current

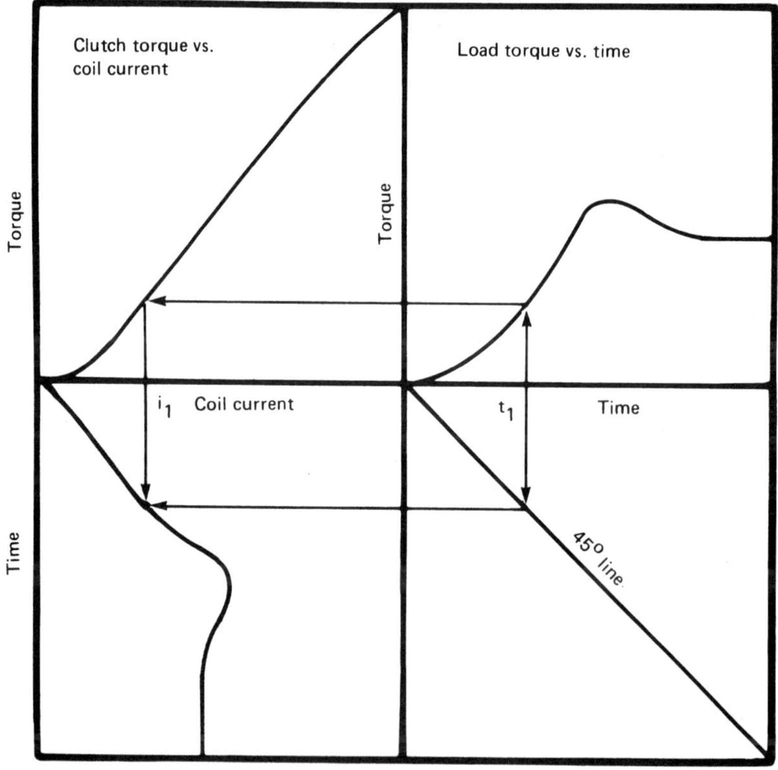

FIGURE 23 Graphical determination of the control current as a function of time to produce a prescribed soft start.

axis locates point (i_1, t_1) on the desired coil current versus time curve. Continuing in this manner for a sequence of points enables us to find sufficient points to complete the coil current versus time curve as shown.

Our program may be written in a parallel manner. After entering tabular data CTCC describing the clutch torque versus coil current and tabular data LTT describing the load torque versus time, we select a sequence of times $t(i)$. For each of these $t(i)$ values we use the LTT data to interpolate to find corresponding torques $TQ(i)$. We then use the CTCC data to interpolate to find the coil current $I(i)$ associated with torque $TQ(i)$. Thus we have tabulated $TQ(i)$ as a function of $I(i)$. These data, if plotted, would yield the coil current versus time curve used to control the soft start.

XIII. NOTATION

a	linear acceleration or deceleration (lt^{-2})
C_p	specific heat at constant pressure
c	damping coefficient (mt^{-2})
d	diameter (l)
E	energy (ml^2t^{-2})
F	force (mlt^{-2})
g	acceleration due to gravity (lt^{-2})
h	height (l)
I	moment of inertia (ml^2)
KE	kinetic energy $(ml^2t\text{-}^2)$
k	integer (l)
m	mass (m)
N	intcgcr (1)
n	revolutions/minute (rpm) (t^{-1})
n_{ij}	speed ratio of gear i relative to gear j (l)
p	pressure $(ml^{-1}\ t^{-2})$
Q	heat $(mt^2\ t^{-2})$
r	radius (l)
r_g	radius of gyration (l)
t	time (t)
v	velocity (lt^{-1})
W	work $(ml^2\ t^{-2})$
w	weight (mlt^{-2})
α	angular acceleration or deceleration (t^{-2})
γ	mass/length (ml^{-1})
Δ	increment of the quantity that follows
Θ	temperature (θ)
θ	angular position (1)

ϕ	angular position (l)
ω	angular velocity (t^{-1})

XIV. FORMULA COLLECTION

Braking time, variable torque

$$t_2 - t_1 = -I \int_{\omega_2}^{\omega_1} \frac{d\omega}{T(\omega) + c\omega}$$

Braking time, constant torque

$$t_2 - t_1 = \frac{I}{c} \ln \frac{T + c\omega_1}{T + c\omega_2}$$

Braking time, full stop, constant torque

$$t = \frac{I}{c} \ln\left(1 + \frac{c}{T}\omega\right)$$

Braking time or clutch, acceleration time, negligible damping, constant torque

$$T \approx \frac{I\omega}{t} = \frac{mr_g^2 n}{10t} (\text{SI units}) = \frac{Wr_g^2 n}{307t} (\text{OE units})$$

Constant braking torque, full stop

$$T = c \frac{\omega}{e^{\frac{c}{I}t} - 1}$$

Constant clutch torque, start from rest

$$T = c \frac{\omega}{1 - e^{-\frac{c}{I}t}}$$

Clutch acceleration time, variable torque

$$t_2 - t_1 = I \int_{\omega_1}^{\omega_2} \frac{d\omega}{T(\omega) + c\omega}$$

Clutch acceleration time, constant torque

$$t_2 - t_1 = \frac{I}{c} \ln \frac{T - c\omega_1}{T - c\omega_2}$$

Clutch acceleration time, constant torque, from rest

$$t = \frac{I}{c} \ln \frac{1}{1 - (c/T)\omega}$$

Clutch, heat dissipated during acceleration

$$W(\omega) = T\,\omega_i\,\tau - I \int_0^{\omega_i} \frac{T(\omega)\,\omega}{T(\omega) - c\omega}\, d\omega$$

9

Centrifugal, One-Way, and Detent Clutches

These are special-purpose clutches that are used in automatic transmissions, in devices for bringing high-speed machinery up to speed, in chain saws, in conveyor drives, and in similar industrial, vehicular, and large- and small-equipment applications. The centrifugal clutches provide a speed-dependent torque which acts only when the rotational speed exceeds a particular value; the one-way, or overrunning, clutches provide a torque that is not speed dependent once they are engaged, but is dependent on the direction of rotation; and the detent clutches provide a torque that cannot exceed a prescribed value.

I. CENTRIFUGAL CLUTCHES

A centrifugal clutch may be described as consisting of an inner cylinder that is attached to the input shaft and an outer housing that is attached to the output shaft, as in Figure 1. Sectors of the inner cylinder are cut out to allow it to be fitted with weights that can slide radially outward as the inner cylinder rotates so that the weights are forced against the outer housing by centrifugal force and thereby transmit torque to the outer housing. Centrifugal clutches designed for lower power transfer may use simpler designs. In some chain saws, for example, it is the weights themselves that are recessed to accept radial guides from the central shaft.

FIGURE 1 Centrifugal clutch. (Courtesy Dana Corp., Inc., Toledo, OH.)

Because of the variety of centrifugal clutch designs, their analysis will be described in general terms. Let A denote the cross-sectional area of each weight in a plane perpendicular to the axis of rotation, written in the form of an annular sector of angle ϕ_o as

$$A = c\phi_o r_o^2(1 - \beta^2) \tag{1-1}$$

where $\beta = r_i/r_o$. Parameters β and c are factors that may be used to express other cross-sectional areas in this form of equation (1-1). When $\beta = 0, c = 1/2$, and $\phi_o = 2\pi$, area A in equation (1-1) becomes that of a disc of radius r_o.

Let w denote the width of each weight, measured in a direction parallel to the axis of rotation, and let γ represent the mass density of the weights. If the static deflection of a retaining spring attached to each mass is δ_s, then its spring constant k is given by

$$k = \gamma A \frac{wg}{\delta_s} \tag{1-2}$$

in which g is the acceleration of gravity, taken to be 9.8067 m/sec^2, or 32.2 ft/sec^2.

Denote the radius to the center of gravity of each weight by r_c. Then the centrifugal force acting on each weight as it rotates at angular velocity ω about the axis of the clutch and moves outward a distance δ is then given by

$$F = \gamma w A (r_c + \delta)\omega^2 - g\kappa \frac{\delta + \delta_s}{\delta_s} \tag{1-3}$$

where the spring constant may be increased by the factor κ to hold each weight more securely against its stop at low rotational speeds.

Consider a prototype weight as being made from a sector of a thick cylinder whose inner radius is r_i and whose outer radius is r_o. Form the sector by cutting the cylinder to length w, which will be the width of the sector, and then cut the cylinder with two radial planes separated by angle ϕ_o. Retain one of the two sectors that subtend angle ϕ_o as the prototype weight shown in later Figure 3(a). The radius of gyration of this weight about the axis of the original cylinder is given by

$$r_c = \frac{r_o}{\sqrt{2}} \sqrt{1 - \beta^2} \tag{1-4}$$

In order to express the radius of gyration of other geometries in this form, let

$$r_c = \lambda r_o \sqrt{1 - \beta^2} \tag{1-5}$$

The torque that can be delivered by N of these weights after they have moved outward a distance δ to make contact with the inner surface of the housing at radius r_o may be written as

$$T = \mu r_o F = \mu \gamma w r_o N A \left[(r_c + \delta)\omega^2 - g\kappa(1 + \eta) \right] \tag{1-6}$$

where $\eta = \delta/\delta_s$. This relation may be solved for the w required for the clutch to transmit torque T at angular speed ω to get

$$w = \frac{T}{N\mu\gamma\phi_o c r_o^3 (1 - \beta^2) \left[(r_c + \delta)\omega^2 - g\kappa(1 + \eta) \right]} \tag{1-7}$$

Maximum pressure on the lining may be found from

$$F = r_o w \int_{-\phi_o/2}^{\phi_o/2} p \cos \phi \omega = r_o w p_{\max} \int_{-\phi_o/2}^{\phi_o/2} \cos(\phi)^2 d\phi$$
$$= \frac{p_{\max}}{2} r_o w(\phi_o + \sin \phi_o) \qquad (1\text{-}8)$$

upon using the pressure distribution from equation (1-2) in Chapter 4. Hence,

$$p_{\max} = \frac{2F}{r_o w(\phi_o + \sin \phi_o)} \qquad (1\text{-}9)$$

The angular velocity of the input shaft when the weights make initial contact with the drum may be found by setting the square bracket in equation (1-6) equal to zero. Substitution of $\omega = 2\pi n/60$, where n is in rpm, followed by solving the resulting expression for n, yields

$$n = \frac{\pi}{30} \sqrt{\kappa \frac{g}{r_c(\beta) + \delta} (1 + \eta)} \qquad (1\text{-}10)$$

The role of parameters β and λ in equation (1-5) upon force F, equation (1-3), and therefore upon $p_{\max}$, equation (1-9) and the speed at which they first contact the drum are shown in Figures 2(a) through (d).

Observe that the variation of the pressure, and hence the force, that each weight exerts against the drum is a linear function of parameter λ and that it becomes a nearly linear function of β, and hence of r_i, for β greater than about 0.3. The dependence of the width of each weight, however, becomes increasingly nonlinear as β increases and as λ decreases. The rotational speed for initial contact is also nearly linear for $\beta < 0.6$, especially for the larger values of λ.

Example

Design a centrifugal clutch to provide a torque of 2400 N-m when the rotational speed reaches 870 rpm using sector weights having the geometry shown in Figure 3(a). Preferred characteristics are that initial contact between weights and drum occur at between 220 and 230 rpm and that the width of the weights be less than 30 cm. Assume a lining coefficient of friction of 0.35 and design for an inside drum radius (minus the lining thickness) of 15 cm, a displacement δ of 3 mm for the segments to contact the drum, and a static deflection of 1 mm. The segments are to be made from an iron alloy having a nominal density of 7880 kg/m^3, and a safety factor of 3.5 is mandated. Hold

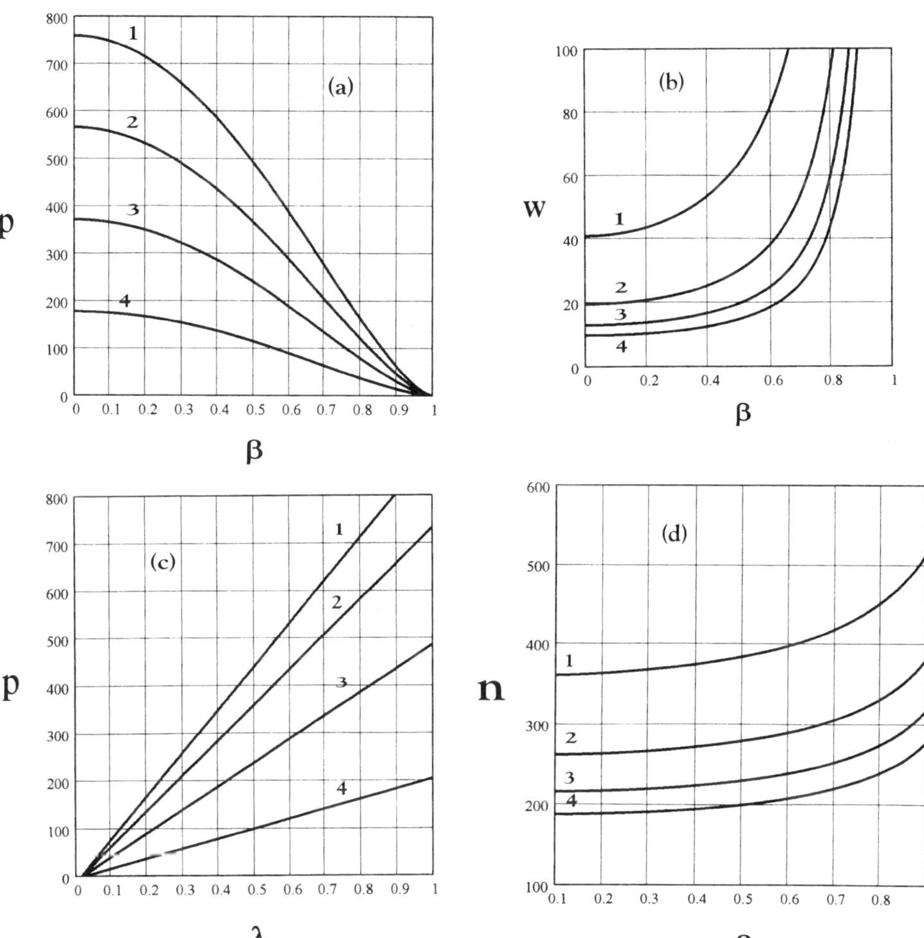

FIGURE 2 (a) Variation of pressure (kPa) with β for $\lambda = 0.2$, 0.4, 0.6, and 0.8 for curves 1, 2, 3, and 4, respectively. (b) Variation of width (cm) with β for $\lambda = 0.2$, 0.4, 0.6, and 0.8 for curves 1, 2, 3, and 4, respectively. (c) Variation of pressure (kPa) with λ for $\beta = 0.2$, 0.4, 0.6, and 0.8, respectively. (d) Variation of contact speed (rpm) with λ for $\beta = 0.2$, 0.4, 0.6, and 0.8 for curves 1, 2, 3, and 4, respectively.

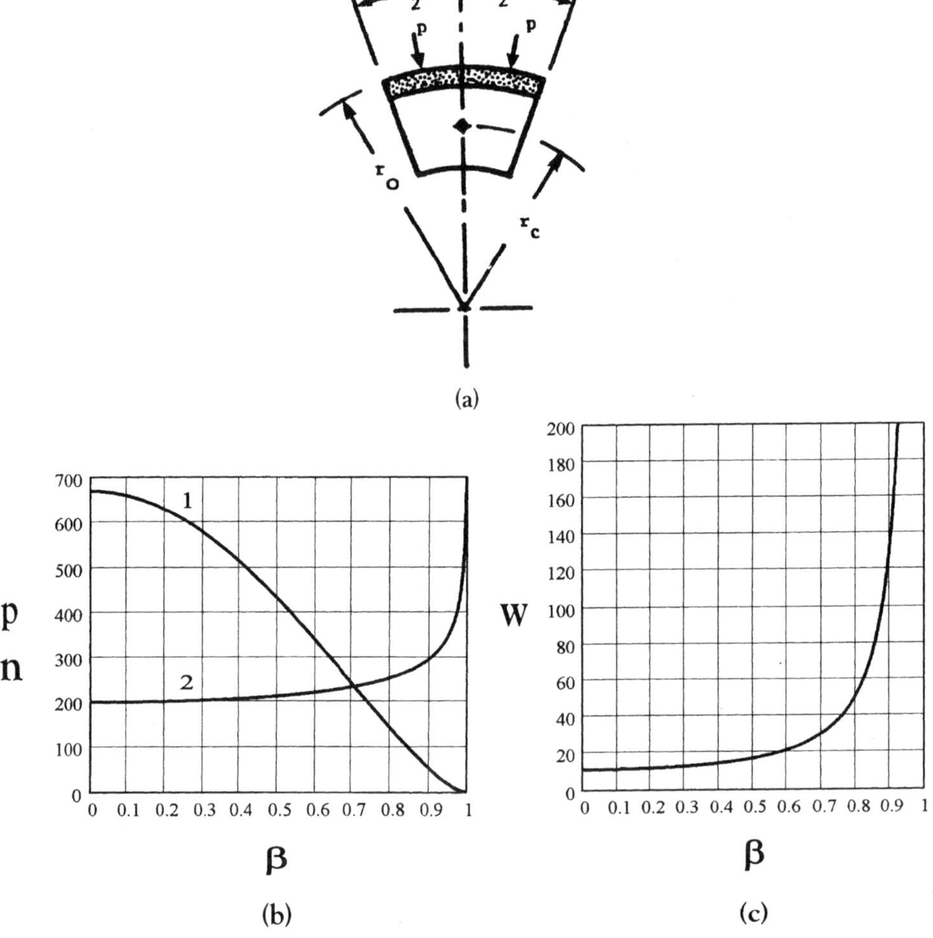

FIGURE 3 (a) Sector cross section. (b) Curve 1: pressure P (kPa) vs. β; Curve 2: initial contact speed n (rpm) vs. β. (c) Sector width w (cm) vs. β.

the weights against their rest position with a force 1.2 times their weight. Maximum lining pressure of less than 440 kPa is preferred.

Because the lining pressure on a segment decreases with angle θ from the centerline of that segment according to equation (4-2), use six weights to get a greater force transfer, each subtending an angle $\phi_n = 42°$.

Begin the design process by plotting pressure p, contact speed n, and width w against β by substituting the following values into equations (1-1) through (1-3), (1-5) through (1-7), and (1-9).

$T = 24,000$ N-m	$r_o = 150$ m	$\gamma = 7880\ \mathrm{kg/m^3}$	$n = 870$ rpm	$c = 0.50$
$\delta = 0.003$ m	$\delta_s = 0.001$ m	$\kappa = 1.2$	$\phi_o = 42°$	$\mu = 0.35$
$\eta = 3$	$N = 6$	$\lambda = \dfrac{1}{\sqrt{2}}$	$g = 9.8067\ \mathrm{m/sec^2}$	

These plots are shown in Figure 3(b) and (c).

Figure 3(b) shows that an initial contact speed between 220 and 230 rpm may be had for β between 0.6 and 0.7, Figure 3(c) shows that the corresponding width of the sector would be less than 30 cm. Substituting $\beta = 0.65$ into equation (1-10) yields $n = 226.59$ rpm, which is within the desired range. This is close enough to the preferred value of 225 rpm for manual iteration of β to find that

$$n = 225.001 \text{ rpm} \qquad \text{at} \qquad \beta = 0.6367$$

The width of each weight and the maximum lining pressure corresponding to $\beta = 0.6357$ are found to be

$$w = 23.8 \text{ cm} \qquad \text{and} \qquad p_{\max} = 304 \text{ kPa}$$

by substitution into equations (1-7) and (1-9), respectively. The required spring constant may be found by substituting from equation (1-1) into equation (1-2) to get

$$k = wc\phi_o r_o^2 \gamma \frac{g}{\delta_s} \kappa (1 - \beta^2) \tag{1-11}$$

Substitution into this expression yields

$$k = 1366 \mathrm{N/mm}$$

II. ONE-WAY CLUTCH: THE SPRING CLUTCH

Wiebusch gave the first description of this clutch, shown in Figure 4, in 1930 [1]. As may be soon from the figure, it consists of a helical spring snugly, wrapped about both the input and output hubs, parts 1 and 3 in Figure 4, but is attached to neither of them. If the input hub tends to turn in the direction that causes the helix to tighten, the increased friction between the spring and hubs tends to resists any further relative rotation. Relative rotation in the

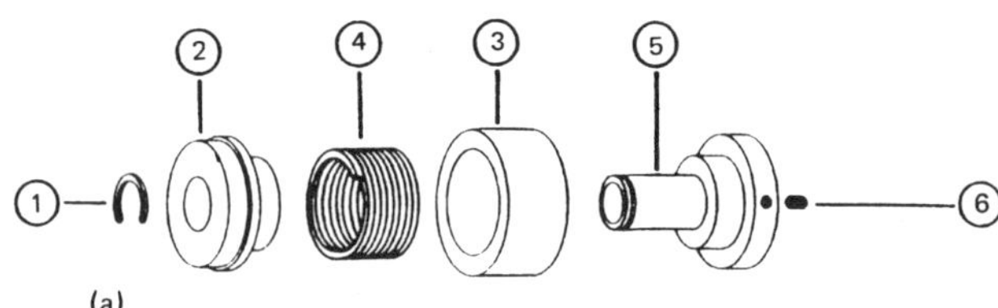

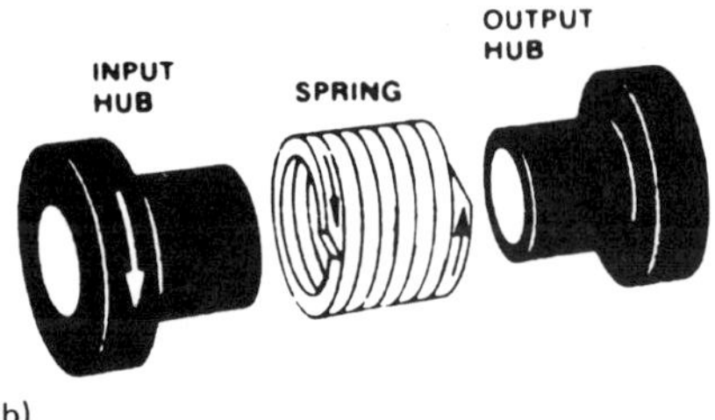

FIGURE 4 Spring clutch and its components. (Courtesy Warner Electric Brake & Clutch Co., South Beloit, IL.)

other direction, however, tends to loosen the helix, and relative rotation may proceed with only a relatively small restraint by the spring clutch.

Although Wahl [2] appears to have derived a more accurate expression for the torque that may be transmitted, T_t, agreement between the Wiebusch theory and experiment seems to be close enough to justify use of the simpler relationship, which is

$$T_t = EIr_h\left(\frac{1}{R_2} - \frac{1}{R_1}\right)\left(e^{2\pi N\mu} - 1\right) \tag{2-1}$$

in terms of the elastic modulus E of the spring material, the moment of area I of the spring wire in bending, the radius R_1 of the neutral surface of the wire in helix 4 in Figure 4 when it is free of external load, the radius R_2 of the wire when the helix is in tight contact with hubs 1 and 5 in the figure, and the number of turns N on one hub if both hubs have the same number of turns. If both hubs do not have the same number of turns, N is the smaller of the two. The friction coefficient is represented by μ, and r_h denotes the hub radius.

Wiebusch found that the torque T_u in the unwinding direction was approximately equal to

$$T_u = EIr_h\left(\frac{1}{R_2} - \frac{1}{R_1}\right)\left(1 - e^{-2\pi N\mu}\right) \tag{2-2}$$

Equation (2-1) obviously holds for a torque less than that which corresponds to the maximum force than can be carried by the spring wire at yield. Kaplan and Marshall [3] have indirectly suggested that the limiting torque satisfies the inequality

$$T_{\max} \leqslant bt^2 \frac{1.05}{2r_h}\left(R_1 - R_2 - \frac{t}{2}\right) \tag{2-3}$$

for rectangular wire whose dimension in the radial direction is t and whose dimension in the axial direction of the helix is b.

III. OVERRUNNING CLUTCHES: THE ROLLER CLUTCH

These clutches are designed to transmit torque from shaft A to shaft B when shaft A tends to rotate faster than shaft B but to disengage when shaft B rotates faster than A. Details of four designs that accomplish this are shown in Figure 5, which shows that the clutch consists of two concentric races, in which one is circular and the other consists of a series of cams, with a roller under, or above, each cam. Relative rotation which wedges the rollers between the narrow portion of the cam and the circular surface of the other race forces both races to rotate together, while relative rotation in the opposite direction frees the rollers and allows the two races to rotate at different angular rates.

In particular, if the cams are cut in the outer race and tapered in the direction shown in Figure 5(a), (b), and (c), rotation of the inner race in the clockwise direction will cause the rollers to wedge themselves between the two races so that the outer race must also rotate in the clockwise direction, that is, when

$$\omega_i > \omega_o$$

If the outer race is then accelerated to a rotational speed greater than that of the inner race so that

$$\omega_o > \omega_i$$

the roller will move to the larger ends of the cam and the outer race is free to accelerate to a speed greater than that of the inner race. The sequence just described is, for example, that used in starting gas turbines with an electric motor to get them up to operating speed, at which point the turbine accelerates under its own power and disengages the starter motor, which is then shut off.

If the cam surface is cut in the inner race and tapered as shown in Figure 5 (d), clockwise rotation of the outer race will drive the inner race whenever

$$\omega_o > \omega_i$$

Acceleration of the inner race in the clockwise direction will cause the clutch to disengage whenever

$$\omega_o < \omega_i$$

as is obvious from the taper geometry. These clutches are said to be free-wheeling or overrunning when the relative rotation of the race is such that no torque is transmitted from one to the other.

From the geometry of Figures 5 it follows that the torque transmitted to a roller and a convex race is limited by the maximum contact stress that can be sustained along the line of contact (actually, a narrow strip after the surfaces have deformed slightly) between the roller and the race with the smaller radius of curvature.

$$\begin{aligned}\sigma_{xx} = {} & \frac{-2F}{\pi^2 a}(a^2 + 2x^2 + 2z^2)\frac{z}{a}\Psi - 2\pi\frac{z}{a} - 3xz\Phi \\ & + \mu(2x^2 - 2a^2 - 3z^2)\Phi + 2\mu\pi\frac{x}{a} + 2\mu(a^2 - x^2 - z^2)\frac{x}{a}\Psi\end{aligned} \tag{3-1a}$$

$$\sigma_{zz} = -\frac{2F}{\pi^2 a}z(a\Psi - x\Phi + \mu z\Phi) \tag{3-1b}$$

$$\sigma_{xz} = -\frac{2F}{\pi^2 a}\left[z^2\Phi + \mu(a^2 + 2x^2 + 2z^2)\frac{z}{a}\Psi - 2\pi\mu\frac{z}{a} - 3\mu xz\Phi\right] \tag{3-1c}$$

away from the contacting surfaces and by

$$\begin{aligned}\sigma_{xx} &= -\mu\frac{4F}{\pi a}\left[\frac{x}{a} - \left(\frac{x^2}{a^2} - 1\right)^{1/2}\right], \quad x \geqslant a \\ &= -\frac{2F}{\pi a}\left[\left(1 - \frac{x^2}{a^2}\right)^{1/2} + 2\mu\frac{x}{a}\right], \quad -a \le x \le a \\ \sigma_{xx} &= -\mu\frac{4F}{\pi a}\left[\frac{x}{a} - \left(\frac{x^2}{a^2} - 1\right)^{1/2}\right], \quad x \leqslant -a\end{aligned} \tag{3-2a}$$

If a finite element analysis program with contact stress capability is not available, the pertinent stress components may be estimated from an analysis by Smith and Liu [4] for the contact (Hertzian) stresses between two parallel

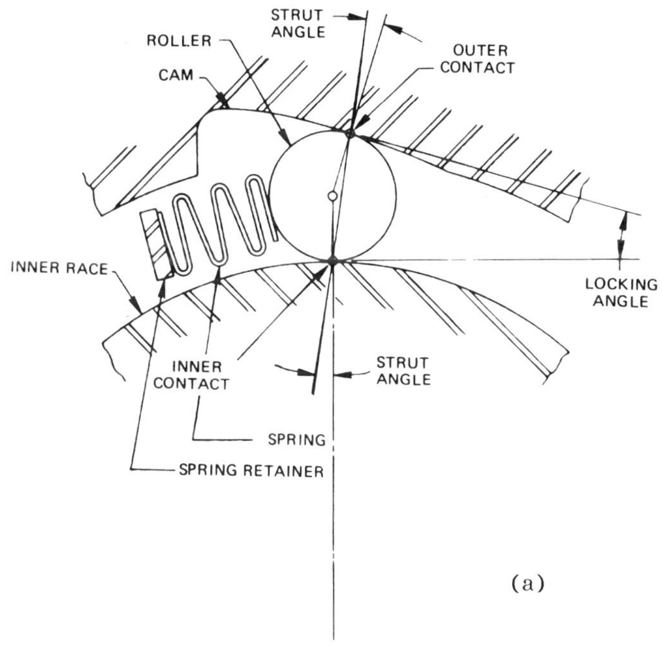

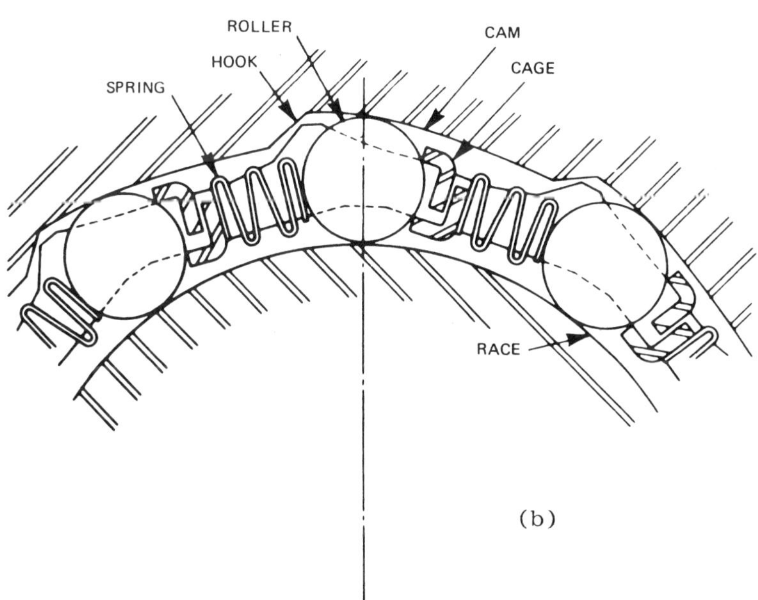

FIGURE 5 Typical roller clutch configurations. (a) Outer cam type of roller one-way clutch diagram. (b) Caged roller type of clutch diagram (hook-type cam). (c) Loose roller type of clutch diagram (leg-type cam). (d) Inner can type of roller one-way clutch diagram. (Reprinted with permission; © 1984 Society of Automotive Engineers, Inc.)

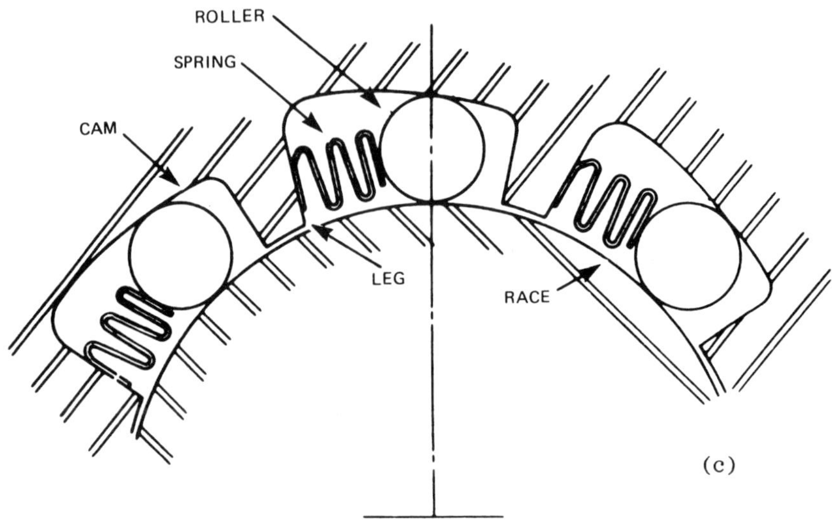

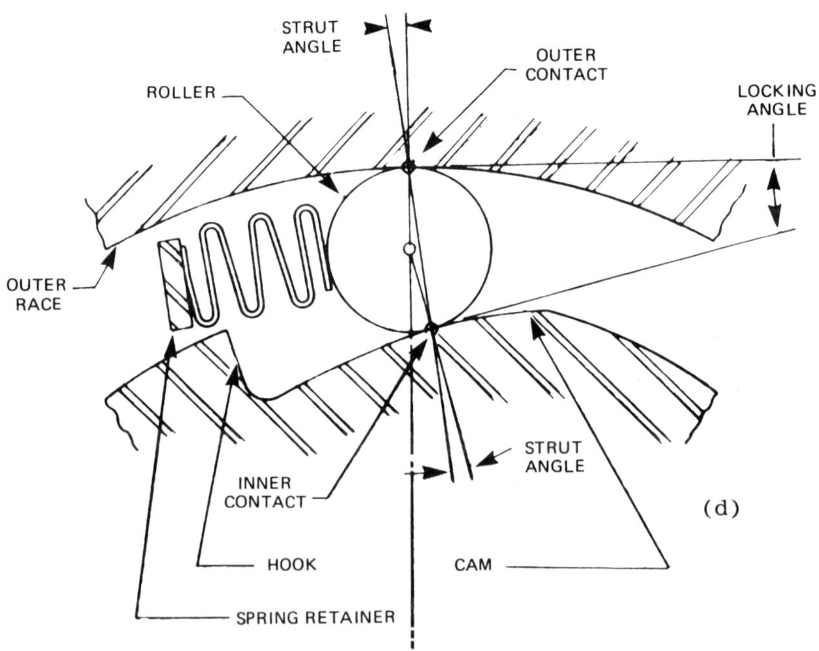

FIGURE 5 Continued.

cylinders, such as between the rollers and the inner races in Figure 6 (a), (b), and (c).

$$\begin{aligned}\sigma_{zz} &= -\frac{2F}{\pi a}\left(1-\frac{x^2}{a^2}\right)^{1/2}, \qquad -a \leqslant x \leqslant a \\ &= 0, \qquad x \leqslant -a,\ x \geqslant a\end{aligned} \tag{3-2b}$$

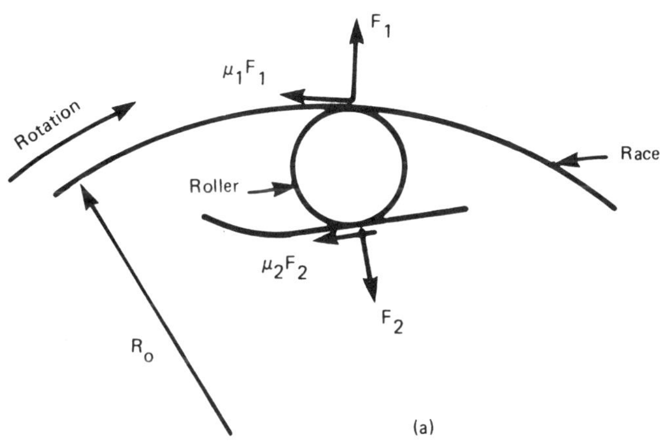

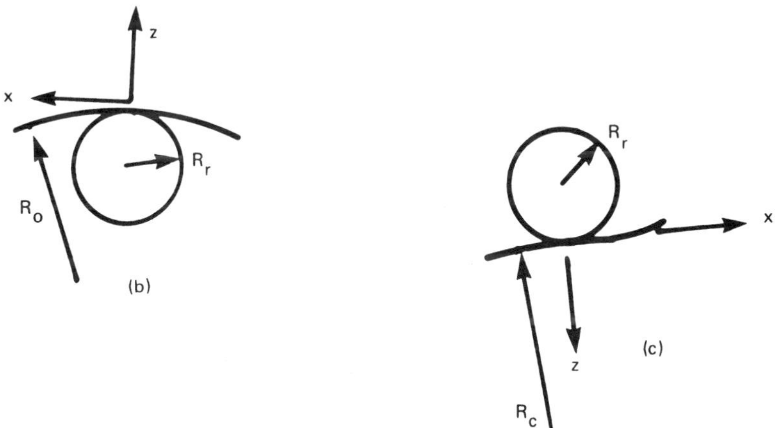

FIGURE 6 Forces on cam and race in an overrunning roller clutch.

$$\sigma_{xz} = -\frac{2F}{\pi a}\mu\left(1-\frac{x^2}{a^2}\right)^{1/2}, \quad -a \leqslant x \leqslant a$$
$$= 0, \quad x \leqslant -a,\ x \geqslant a \tag{3-2c}$$

on the contacting surface. For larger values of μ the maximum stress is on the surface and for smaller values it lies below the surface. Quantities Φ and Ψ in equation (3-1) are defined by

$$\Phi = \frac{\pi}{k_1\phi}\left[1-\left(\frac{k_2}{k_1}\right)^{1/2}\right] \qquad \Psi = \frac{\pi}{k_1\phi}\left[1+\left(\frac{k_2}{k_1}\right)^{1/2}\right]$$

$$\phi = \left(2\frac{k_2}{k_1}\right)^{1/2}\left[\left(\frac{k_2}{k_1}\right)^{1/2} + \frac{k_1+k_2-4a^2}{2k_1}\right]^{1/2}$$

where

$$k_1 = (a+x)^2 + z^2 \qquad k_2 = (a-x)^2 + z^2$$

$$a = 2\left(\frac{E}{\pi}\frac{\dfrac{1-\nu_1^2}{E_1}+\dfrac{1-\nu_2^2}{E_2}}{\dfrac{1}{r_1}+\dfrac{1}{r_2}}\right)^{1/2}$$

Quantities ν_1, ν_2, r_1, r_2, E_1, and E_2 refer to the Poisson ratios, radii, and Young's moduli of the components in contact, i.e., a roller and the outer race or a roller and the inner race. Since the trios of quantities ν_1, r_1, E_1 and ν_2, r_2, E_2 enter symmetrically into the expression for a, either trio may be associated with a roller and the other trio associated with the inner race. Coordinates x and z lie in the circumferential and radial directions, respectively, and quantity a represents the half-width of the contact area measured in the circumferential direction, hence the inclusion of F in the definition of a. Loading force per unit length in the axial direction between a race and a roller is denoted by F. One misprint and one apparent misprint have been corrected by the author in reproducing equations 4 and 12 in Ref. 4 in the preceding set of equations.

Maximum compressive stress at the surface between a roller and a concave race is given by

$$\sigma_{zz} = \frac{-1}{\pi}\left[F\frac{\left(\dfrac{1}{r_1}-\dfrac{1}{r_2}\right)}{\left(\dfrac{1-\nu_1^2}{\pi E_1}+\dfrac{1-\nu_2^2}{\pi E_2}\right)}\right]^{1/2} \tag{3-3}$$

from Ref. 5.

Insertion of the force F used to calculate the previous stresses into equation (3-3) will give the torque transferred to the outer ring as

$$T = FNl[r_1 + r_3(l + \cos\alpha)](\sin\alpha + \mu\cos\alpha) \tag{3-4}$$

in terms of the angle α between F_1 and the direction of F_2, the length/of each of the rollers, the number of rollers, N, the radius r_1, and the radius of a roller, r_3.

Forces acting on a roller as it is wedged between the inner and outer races are shown in Figure 7. Summing forces in the direction of F_1 yields

$$F_2\cos\alpha + \mu_2 F_2 \sin\alpha = F_1 \tag{3-5}$$

and summing forces perpendicular to the direction of F_2 yields

$$F_2\sin\alpha = \mu_2 F_2 \sin\alpha + \mu_1 F_1 \tag{3-6}$$

From Figure 7 we note that the magnitude of the force transmitted to the outer race is given by $F_2(\sin\alpha - \mu_2\cos\alpha)$. This force would reach its maximum and the shear stress on the roller and outer race would vanish if only a normal force acted between a roller and the outer race or, more precisely, the

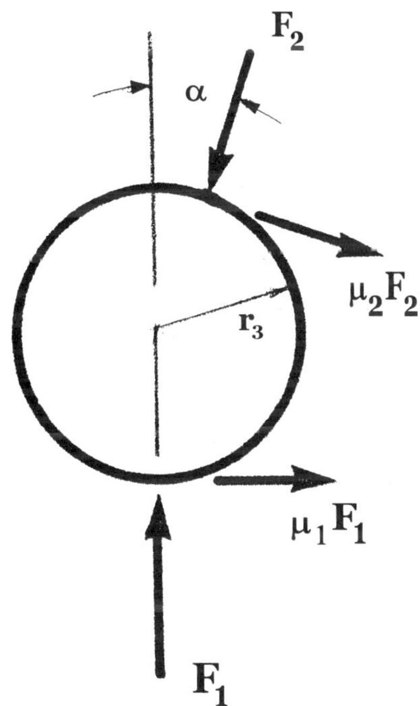

FIGURE 7 Forces acting on a roller wedged between an outer cam and and the inner race.

internal cams on the outer ring. To find the requirement for this condition to be true, set $\mu_2 = 0$ in equations (3-5) and (3-6) to get the equations that hold when only a normal force. F_2, acts between a cam and a roller, as in Figure 7. Substitute the value of F_1 from equation (3-5) into equation (3-6). The result is $\sin \alpha = \mu_1 \cos \alpha$, or

$$\tan \alpha = \mu_1 \tag{3-8}$$

Therefore lubrication of the races (races and cams) and rollers not only reduces wear on the contacting surfaces, but also reduces the angle of the outer cams, for which equation (3-8) is satisfied.

Example

Is it possible to produce a roller clutch to provide a torque of 950 ft-lb that has an inner race no more than 5 in. in diameter and that has a roller length equal to or less than 1.8 in.? Assume that the roller and races are made from a material whose working stress should be no more than 100,000 psi in either tension or shear and that its Young's modulus is that of steel.

Assume Poisson's ratio of 0.3, a friction coefficient of 0.34, and a Young's modulus of 3×10^7 psi. Since the circumference of a 5-in. diameter race is 15.71 in., initially select 10 rollers with 0.50-in. diameters and assume a strut angle of 9°. Let the initial trial contact length of each roller be 1.8 in.

First solve equation (3-5) for F to find $F = 36.776$ lb/in. Substitute $F = 36.776$ lb/in. into equations (3-1) and (3-2) and the definitions of the parameters to get $a = 7.925 \times 10^{-4}$ in., so the contact width is from $x = -0.0079$ in. to $x = 0.00079$ in. Values of k_1 and k_2 are $k_1 = 2.389 \times 10^{-6}$ in.^{2} and $k_2 = 1.676 \times 10^{-6}$ in.^{2}, and parameters Φ and Ψ are given by $\Phi = 1.610 \times 105$ in.$^{-2}$ and $\Psi = 1.826 \times 10^6$ in.$^{-2}$. After calculating the foregoing stress component at the surface and at 0.001 in. below the surface, we find that the largest tensile stress for the points calculated on the surface was 78,060 psi at $x = -0.001$, the largest shear stress was 43,560 psi at the center of the contact area and the largest compressive stress was $-78{,}060$ psi at $x = 0.001$ in. At $z = 0.001$ in. below the surface, all of the direct stresses calculated were compressive, the largest being $\sigma_{xx} = 1{,}642{,}000$ psi at $x = -0.001$ in. The largest stress found was $\sigma_{xx} = 98{,}140$ psi at $x = -0.00018$ in., just outside of the contact region at $z = 0.001$ in. Equation (3-3) gave a compressive stress $\sigma_{xx} = -26{,}530$ psi at the surface of the cam on the outer ring.

IV. OVERRUNNING CLUTCHES: THE SPRAG CLUTCH

A representative sprag clutch is shown in Figure 8. These clutches are also direction depenent, but they differ from the roller clutches in that sprags are used rather than circular cylindrical rollers. Sprags are cylinders whose cross section, as shown in Figure 9, is designed to allow them (1) to engage and

FIGURE 8 Overrunning sprag clutch. (Courtesy Dana Corp., Inc., Toledo, OH.)

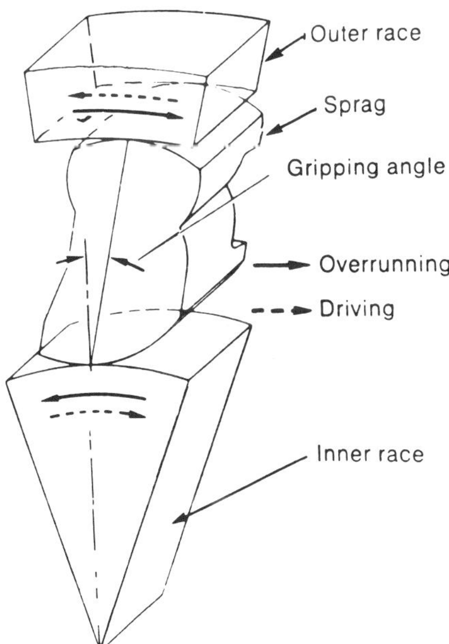

FIGURE 9 Conventional sprag and a sector of its inner and outer races.

disengage in a fraction of a turn and (2) to provide a larger radius of curvature along the contact line between the sprag and the races than would be possible if complete circular cylinders were used. Placing more sprags between the inner and outer races increases the torque by increasing N in equation (3-5) while increasing the radius of curvature reduces the $(1/R_1 - 1/R_2)$ term in equations (3-1) and (3-2) and thereby reduces the contact stress for a given value of F, which permits an increase in the magnitude of F in equation (3-5) for a given stress level in each race. These comments also apply to sprags whose cylindrical surface has been cut by intersecting planes, as in Figure 10, to further increase N, the number of contacting sprags.

Some sprags are designed to respond to centrifugal forces as well as frictional forces, so that as the rotational speed of the faster race increases, the sprags rotate under the influence of the centrifugal force and break contact with one of the races, thereby reducing wear and drag.

Sprags designed to respond only to friction are often termed conventional sprags, while those which are designed to respond to both friction and to an increase in speed are termed either throw-out or throw-in sprags, depending on whether they disengage from the inner or outer race at the lift-off speed.

Conventional sprags and three installation variations are shown in Figure 11. The energizing spring in all three configurations is to hold the sprags in a position to become engaged as soon as conditions permit, while the cages, shown in Figure 11(a) and (c), are to space the sprags apart to reduce

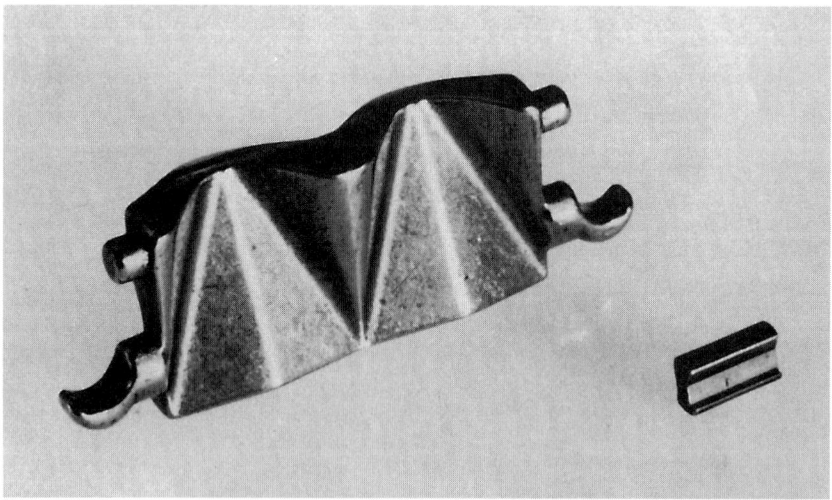

FIGURE 10 Modified sprag design for closer packing. (Courtesy Georg Muller of America, Inc., Schaumburg IL.)

FIGURE 10 Continued.

friction between them and permit faster engagement. Double cages are used to allow the first few sprags that make contact to force the remaining sprags into contact as these first ones assume a more radial orientation and thus cause angular rotation of one cage relative to the other. They may be used with external connections to force disengagement before the driven race reaches a speed greater than that of the driving race. Full complement configurations, represented by Figure 11(b), are used for larger torsional loads, while the single- and double-cage configurations are for smaller loads and faster response.

One version of an overrunning throw-out sprag is shown in Figure 12, in which the cage rotates with the outer race. The small projection at the left of each sprag in this figure not only aids in moving the center of gravity to provide a rotational moment due to the centrifugal force, but acts as a stop

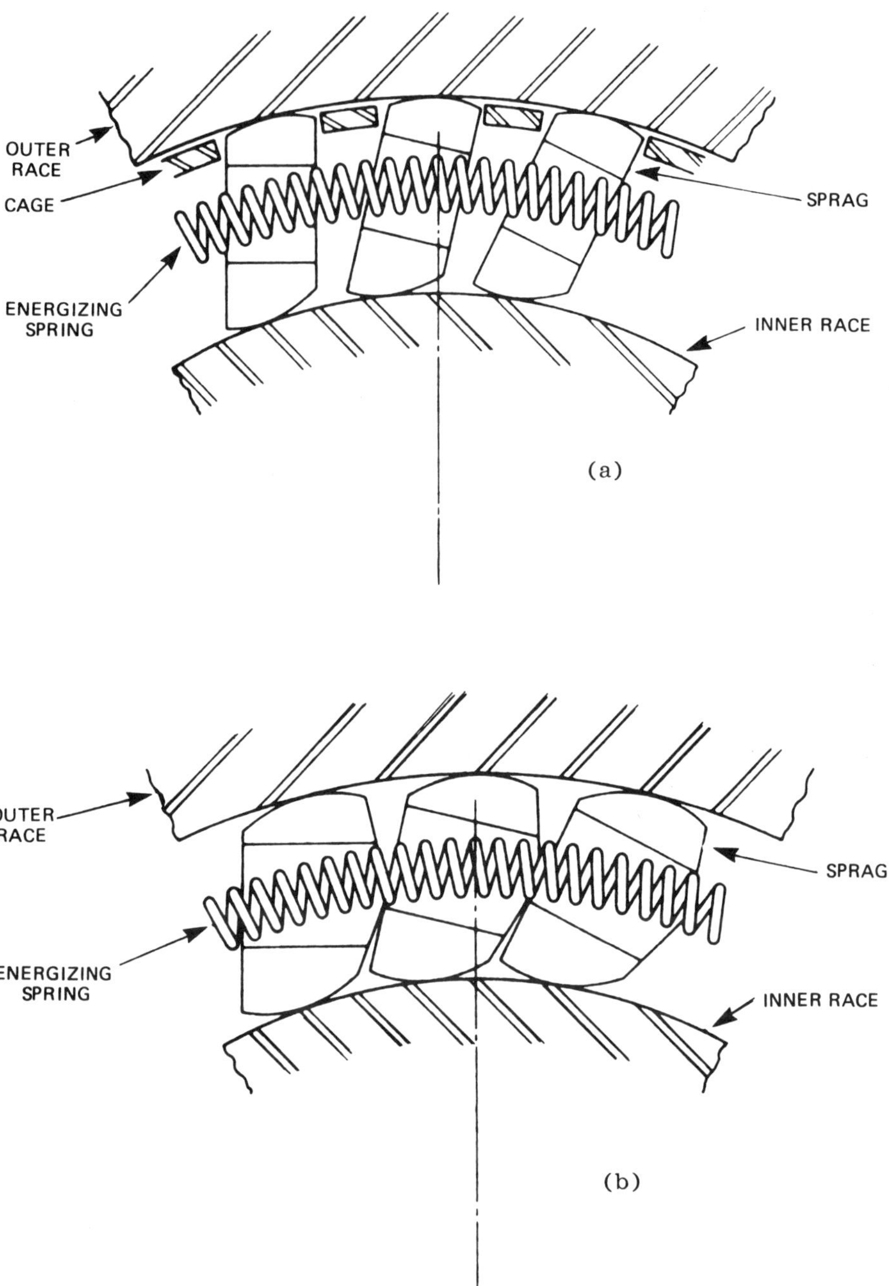

Figure 11 Conventional sprags with their retaining springs and cages. (a) Typical single-cage one-way clutch diagram. (b) Typical full-complement sprag one-way clutch diagram. (c) Typical double-cage sprag one-way clutch diagram. (d) Sprag one-way clutch diagram. (Reprinted with permission; © 1984 Society of Automotive Engineers, Inc.)

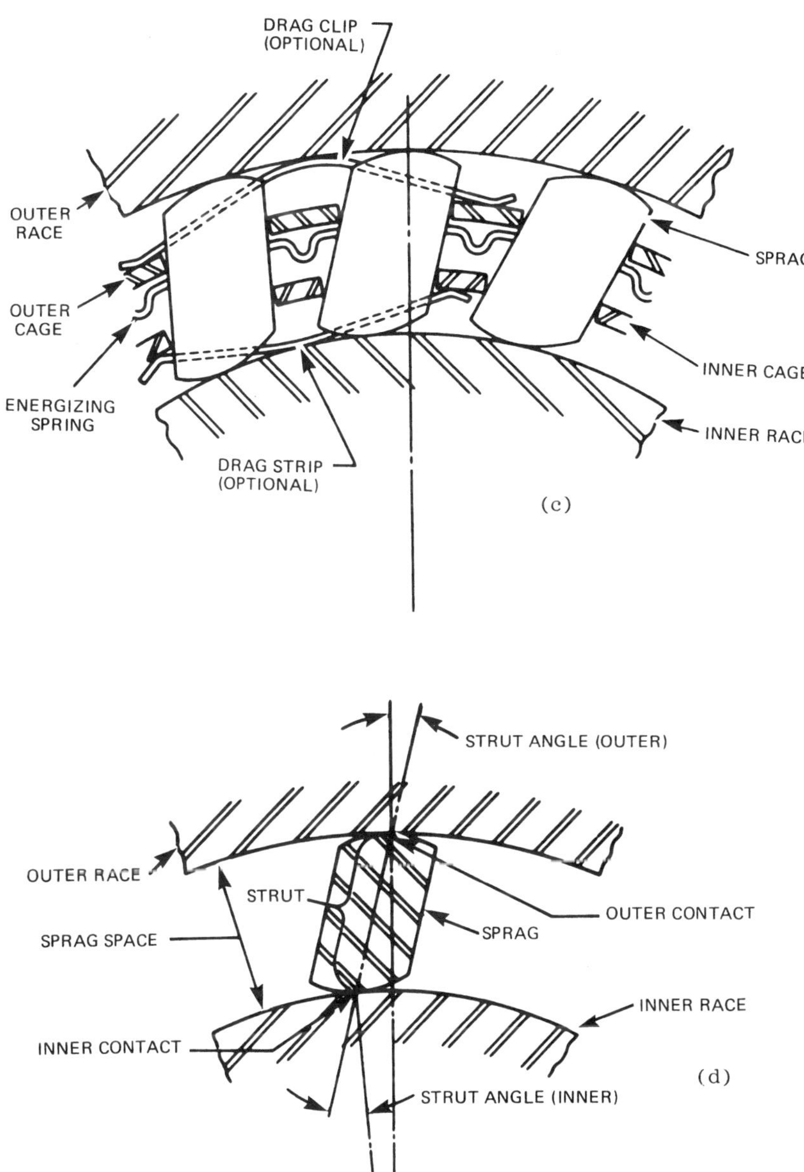

FIGURE 11 Continued.

to prevent sprag rollover under clutch overload conditions, as pictured in Figure 12(c). Springs between the cage and the sprags, not shown in these figures, may be selected to control the lift-off speed.

One manufacturer of throw-in clutches uses a sprag and spring design, as shown in Figure 13, where the sprag retainer, or cage, moves with the inner race so that as the speed of that race increases, the centrifugal force acting on the center of mass of the sprag, which lies to the right of the pivot, causes the

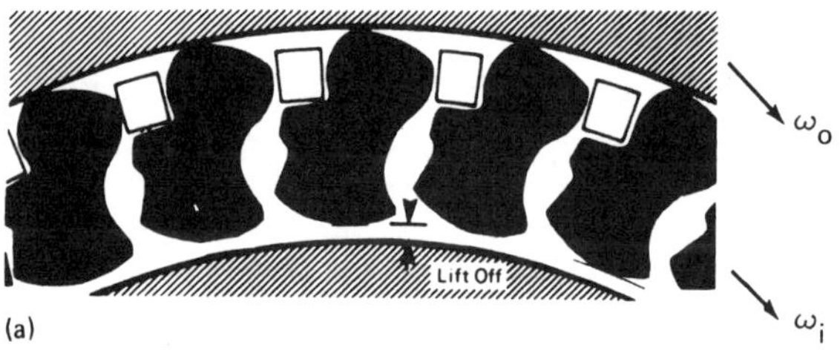

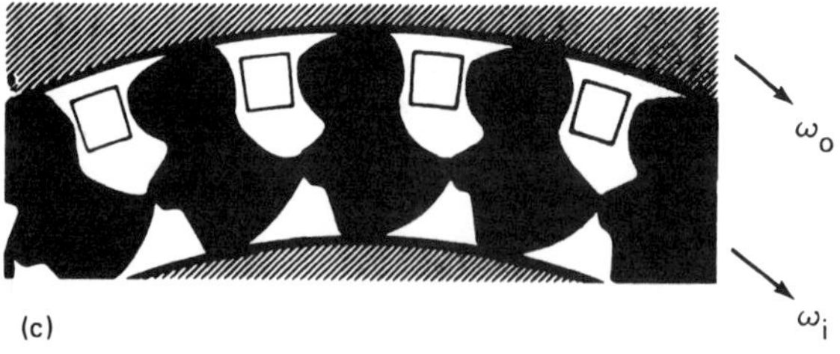

Figure 12 Throw-out sprags with antiroller rails; *C/T* designates centrifugal throw-out: (a) High RPM–*C/T* overrunning: $\omega_o > \omega_i$. (b) Regular engagement condition: $\omega_o = \omega_i$. (c) Overload-imposed conditions: $\omega_o = \omega_i$. (Courtesy Dana Corp., Toledo, OH.)

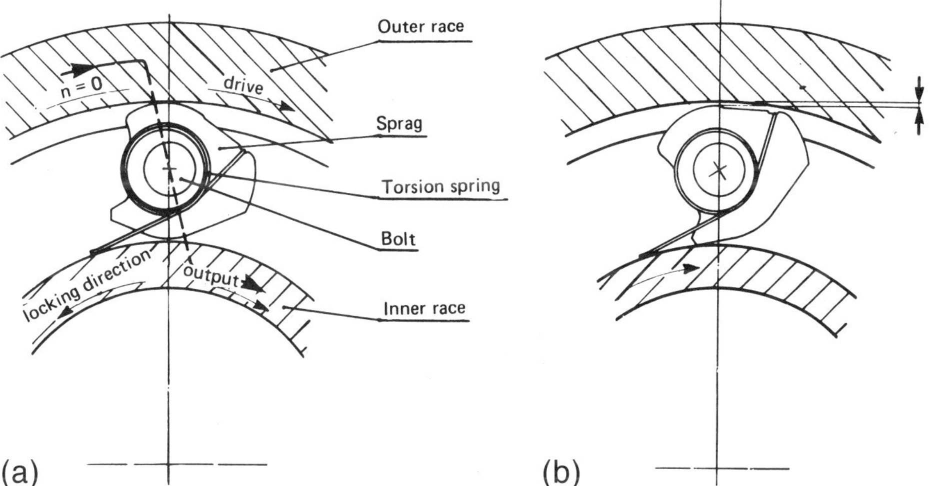

FIGURE 13 Throw-in sprag configuration and spring. (a) Regular engagement. (b) Centrifugal throw-in engagement.

sprag to rotate in a counterclockwise direction and lose contact with the outer race. As the speed of the inner race decreases, the spring causes the sprag to rotate in the clockwise direction so that it will again make contact with the outer race.

Returning now to the conventional sprag profile, let A and B denote the contact points on the profile of a sprag, as shown in Figure 14, let the line between A and B termed the *strut*, and let α represent the angle subtended by the strut at the center of the clutch. Let r_o and r_i represent the radii of the outer and inner races, respectively, let μ_o and μ_i denote the corresponding coefficients of friction, and let F_o and F_i refer to the associated normal forces. In these terms,

$$\mu_o F_o r_o - \mu_i F_i r_i = 0 \tag{4-1}$$

is the moment equilibrium equation of the sprag about the axis of rotation of the clutch. Summing forces in the direction of the friction force at A gives

$$\mu_o F_o + F_i \sin\alpha - \mu_i F_i \cos\alpha = 0 \tag{4-2}$$

as the equilibrium condition in that direction. Substitution for F_o in equation (4-2) from equation (4-1) yields

$$\frac{r_i}{r_o} = \cos\alpha - \frac{1}{\mu_i}\sin\alpha \tag{4-3}$$

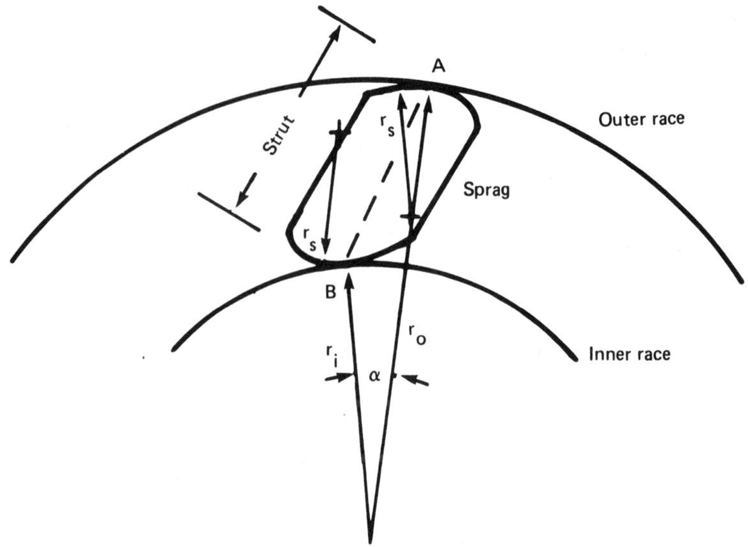

FIGURE 14 Sprag and race geometry: r_i = radius of inner race; r_o = radius of outer race; r_s = radius of sprag contact surface (commonly the arc of a circle or of a logarithmic spiral).

as a guide in selecting angle α. The value of sin α may be found by substituting for cos α from equation (4-3) and substituting into the trigonometric identity

$$\sin^2 \alpha + \cos^2 \alpha = 1$$

and then solving for sin α from the quadratic formula to give

$$\sin \alpha = \frac{r_i}{r_o} \frac{\mu_i}{1+\mu_i^2} \left\{ \left[1 + (1+\mu_i^2)\left(\frac{r_o}{r_1} - 1\right) \right]^{1/2} - 1 \right\} \tag{4-4}$$

Forces F_o and F_i are related to the torque according to

$$F_i \leq \frac{T}{\mu_i - N - r_i} \tag{4-5}$$

and, from the equilibrium equations for the sprag,

$$F_o = \frac{\mu_i \frac{r_o}{l} \sin(\alpha) - \cos\left(\alpha \sin\left(\frac{r_o}{l}\sin(\alpha)\right)\right)}{\mu_o \sin\left(\alpha + \alpha \sin\left(\frac{r_o}{l}\sin(\alpha)\right)\right) - \cos\left(\alpha + \alpha \sin\left(\frac{r_o}{l}\sin(\alpha)\right)\right)} \tag{4-6}$$

$$> \frac{T}{\mu_o N r_o}$$

in which l is the length from A to B in Figure 14. These values of r_s and r_i may be substituted into equations (3-1) and (3-2) and values of r_s and r_o substituted into equation (3-3) or into a finite element program for contact stresses at the inner and outer radii, to find the minimum radius of curvature r_s, shown in Figure 14, that will give a permissible stress for these sprag profiles for the inner race (IR) and for the outer races (OR). Different radii may be selected for contact stresses at the IR and OR for the sprag configurations shown in Figure 11.

Sprag overrunning clutches have speed envelopes within which they can operate as designed. Although these envelopes have the same general shape, the nature of the envelope in the third quadrant (that to the left of the vertical axis and below the horizontal axis) may vary as shown in Figures 15 and 16 for sprag clutches. Recommended operating speeds lie between the upper curve, which is the upper boundary of the envelope, and the 45° diagonal, which is the lower boundary of the envelope. In both figures the upper curved arrow in each quadrant depicts the direction of rotation of the outer race in that quadrant, the lower curved arrow in each quadrant indicates the direction of rotation of the inner race in that quadrant, and the inclined line between the

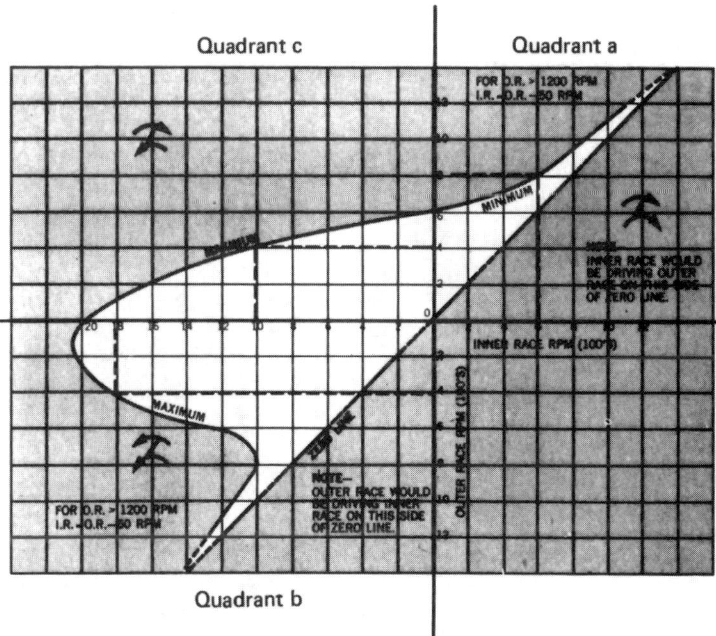

FIGURE 15 Relative overrun speed envelope showing the directions of rotation and the strut angle. (Courtesy Dana Corp., Toledo, OH.)

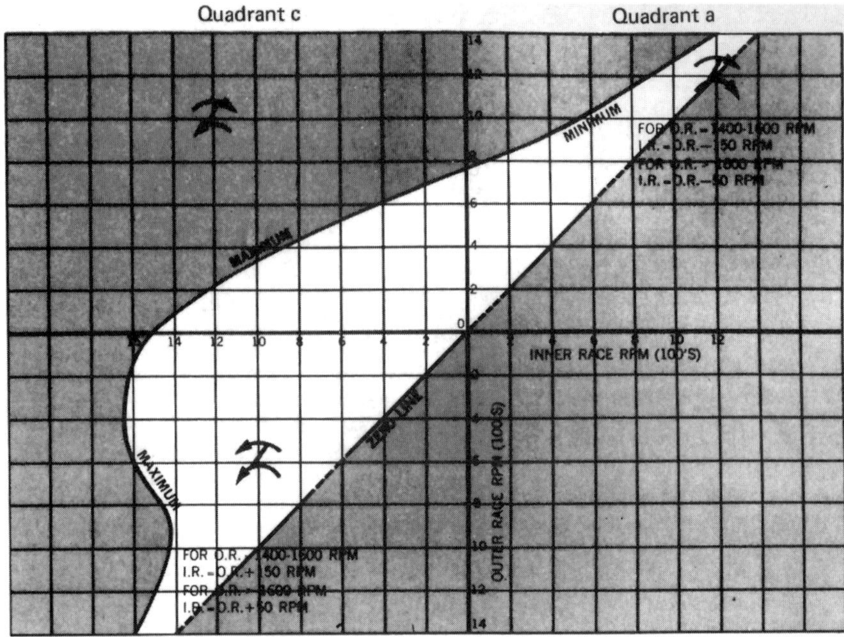

FIGURE 16 Relative overrun speed envelope showing representative variation between models. (Courtesy Dana Corp., Toledo, OH.)

curved arrows indicates the direction of the strut. In all cases clockwise rotation is taken as positive.

No curved arrows are shown in the fourth quadrant because in that quadrant the IR rotation is positive, the OR rotation is negative, and the strut thrust angle is such that one will always drive the other, so that no overrunning is possible. No overrunning will occur at points in the first quadrant below the diagonal because in this region the OR rotates more slowly than the IR but the strut angle is such that the IR cannot overrun the OR. Similar reasoning regarding the third quadrant will show that the OR cannot override the IR.

Rotational combinations corresponding to points above and/or to the left of the envelope in the first and second quadrants are not recommended even though overrunning is possible in these regions because at these higher rotational speeds of the OR it tends to accelerate the IR in the first quadrant and decelerate it in the second quadrant. Similarly, rotational speeds corresponding to points below and/or to the left of the envelope in the third quadrant, where overrunning in possible, are not recommended because these IR speeds tend to accelerate the OR. As noted in Figure 15, at OR speed in excess of 1200 rpm in the first and third quadrants, the IR rotational speed tends to be only 50 rpm less than the OR speed.

V. TORQUE LIMITING CLUTCH: TOOTH AND DETENT TYPES

Although tooth clutches, as pictured in Figure 17, are usually used for positioning one shaft relative to the other, they may, in an emergency, also serve as overload detent clutches, because their torque is limited by the axial force

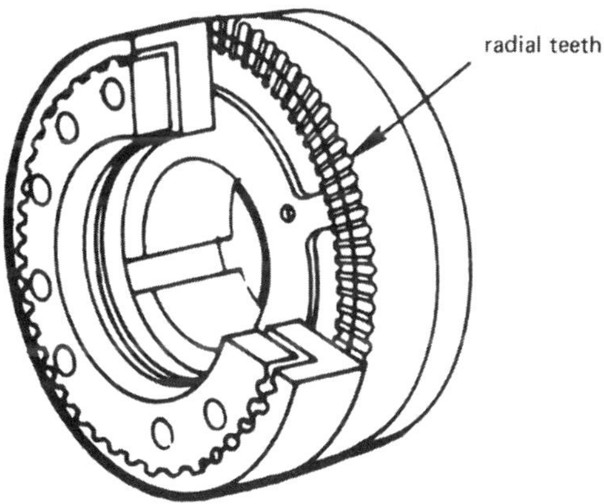

FIGURE 17 Tooth clutch for shaft positioning.

holding the toothed jaws together. From Figure 18 we find that the tangential, F_t, and normal, F_n, forces are given by

$$F_t = N \sin \zeta + \mu N \cos \zeta \tag{5-1}$$

and

$$F_n = N \cos \zeta - \mu N \sin \zeta \tag{5-2}$$

from which it follows that the ratio F_t/F_n, the ratio of the tangential load to the axial load, becomes

$$\frac{F_t}{F_n} = \frac{\sin \zeta + \mu \cos \zeta}{\cos \zeta - \mu \sin \zeta} \tag{5-3}$$

which may be simplified to read

$$\frac{F_t}{F_n} = \tan(\zeta + \beta) \tag{5-4}$$

if β is defined to be

$$\beta = \tan^{-1} \mu \tag{5-5}$$

The maximum torque that a tooth clutch with wedge-shaped teeth can transmit may be estimated from

$$T = N r_k F_n \tan(\zeta + \beta) \tag{5-6}$$

where N is the number of teeth and r_k is the radius from the axis of the clutch to the circle that passes through the center of the teeth in Figure 17.

Clutches designed specifically as overload release clutches remain engaged only if the transmitted torque is less than a prescribed critical value. Once that value is exceeded, the clutch is disengaged and remains disengaged until it is manually reset. Several versions will be considered here.

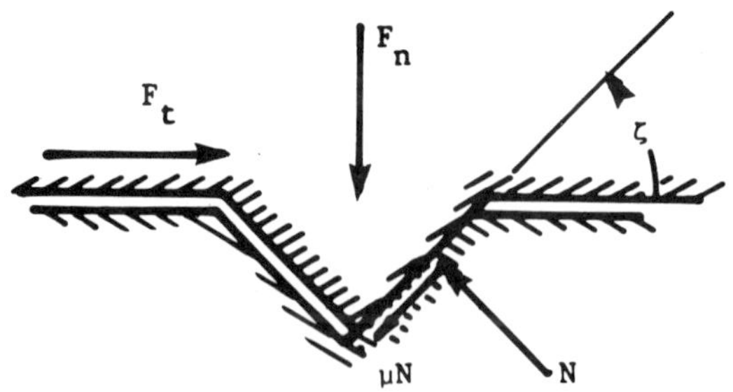

FIGURE 18 Forces acting on a single wedge tooth. (Courtesy Horton Mfg. Co., Minneapolis, MN, and Machine Design, Penton Press, Cleveland, OH.)

FIGURE 19 Overload release clutch using clutch plates and wedge-shaped release cam. (Courtesy of Carlyle Johnson Machine Co., Manchester, CT.)

The first is shown in Figure 19. Torque is transferred by means of clutch plates that are alternately keyed to the driver ring on the left and to the clutch body that is concealed by the clutch plates and is enclosed by the sleeve on the right in this figure. A collar on the input shaft (not shown) is bolted to the driver ring and the clutch body is keyed to the output shaft. Pressure between clutch plates is exerted by the adjusting ring that is shown just to the right of the clutch plates. A trio of levers that lie between that clutch body and the outer sleeve that extends to the right of the adjusting ring hold the adjusting ring in place when the race on the sleeve engages the cam on the driver ring. An overload causes the clutch plates to slip, which in turn allows the cam on the driver ring to push against the race on the sleeve and cause it to move axially to the right to disengage internal levers that maintain clamping pressure on the clutch plates. Until the clutch is reset the torque transfer drops to 1% of the rated torque with no ratcheting. Rated torque capability for clutches of this type from this manufacturer range from 20 lbs ft. to 2400 lbs ft., depending upon size.

A second style of overload clutch may employ spring-loaded rollers (or balls) held in sockets attached to one plate such that the rollers rest in pockets in the other, as shown in Figure 20. These rollers will remain in the pockets as long as the tangential force between plates is satisfies

$$F_t \leqslant F_k \frac{\cos\zeta + \mu(1 + \sin\zeta)}{\sin\zeta - \mu(1 + \cos\zeta)} \tag{5-7}$$

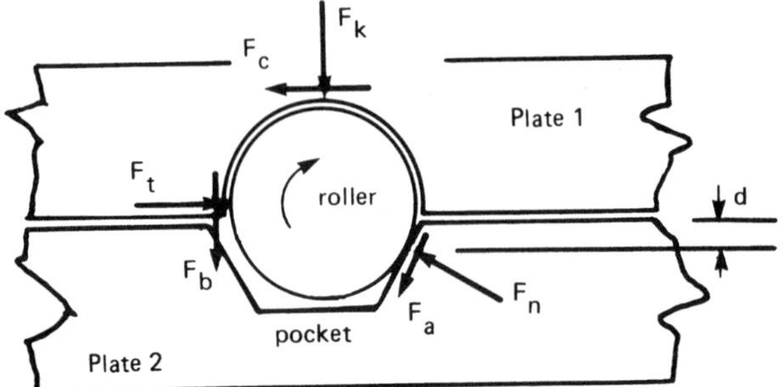

FIGURE 20 Forces on a typical roller or ball in a detent clutch using one or more such elements. The spring-loading mechanism is replaced by F_k and F_c.

where F_k is the spring force on the ball, F_t is the lateral force on the ball, μ is the coefficient of friction, and ς is the angle between the detent wall and the vertical. This relation may be derived by taking moments about the instantaneous center at the contact between the roller and the pocket in Figure 20.

Torque transmitted by the clutch may be written as

$$T = F_t RN \tag{5-8}$$

where R is the radius from the center of the balls to the axis of rotation of the disk on which they are mounted and N is the number of spring-and-ball assemblies on the disk. When equality holds in equation (5-7), substitution for F_t from equation (5-8) into equation (5-7) yields that the spring force must satisfy

$$F_k = \frac{T}{NR} \frac{\sin \zeta - \mu(1 + \cos \zeta)}{\cos \zeta + \mu(1 + \sin \zeta)} \tag{5-9}$$

Most, if not all, detent overload clutches use something similar to the geometry shown schematically in Figure 20, in which the detents are in one plate and the spring-loaded balls and their retainers are mounted on the other plate. Immediately after an overload occurs, the balls are pushed from their detents and pop into and out of adjacent detents until either the rotation is stopped or the overload is removed. Consequently it is often recommended that they be used on shafts that rotate at less than 500 rpm to reduce damage to both the balls and the detents. The transmitted torque after the balls are pushed from their detents may be about 5% of the rated torque of the clutch. Ball and detent arrangements in clutches where indexing (maintaining a constant angular relation between input and output shafts) after an overload is not required usually are arranged in axial symmetry in order to reduce shaft vibration and noise.

Many manufacturers of ball and detent overload clutches designed for applications where indexing is important are reluctant to display the particular designs used to achieve automatic indexing after an overload is removed. The following description of a detent arrangement is offered, therefore, only to show that a simple detent layout is possible that can provide automatic indexing. It is based upon the observation that without axial symmetry of the detent positions there should be only one relative position between mating disks where all of the balls in one disk fit into all of the detents in the other disk so that the clutch can transmit its rated torque.

An example of one possible configuration is that shown in Figure 21, in which each detent subtends an angle of 10° from the center of the disk and all eight detents lie in a circle about the center of the disk so that they all contribute equally to the total torque.

Three disadvantages of this arrangement are: (1) the balls slam into and out of the detents when the disks rotate relative to one another in the overload condition; (2) the plates must have masses either added or removed to establish dynamic balance; and (3) shaft speed should usually be no more than 500 rpm. During overloading, this type of clutch may transmit a small fluctuating torque that could be of the order of 5–13% of the rated torque until the overload is removed. The compensating advantage is that after the overload is removed, the clutch automatically reindexes to the proper position.

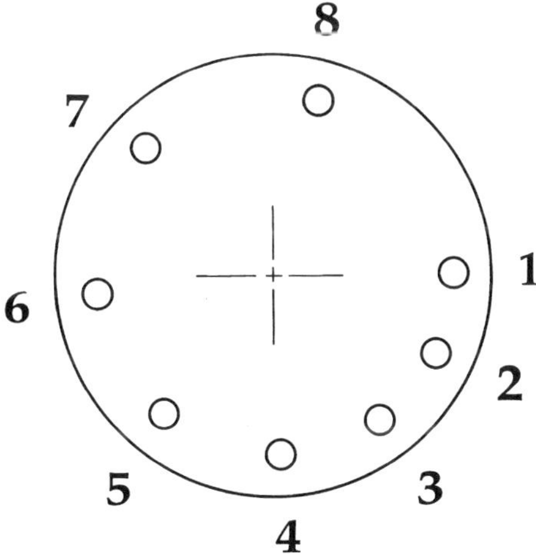

FIGURE 21 Detent positions for an indexing ball and detent overload clutch.

Detent center locations, and that of the mating detent balls, shown in Figure 21, are arranged according to the angular positions shown in the following table. Addition of $\Delta\theta^\circ$ shown in the top row above detent number 1 to the value of θ° just above detent number 1 gives θ° for detent number 2, and so on.

$\Delta\theta^\circ$	25	30	35	40	45	50	60	75	
θ°	0	25	55	90	130	175	225	285	360
Detent number	1	2	3	4	5	6	7	8	1

This detent arrangement could be improved. That is because in each rotation of the disk fitted with the spring-loaded ball assembly relative to the disk in which the detents are cut there are three relative orientations where two ball-and-detent pairs engage. In those three instances the torque may momentarily jump to 25% of the rated torque rather than to the 12.5% that may occur when only one ball-and-detent pair engages.

To elaborate, if the ball-and-detent pairs are numbered in the clockwise direction when viewed from the driving plate to the driven plate, as in Figure 21, we find that during clockwise rotation of the driving plate relative to the driven plate three instances occur wherein two ball-and-detent pairs are engaged before the plates reindex. The first instance occurs when balls 8 and 3 engage detents 1 and 5. This is because the angular separation between detents 8 and 3 is the same as that between detents 1 and 5. In particular, from the preceding table of the angular positions of the ball-and-detent pairs and the angular separation between centers we find

75° between detents 8 and 1
25° between detents 1 and 2
30° between detents 2 and 3
130° between detents 8 and 3

and that the angular separation between detent centers 1 and 5 is given by

25° between detents 1 and 2
30° between detents 2 and 3
35° between detents 3 and 4
40° between detents 4 and 5
130° between detents 1 and 5

The second instance occurs when balls 8 and 1 engage detents 1 and 5, where ball centers 8 and 1 are separated by 75° and detent centers 3 and 4 are separated by 35° and 4 and 5 are separated by 40° for a combined separation of 75°. The third and last instance is when balls 7 and 1 engage detents 4 and 7. Placing ball-detent pairs at different radii eliminates engagement except at the index position, but those ball and detent locations at smaller radii transmit less torque.

VI. TORQUE LIMITING CLUTCH: FRICTION TYPE

Torque limiting friction clutches are another version of overload clutches. They differ from those considered in the previous sections in that the transmitted torque does not drop sharply from the rated torque. Instead, the

FIGURE 22 Torque limiting clutch. From the Carlyle Johnson Machine Co., Bolton, CT, Web page on the Thomas Register Web site.

transmitted torque does not exceed a preset value regardless of the speed or of the torque imposed by the driven unit. The clutch simply slips when its preset torque is exceeded. A torque limiting friction clutch, as shown in Figure 22, consists of a series of spring-loaded clutch plates in which greased alternate steel and bronze plates are keyed to the input and output sections of the clutch. Spring loading to set the torque limit is accomplished by controlling the spring force on the plates by means of the adjusting nut at the right-band end of the clutch hub. Grease sealed within the hubs provides a lubricant between the clutch plates, and it is the viscous characteristics of the grease that are used to determine the torque characteristics of the clutch, i.e., the slope of the curves in Figure 23. Consequently, a variety of torque characteristics are available. Torque limits are a linear function of the spring compression, also as illustrated in Figure 23 for a particular grease.

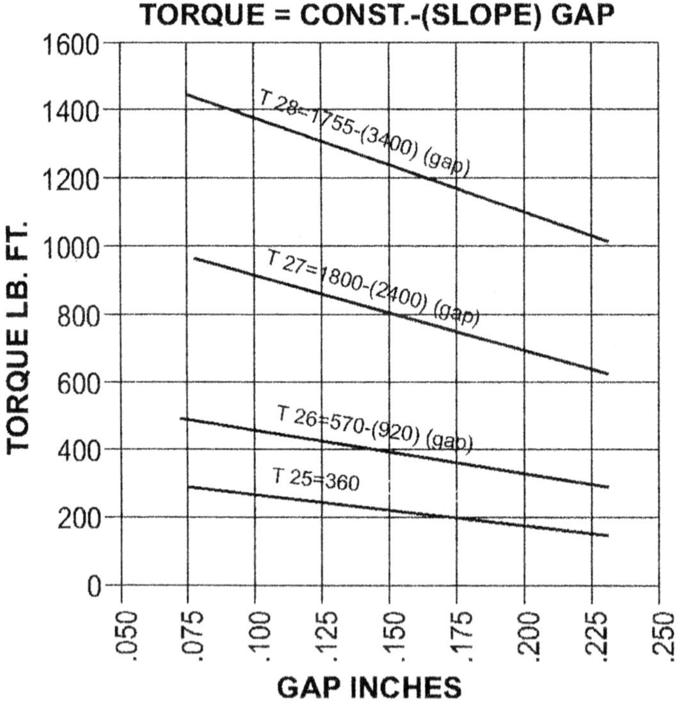

FIGURE 23 Torque as a function of the adjustable gap for the models noted. From the Carlyle Johnson Machine Co., Bolton, CT, Web page on the Thomas Register Web site.

Torque capabilities of models similar to that shown in Figure 22 range from 0 to 10 lb-ft up to 0 to slightly over 1400 lb-ft.

VII. NOTATION

A	area (l^2)
a	factor in Hertzian (contact) stress formulas (l)
b	width, rectangular wire (l)
c	correction factor (1)
E	elastic modulus (Young's Modulus) ($ml^{-1}t^{-2}$)
F	force (mlt^{-2})
g	acceleration due to gravity (lt^{-2})
I	moment of area (l^4)
k	spring constant (mt^{-2})
m	mass (m)
N	number, an integer (1)
n	angular velocity in revolutions per minute, rpm (t^{-1})
p	pressure ($ml^{-1}t^{-2}$)
R,r	radius (l)
r_o	radius to a centroid (l)
T	torque (ml^2t^{-2})
t	time (t) or wire thickness (l)
W	weight (mlt^{-2})
w	width (l)
x,z	cartesian coordinates (l)
α	angle (1)
β	ratio of radii (1)
γ	mass density (ml^{-3})
δ	displacement (l)
ζ	safety factor (1)
η	displacement ratio (1)
θ	angle (1)
κ	spring multiplication factor (1)
λ	centroid parameter (1)
μ	friction coefficient (1)
ν	Poisson's ratio (1)
σ	stress ($ml^{-1}t^{-2}$)
Φ	factor in Hertzian (contact) stress formulas (l^{-2})
ϕ	angle (1) and intermediate parameter in Hertzian stress formulas
Ψ	factor in Hertzian (contact) stress formulas (l^{-2})
γ	angular velocity in radians (t^{-1})

VIII. FORMULA COLLECTION

Maximum pressure, centrifugal clutch:

$$p_{\max} = \frac{2F}{r_o w(\phi_o + \sin\phi_o)}$$

Torque, centrifugal clutch:

$$T = \gamma r_o F = \mu\gamma w r_o \, NA\left[(r_o + \delta)\omega^2 - g\kappa(1+\eta)\right] \quad \eta = \delta/\delta_s$$

Torque, spring clutch, winding direction:

$$T_t = Elr_h\left(\frac{1}{R_2} - \frac{1}{R_1}\right)(e^{2\pi\mu N} - 1)$$

Torque, spring clutch, unwinding direction:

$$T_u = Elr_h\left(\frac{1}{R_2} - \frac{1}{R_1}\right)(1 - e^{-2\pi\mu N})$$

Maximum torque, spring clutch, based on wire dimensions:

$$T_{\max} \leqslant bt^2 \frac{1.05}{2r_h}\left(R_1 - R_2 - \frac{t}{2}\right)$$

Normal plus tangential contact stress, away from surface:

$$\sigma_{xx} = -\frac{2F}{\pi^2 a}\left[(a^2 + 2x^2 + 2z^2)\frac{z}{a}\Psi - 2\pi\frac{z}{a} - 3xz\Phi + 2\pi\mu\frac{x}{a} + \mu(2x^2 - 2a^2 - 3z^2)\Phi + 2\mu(a^2 - x^2 - z^2)\frac{x}{a}\Psi\right]$$

$$\sigma_{zz} = -\frac{2F}{\pi^2 a} z(a\Psi - x\Phi + \mu z\Phi)$$

$$\sigma_{xz} = -\frac{2F}{\pi^2 a}\left[z^2\phi + \mu(a^2 + 2x^2 + 2z^2)\frac{z}{a}\Psi - 2\mu\pi\frac{z}{a} - 3\mu xz\Phi\right]$$

Normal plus tangential contact stress, on surface:

$$\sigma_{xx} = -\mu\frac{4F}{\pi a}\left[\frac{x}{a} - \left(\frac{x^2}{a^2} - 1\right)^{1/2}\right] \quad \text{for } x \geqslant a$$

$$= -\frac{2F}{\pi a}\left[\left(1 - \frac{x^2}{a^2}\right)^{1/2} + 2\mu\frac{x}{a}\right] \quad \text{for } -a \leqslant x \leqslant a$$

$$\sigma_{xx} = -\mu\frac{4F}{\pi a}\left[\frac{x}{a} + \left(\frac{x^2}{a^2} - 1\right)^{2}\right] \quad x \leqslant -a$$

$$\sigma_{zz} = -\frac{2F}{\pi a}\left(1 - \frac{x^2}{a^2}\right)^{1/2} \qquad \text{for } -a \leqslant x \leqslant a$$

$$= 0 \qquad \text{for } x \leqslant -a,\ x \geqslant a$$

where

$$\Phi = \frac{\pi}{k_1 \phi}\left[1 - \left(\frac{k_2}{k_1}\right)^{1/2}\right] \qquad \Psi = \frac{\pi}{k_1 \phi}\left[1 + \left(\frac{k_2}{k_1}\right)^{1/2}\right]$$

$$\phi = \left(2\frac{k_2}{k_1}\right)^{1/2}\left[\left(\frac{k_2}{k_1}\right)^{1/2} + \frac{k_1 + k_2 - 4a^2}{2k_1}\right]^{1/2}$$

$$k_1 = (a + x)^2 + z^2 \qquad k_2 = (a - x)^2 + z^2$$

$$a = 2\left(\frac{F_r}{\pi}\,\frac{\dfrac{1 - v_1^2}{E_1} - \dfrac{1 - v_2^2}{E_2}}{\dfrac{1}{r_1} + \dfrac{1}{r_2}}\right)^{1/2}$$

for all previous normal and tangential contact stresses (i.e., modified Hertzian stresses).

Hertzian stress, outer ring, cam:

$$\sigma_{zz} = \frac{-1}{\pi}\left[F\frac{\dfrac{1}{r_1} - \dfrac{1}{r_2}}{\left(\dfrac{1 - v_1^2}{\pi E_1} + \dfrac{1 - v_2^2}{\pi E_2}\right)}\right]^{1/2}$$

Radial force, centrifugal clutch:

$$F = \gamma w A\left[(r_c + \delta) - g\kappa\frac{\delta + \delta_s}{\delta_s}\right]$$

Torque, overload detent clutch:

$$F_t = \frac{T}{NR}\,\frac{\sin\zeta - \mu(1 + \cos\zeta)}{\cos\zeta + \mu(1 + \sin\zeta)}$$

Spring constant, centrifugal clutch:

$$k = wc\phi_o^2 r_o^2 \gamma \frac{g}{\delta_s}\kappa(1 - \beta^2)$$

REFERENCES

1. Wiebusch, C. F. (1939). The spring clutch. *Journal of Applied Mechanics*. Vol. 6:A103–A108.
2. Wahl, A. M. (1940). Discussion of the spring clutch. *Journal of Applied Mechanics*. Vol. 7:A89–A91.
3. Kaplan, J., Marshall, D. (1956). Spring clutches. *Machine Design*. Vol. 28:107–111.
4. Smith, J. O., Liu, C. K. (1953). Stresses due to tangential and normal loads on an elastic solid with application to some contact stress problems. *Journal of Applied Mechanics*. Vol. 20:157–168.
5. Timoshenko, S. P., Goodier, J. N. (1970). *Theory of Elasticity*. NY: McGraw-Hill Book Co., pp. 417–418.
6. Poritsky, H. (1950). *Journal of Applied Mechanics*. Vol. 17:191.

10

Friction Drives with Clutch Capability

Friction drives that also have clutch capabilities are attractive because they are relatively simple and inexpensive. However, they have been inherently limited to relatively low-power applications because of their dependence upon a coefficient of friction that is usually less than 0.6 between the contacting materials. A friction drive was used in an early automobile, but it was discontinued because of its power limitation.

Friction drives recently have been given new life with the development of elastohydrodynamic fluids that become solid under pressure and can change from solid to liquid and back within microseconds. The fluids provide an effective friction coefficient that may be 1.0 or greater as long as the tangential forces impose a shear stress that is less than the ultimate shear stress of the solid-state form of the elastohydrodynamic fluid. Hence, these drives, which feature metal-to-fluid/solid-to-metal contact, can transmit sufficient power to find industrial and automotive applications that benefit from their ability to easily and simply provide continuously variable speeds. At this time they are relatively expensive because of the structure needed to support the large contact forces that induce the fluid-to-solid transformation. They are presently known as *traction drives*. At this time, however, no known traction drives in production include a clutch capability; consequently they will not be included in this chapter.

Several formulas presented in this chapter may be written in nondimensional form for three reasons: (1) the nondimensional form indicates the relative significance of the ratios selected; (2) it allows drive designs to easily be scaled up or down for various applications; and (3) it allows any consistent set of units to be used for each ratio, and the resulting ratios are independent of the units used.

Relatively broad curves are shown in the following computer-generated graphs for easy reading to show characteristic behavior and to provide contrast against the grid lines. Associated routines, such as Mathcad Trace, appear to read them from the originating data, thereby eliminating the reading errors associated with trace widths.

I. BELT DRIVES

Equipment using nonmetallic belt drives may include the clutch capability by mounting the motor (because it is usually smaller than the driven machine) either upon a hinged base or upon a sliding base fitted with a lever or a linkage that permits the motor to be moved to and from the driven machine in order to apply and relieve the belt tension and thereby give clutching (applying belt tension) and declutching (relieving belt tension) capability.

These designs eliminate the need for a mechanical clutch. Their simplicity is achieved, however, at the risk of introducing the possibility that frictional heating of the belt during idling, when the belt (or belts) may rest on the motor's rotating sheave (pulley). That may generate enough heat to cause belting materials to slowly shrink. This reduction in the center distance between the driving and driven pulleys, or sheaves, may be great enough to cause an unintended re-engagement of the motor and the driven machine. It may also inhibit their disengagement. Consequently, some belt manufacturers produce belts that resist shrinkage due to heating for use in these clutching and declutching applications.

Torque capability for these drives is a separate calculation to be performed according to the procedures given by the belt manufacturers. Therefore, it will not be considered in the following discussion.

A. Hinged Base

At first glance it may appear that moving a motor by mounting it either on a hinged base or on a sliding base is so simple that no analysis is necessary. An analysis, however, does bring forth several considerations that may be missed in selecting the dimensions of the base plate, in locating the position of the base plate hinge, or in designing the linkage for the sliding base plate.

Two similar, but distinct, mounting designs for hinged bases will be considered. In these configurations it is the weight of the motor alone that provides the belt tension. The tension vector shown in Figure 1(a) and (b) acting at the center of the motor shaft represents the sum of the tension acting through the upper and lower belts.

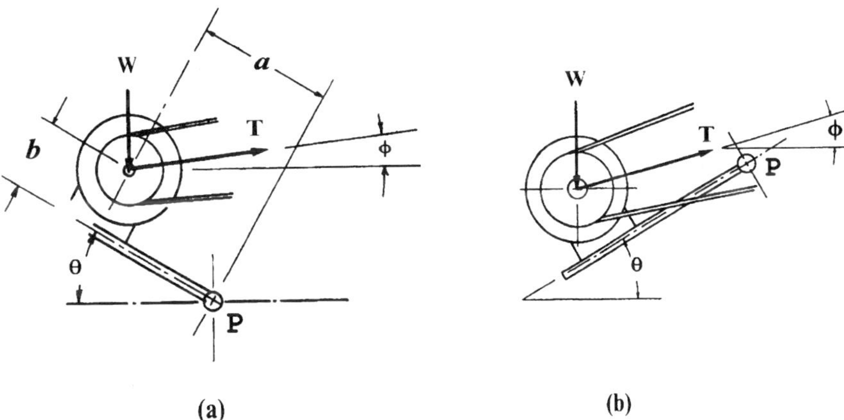

FIGURE 1 Hinged base, belt drive.

Analysis of the first of the two is based upon the configuration shown in Figure 1(a). Upon taking moments about the hinge point P in Figure 1(a) we have

$$W(a\cos\theta - b\sin\theta) = T[a\sin(\theta+\varphi) + b\cos(\theta+\varphi)]$$

After letting $\xi = b/a$, this equation may be written as

$$\frac{T}{W} = \frac{\cos\theta - \xi\sin\theta}{\sin(\theta+\phi) + \xi\cos(\theta+\phi)} \tag{1-1}$$

where θ is positive in the clockwise direction from a horizontal plane through point P and ϕ is positive counterclockwise from a horizontal plane either through or parallel to the motor's axis of symmetry.

Figure 2(a) and (b) show that the weight-to-tension ratio $W/T = 1/(T/W)$ decreases with angle θ when $\phi = 0$. In other words, since W is constant, a decreasing W/T ratio means that tension T increases as θ decreases until θ becomes negative enough for the tension vector **T** to pass through the hinge line that passes through point P. That occurs at the point where $W/T = 0$ on the two lower curves in Figure 2(a). Tension T goes to infinity in Figure 2(b) at those points that are at approximately $\theta = -30.8^\circ$ for $\xi = 0.6$ on the middle curve and at approximately $\theta = -16.6^\circ$ for $\xi = 0.3$ on the lower curve, as determined either by using Mathcad's x – y Trace feature or by interpolation. By comparing the curves in Figure 2(a) it is evident that the W/T ratio also increases as ξ increases for $\phi = 0$ and $\theta = 20^\circ$.

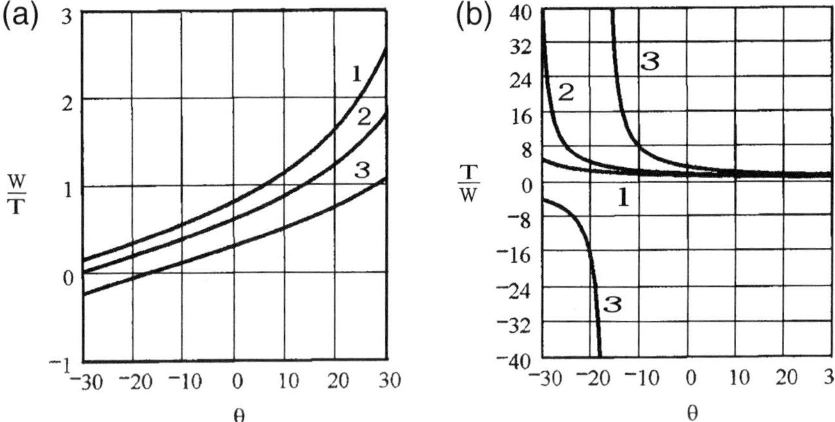

FIGURE 2 Variation of weight-to-tension W/T ratio and tension-to-weight ratio T/W with angle. (a) Plot of W/T, in which $\phi = 0$ for all curves; curve 1, $\xi = 0.8$; curve 2, $\xi = 0.6$; curve 3, $\xi = 0.3$. (b) Plot of T/W, in which $\phi = 0$ for all curves; curve 1, $\xi = 0.3$; curve 2, $\xi = 0.6$; curve 3, $\xi = 0.8$.

Upon turning to Figure 2(b) and recalling that nonmetallic belts under tension stretch over time, it is clear that whenever these belts are used, the tension on them will increase as the angle θ decreases due to the belt's stretching. Hence, the motor must be moved periodically if the tension is to remain within narrow limits.

The rapid increase in tension for negative values of θ in Figure 2(b) emphasizes that the configuration shown in Figure 1(b) should be avoided whenever possible.

A second hinged configuration, shown in Figure 3, differs from the first because the motor base must be supported in the operating, or clutched, position and then lowered for declutching. A cam is shown in Figure 3 as one of several means for lowering the motor for declutching. Some provision must be made, however, to maintain belt tension as the belt stretches.

Upon taking moments about the hinge and letting l represent the distance from the hinge to the support point (from the hinge to the contact between the cam and base plate in Figure 3) we have that

$$Fl = W(a \cos \theta + b \sin \theta) + T[a \sin(\theta + \phi) - b \cos(\theta + \phi)] \qquad (1\text{-}2)$$

Let

$$\eta = \frac{a}{l} \qquad \Lambda = \frac{W}{T} \qquad \zeta = \frac{b}{a}$$

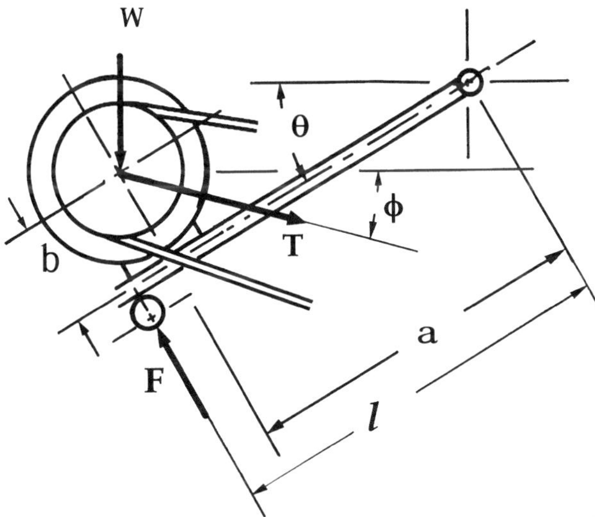

FIGURE 3 Second hinged base configuration, belt drive.

So equation (1-2) may be written in dimensionless form as

$$\frac{F}{T} = \eta[\sin(\theta + \phi) - \zeta \cos(\theta + \phi) + \Lambda(\cos \theta + \zeta \sin \theta)] \tag{1-3}$$

B. Sliding Base

A third mechanism for clutching and declutching involves placing the motor on a sliding base, as shown in the upper drawing in Figure 4, in which the motor base may be both moved back and forth and locked in place by a pair of linkages, one on each side of the sliding base, as shown in the lower drawing in Figure 4. It is locked in place by moving the linkage to a stop below the plane of the slide, as pictured in the lower drawing in Figure 4. This geometry provides a feature not found in the previous two designs: a detent effect on the clutching and declutching force in which the links a and r will snap into the clutched, or engaged, position after a force maximum is reached. This occurs because the belt is stretched slightly beyond its operating length as the motor base moves back and forth from the declutched to the clutched position of the base.

By summing forces in the horizontal direction acting on the slide upon which the motor is mounted, and assuming that the slide is lubricated so that that the small friction force between sliding surfaces may be ignored in com-

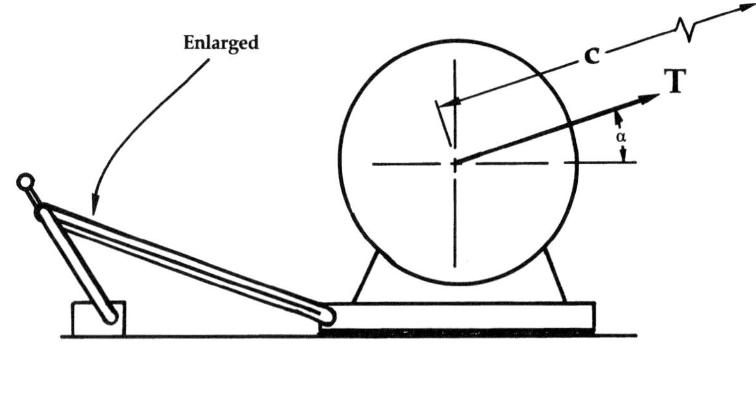

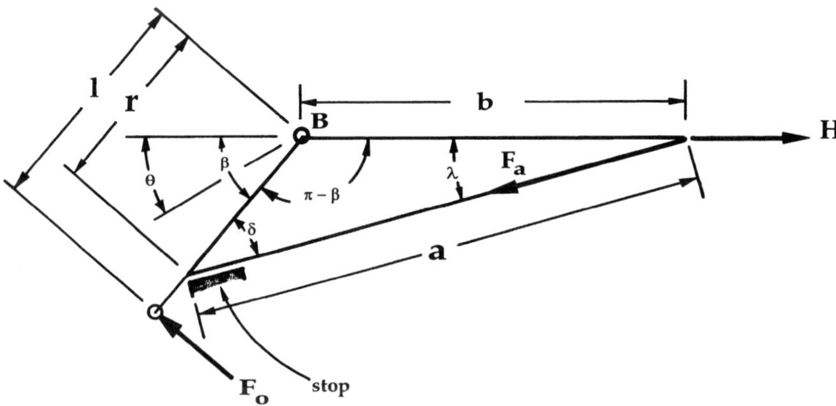

FIGURE 4 Upper drawing: enlarged sketch of sliding motor mount for a belt drive. Lower drawing: linkage geometry.

parison with the belt tension, we find from Figure 4 that the force F_a that acts through link a is related to the horizontal force H acting on the base according to

$$F_a \cos \lambda = H \tag{1-4}$$

where from Figure 4 we also find that

$$H = T \cos \alpha \tag{1-5}$$

The change in angle α as the slide moves is assumed to be small enough relative to changes in angles θ and λ that it may be ignored. Upon taking moments about pivot B in Figure 4(b) we obtain

$$F_o l = F_a r \sin \delta \tag{1-6}$$

where

$$\left.\begin{aligned} \delta &= \beta - \lambda \quad \text{in the clutched, or operating, position} \\ \delta &= \theta - \lambda \quad \text{during de-clutching,} \end{aligned}\right. \tag{1-7}$$

when belt tension is relaxed and where F_o is the force that either the operator or the actuator exerts at the left-hand end of link r. From the law of sines and the geometry in Figure 4, λ is related to θ according to

$$\left.\begin{aligned} a \sin \lambda &= r \sin \theta \quad \text{in the de-clutched position} \\ a \sin \lambda &= r \sin \beta \quad \text{in the clutched (operating) position.} \end{aligned}\right. \tag{1-8}$$

After substituting for δ from the second of equation (1-7) into equation (1-6) and then solving for γ from the first of equations (1-8), equation (1-6) may be rewritten as

$$F_o l = F_a r \sin\left[\theta - \sin^{-1}\left(\frac{r}{a} \sin \theta\right)\right] \tag{1-9}$$

Moving the motor away from the driven machine to begin declutching causes the belt to stretch an amount Δc. The corresponding change in length b is given by

$$\Delta b = \Delta c \cos \alpha \tag{1-10}$$

according to the geometry shown in Figure 4.

The force actıng on the sliding base during the initial declutching motion as the linkage moves to increase the distance b may be written as

$$H + \Delta H = (T + k\, \Delta c)\cos \alpha = T \cos \alpha + k\, \Delta b \tag{1-11}$$

upon using relation (1-10). In equation (1-11), constant k is the spring for the belt, which is defined by k = force/elongation, hence the force required to strech the belt, which is given by $k\, \Delta c$.

Length Δb may be calculated from the law of cosines, by which the length b may be written in terms of the lengths of links r, a and included angle δ as

$$b^2 = r^2 + a^2 - 2ar \cos \delta$$

Substitution from $\delta = \beta - \lambda$ and from the first of equations (1-7) gives

$$b_o = \left\{r^2 + a^2 - 2ar \cos\left[\beta - \sin^{-1}((r/a)\sin \beta)\right]\right\}^{1/2}$$

at the locked, or clutched, position, and substitution of $\delta = \beta - \lambda$ from the second of equations (1-7) gives

$$b = \{r^2 + a^2 - 2ar\cos[\theta - \sin^{-1}((r/a)\sin\theta)]\}^{1/2}$$

Recall that $\theta \leq \beta$ during declutching, and note that both β and θ are positive in the counterclockwise direction from the horizontal plane.

By subtracting b from b_o we have

$$\begin{aligned}\Delta b = &\left\{r^2 + a^2 - 2ar\cos\left[\beta - \sin^{-1}\left(\frac{r}{a}\sin\beta\right)\right]\right\}^{1/2} \\ &- \left\{r^2 + a^2 - 2ar\cos\left[\theta\,\sin^{-1}\left(\frac{r}{a}\sin\theta\right)\right]\right\}^{1/2}\end{aligned} \tag{1-12}$$

which may be rewritten as

$$\begin{aligned}\Delta b = &\, a\{1 + \Gamma^2 - 2\Gamma\cos[\beta - \sin^{-1}(\Gamma\sin\beta)]\}^{1/2} \\ &- a\{1 + \Gamma^2 - 2\Gamma\cos[\theta\,\sin^{-1}(\Gamma\sin\theta)]\}^{1/2}\end{aligned} \tag{1-15}$$

where $\Gamma = r/a$ and β is the limiting value of θ at the operating position when link a rests against a stop as shown in Figure 4(b). Preparatory to the next substitution, note that the belt's effective spring constant k may be written as $k = T/\varepsilon$, where ε is the elongation of the belt due to tension T.

Substitution from equation (1-12) into equation (1-11) and then into equations (1-4) and (1-9) yields

$$\begin{aligned}\frac{F_o}{T} = \kappa\Big\{&\cos\alpha + \gamma[1 + \rho^2 - 2\rho\cos(\beta - \sin^{-1}(\rho\sin\beta))]^{1/2} \\ &- \gamma[1 + \rho^2 - 2\rho\cos(\theta - \sin^{-1}(\rho\sin\theta))]^{1/2}\Big\} \\ &\times \frac{\sin[\theta - \sin^{-1}(\rho\sin\theta)]}{\cos[\sin^{-1}(\rho\sin\theta)]}\end{aligned} \tag{1-13}$$

upon substituting for ka/T according to $ka/T = a/\varepsilon$. Parameters γ and κ are defined by

$$\gamma = \frac{a}{\varepsilon} \qquad \kappa = \frac{r}{l} \qquad \rho = \frac{r}{a} \tag{1-14}$$

By measuring angles in the counterclockwise direction, the force F_o will be positive upward when links a and r are below the horizontal and negative when they are above, indicative of the directions of the initial force to declutch and of the force necessary to keep the linkage in equilibrium when θ goes negative as belt tension is relieved.

Examination of equation (1-13) reveals that κ is a multiplicative constant that decreases the belt tension with increasing lever arm l relative to link r and that γ is a parameter that introduces the effect of belt elasticity.

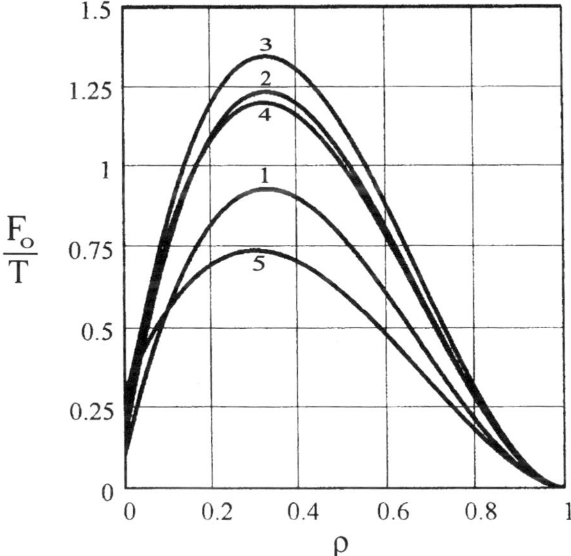

FIGURE 5 Variation of the ratio of operator force to belt tension, F_o/T, with ρ. For all curves, $\kappa = 1$, $\gamma = 1000$, $\alpha = 14°$, and $\beta = 20°$. Angles θ are as follows: curve 1, 6°; curve 2, 9°; curve 3, 12°; curve 4, 15°, and curve 5, 20°.

The plot of F_o/T as a function of ρ in Figure 5 shows that there is an optimum value of ρ that gives the largest detent effect. When $\rho = 0$ there is obviously no belt stretching because $r = 0$ for all finite l. When $\rho = 1$, length r is the same as length a, which implies that they have common pivot points, again making belt stretch impossible. Notice that although the maxima in Figure 5 vary slightly with θ, they lie in the vicinity of $\rho = 0.3$ for the parameters shown.

By plotting F_o/T as a function of θ in Figure 6 we find that the maximum lies at at or close to 12° for $\beta = 20°$. It is also clear that for these values of κ, γ, and β that the choice of β ($\beta \geq \theta$) is important if a detent effect is to be had.

II. FRICTION WHEEL DRIVE

This type of drive, shown in Figure 7, provides both clutch capability and speed variation functions in one pair of discs. This type of friction drive is limited to relatively low-power applications, such as the smaller riding lawnmowers for residential use, because power transfer between discs is limited by the contact force, the friction coefficient, and the shear strength of the tire on

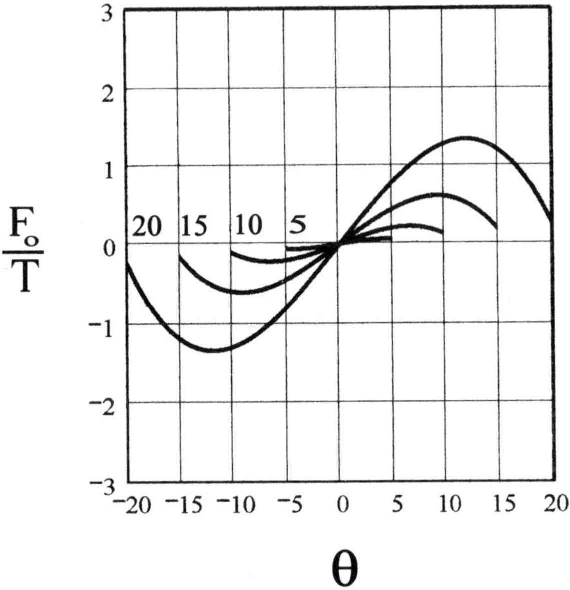

FIGURE 6 Dependence of the ratio of operator force to belt tension, F_o/T, on angle θ. For all curves, $\kappa = 0.001$, $\rho = 0.336$, $\gamma = 1$. Curve 1, $\beta = 20°$; curve 2, $\beta = 15°$; curve 3, $\beta = 10°$; and curve 4, $\beta = 5°$.

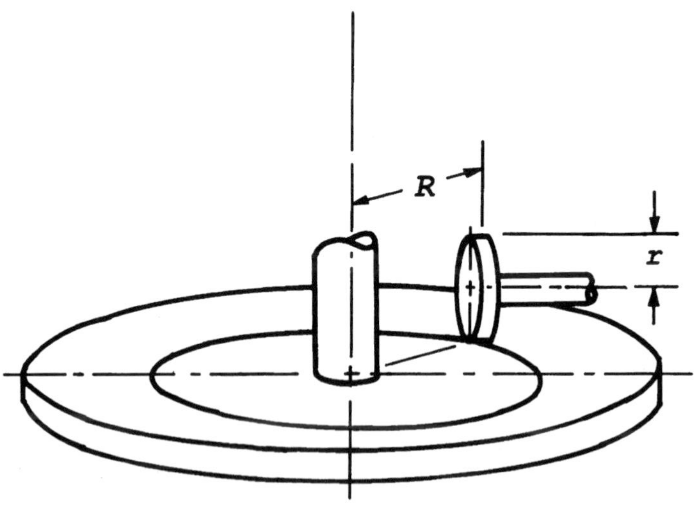

FIGURE 7 Friction drive.

the driven disc. It is an inexpensive alternate to a clutch, bevel gears, and a transmission.

From Figure 7 it follows that the maximum input torque is given by $T_0 = \mu N r$, which is limited by the normal force N and the coefficient of friction μ between the disks. If we let ω_0 and ω_1 represent the angular velocities of the small driver disk and the large driven disk, respectively, it is evident from Figure 7 that the output angular velocity and the maximum output torque T_1 will be given by the following relations if no power is lost due to slippage between the disks; namely,

$$\omega_1 = \omega_0 \frac{r}{R} \quad \text{and} \quad T_1 = T_0 \frac{R}{r} = \mu N R. \tag{2-1}$$

Two possible modes of torque transfer appear possible. In one there may be momentary no-slip contact between the driving and driven discs at some point at or between radii $R - w/2$ and $R + w/2$, where w is the width of the tire on the driven disc. Since the location of this point may change from moment to moment, the driven angular velocity may vary between

$$\omega_{1-} = \omega_0 \frac{R - w/2}{r} \qquad \text{and} \qquad \omega_{1+} = \omega_0 \frac{R - w/2}{r} \tag{2-2}$$

Consequently, the tire may slide over the driver disc except at some point along a line in the contact region. In the other possible mode there may be slip everywhere over the contact region. In the first case, the rotational speed of the driven disc may be found from equation (2-1), and in the second case it will not exceed that given by equation (2-1). Torque transfer may be calculated using the dynamic rather than the static coefficient of friction for the materials involved.

Next, let T_0 denote the torque supplied by the driver disk and T_1 denote the torque transmitted from the driven disk. In terms of the magnitude of the tangential forces f_{max} or f_{min} that act between the surface of the driver disk and the tire of the driven disk at their region of contact, we have

$$\begin{aligned} f_{max}\left(R - \frac{w}{2}\right) &= T_0 \\ f_{min}\left(R - \frac{w}{2}\right) &= T_0 \end{aligned} \tag{2-3}$$

where $\mu N \geq f_{max} > f_{min}$, in which μ is the dynamic coefficient of friction for the materials involved and N is the normal force that presses the driven wheel against the surface of the driving disc. Thus, if the driven wheel is driven at its outer edge,

$$T_{1min} = r f_{min} = T_0 \frac{r}{R + w/2} \tag{2-4}$$

and if the driven wheel is driven at its inner edge,

$$T_{1\max} = rf_{\max} = T_0 \frac{r}{R - w/2} \tag{2-5}$$

hence

$$\frac{T_{\max}}{T_{\min}} = \frac{R + w/2}{R - w/2} = \frac{2 + w/R}{2 - w/R} \tag{2-6}$$

If $w/R = 0.1$, then $T_{\max}/T_{\min} = 2.1/1.9 = 1.11$, which is to say that the torque may vary by slightly more than 10%.

Because of the variations in both speed and torque given by equations (2-2) and (2-5), friction drives of this design also may be limited to those systems where the inertia of the driven elements are large enough to effectively average, and thereby smooth, the speed and torque output of the driven unit.

Clutch action is had by raising and lowering the driven disk from and to the driver disk. Speed control is achieved by moving the driven disk in or out to change the value of R in equation (2-1). This type of relatively inexpensive, easily maintained, drive is used to send power to the rear wheels on one manufacturer's line of small riding lawnmowers designed for residential use. Normal force N may be applied by a spring (cantilever, leaf, or coil), and power may be transferred from the small disk by means of a chain or belt arranged to accommodate the changing positions of the small disc relative to the position of the driven component.

III. FRICTION CONE DRIVE

These friction drives are likewise suited for relatively low-power applications and are employed by one manufacturer of zero-turning-radius residential lawnmowers. Contacting components for this drive are shown in Figure 8, where their axes of symmetry are mutually perpendicular and where each cone rotates about its own axis of symmetry. They are sketched in the declutched configuration, in which there is no contact between cones. The driving element is the central double cone having a vertical centerline, and and the driven elements are the individual single cones, one on either side of the driver cone, having horizontal centerlines. Cones having a horizontal centerline are close enough to the driving cone that clutching and declutching is accomplished simply by moving them up or down to contact the driver cone. Directional control of the rotation of the driven cones is selected by moving them to contact either the upper cone or the lower cone of the double-cone driver.

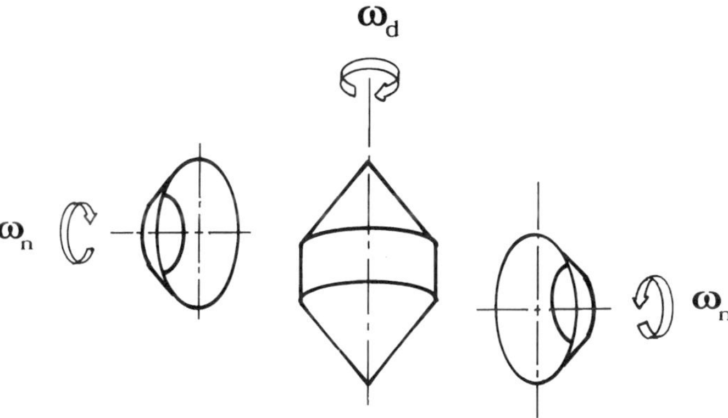

FIGURE 8 Cone friction drive schematic. Central double cone drives output cones on either side.

This type of drive is an inexpensive alternate to bevel gears for a right-angle drive, a clutch, and a reversing mechanism.

When the right-hand cone in Figure 8 is moved downward to contact the upper half of the double cone and the left-band cone is moved upward to contact the lower half of the double cone, both the left- and right-hand cones rotate in the same direction. If the right- and left-hand cones drive the right- and left-hand wheels of a lawnmower, the mower moves forward. If the right-hand cone is moved downward and the left-hand remains downward, the wheels they drive turn in opposite directions and the mower rotates in its own length to provide the zero turning radius. (The unpowered front wheels are on casters.) Finally, if the right-hand cone remains downward and the left-hand cone is moved upward, the mower moves in reverse.

Contact between cones would be along a line where the generators of each cone are in contact if the cones were absolutely rigid. However, the elasticity of the relatively softer cone linings form a contact strip centerd along what would have been the contact line.

Again there are two possible modes of torque transfer; one with no slip at some point or transverse line within the contact region and slip elsewhere, and the other with slip everywhere within the contact area. As illustrated in Figure 9, if slippage at all but one point or line is assumed to occur when two rotating cones are in contact, then the speed of the driven cone will depend upon the location of the no-slip point or transverse line.

From Figure 9 it is evident that with either point or line contact, the angular velocity of the output cones may fall somewhere between the limits determined by the location of that point, or line, within the contact strip where

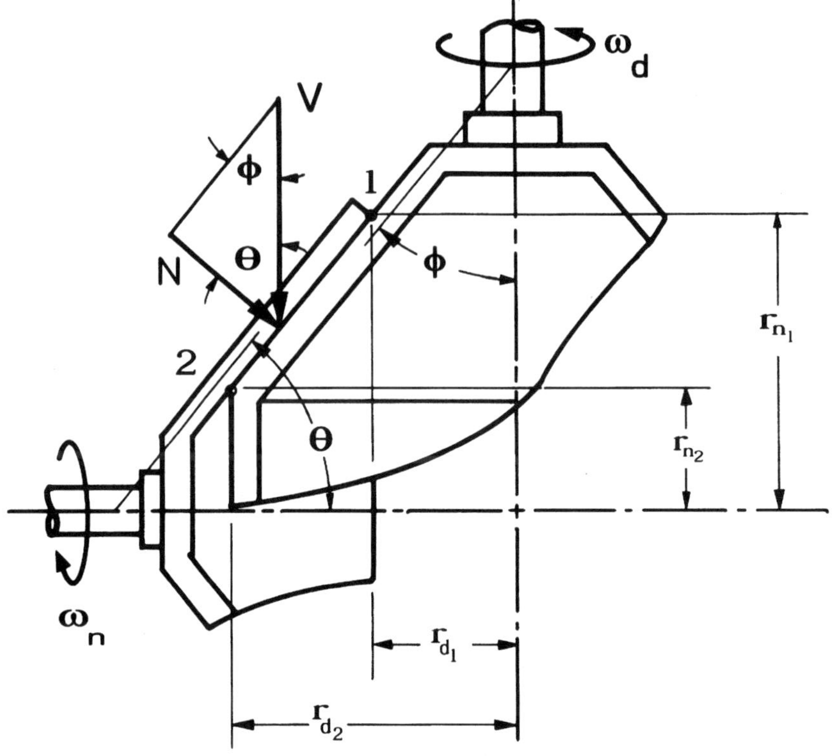

FIGURE 9 Contact between the driver cone and a driven cone.

the two cones may momentarily make contact without slipping. Therefore the angular velocity of an output cone may vary from ω_{nl} to ω_{n2} as given by

$$\omega_{n1} = \frac{\omega_d r_{d1}}{r_{n1}} = \omega_{d1} \frac{l_{d1}}{l_{n1}} \tan^2 \phi \qquad \text{and} \qquad \omega_{n2} = \frac{\omega_d r_{d2}}{r_{n2}} = \omega_d \frac{l_{d2}}{l_{n2}} \tan^2 \phi \tag{3-1}$$

where r_{d1} and r_{nl} are the radii of the driver and driven cones, respectively, at point 1, r_{d2} and r_{n2} are driver and driven radii at point 2, l_{d1}, l_{n1}, l_{d2}, and l_{n2} are the corresponding generator lengths, and $\theta = \pi/2 - \phi$, so that $\tan\theta = \cot\phi$. Angular velocity ω_d is that of the driver cone, and ω_{n1} and ω_{n2} are the angular velocities of a driven cone when driven by contact at points or transverse lines at location 1 or 2, respectively. If there is slip throughout the contact strip, the driven angular velocity will be between the two values for ω_n given in equations (3-1).

It is tacitly assumed that the cones roll smoothly when in contact along a cone generator. This assumption will be valid only if the driving cone fits within the driven cone without interference. Design of cones that will meet this requirement may be begun by returning to the analytical geometry of cones and recalling that the shortest line on the surface of a cone from its apex to it base is called a *generator* of the conical surface and by also recalling that any two-dimensional surface has two principal directions. The radius of curvature is maximum in a plane perpendicular to the surface through one of these principal directions and is minimum in a similar plane through the other principal direction. A generator on a conical surface is the principal direction that has the maximum radius of curvature, infinite, and the minimum radius of curvature of a conical surface at a particular point lies in a plane perpendicular to the generator at that point.

For simplicity in the following discussion, let the minimum radius of curvature of a conical surface be referred to as just the radius of curvature.

From Figure 9 it is evident that for the two cones to fit together without interference, the largest radius of curvature of the driver cone must be equal to, or smaller than, the smallest radius of curvature of the driven cone along their lines of contact. Calculation of the principal radius of curvature in a plane normal to the generator of a cone requires that an expression be obtained for the curve formed by the intersection of the conical surface, as shown in Figure 10(a), and a plane perpendicular to a generator. The equation of a conical surface in the XYZ system shown in Figure 10(b) is

$$X^2 + Y^2 = (Z \tan \phi)^2 \tag{3-2}$$

Substitution for X, Y, and Z in terms of x, y, and z from the coordinate transformation relations corresponding to Figure 10(b) yields

$$\begin{aligned} X &= x \\ Y &= y \cos \theta - z \sin \theta \\ Z &= y \sin \theta + z \cos \theta \end{aligned} \tag{3-3}$$

Equation (3-2) may be written in the x, y, z system by substituting from equations (3-3) into equation (3-2) to get

$$x^2 + (y \cos \theta - z \sin \theta)^2 = (y \sin \theta + z \cos \theta)^2 \tan^2 \phi \tag{3-4}$$

as the equation of a conical surface having a vertex half-angle ϕ whose axis of symmetry lies in the yz-plane and makes an angle θ with the positve z-axis, as shown in Figure 10(a). The equation of the curve formed by the intersection of this conical surface and the plane $z = h$ is found by simply setting $z = h$ in equation (3-4). After this substitution equation (3-4) becomes either the

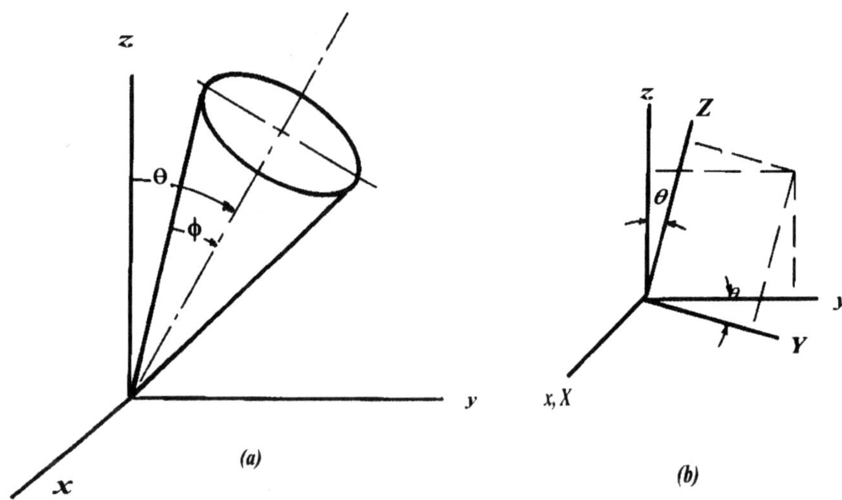

FIGURE 10 Cone geometry and its related coordinates. (a) Cone with vertex at the origin showing radii and angles. (b) Relation between coordinates X, Y, Z and x, y, z.

equation of an ellipse or the equation of a parabola, depending upon the values of θ and ϕ.

Inasmuch as the equation for the radius of curvature ψ of a curve in the xy-plane is given by

$$\psi = \frac{\left[1 + \left(\frac{dy}{dx}\right)^2\right]^{3/2}}{\left(\frac{d^2y}{dx}\right)} \tag{3-5}$$

it is necessary to calculate the first and second derivatives of y with respect to x from equations (3-3). The result is

$$\frac{dy}{dx} = \frac{x}{u} \quad \text{and} \quad \frac{d^2y}{dx^2} = \frac{1}{u} - \frac{x}{u^2}\frac{dy}{dx}\left(\tan^2\phi\,\sin^2\theta - \cos^2\theta\right) \tag{3-6}$$

where

$$u = y\left(\tan^2\phi\,\sin^2\theta - \cos^2\theta\right) + h\left(1 + \tan^2\theta\right)\sin\theta\,\cos\theta. \tag{3-7}$$

Let the line of intersection between the driver and driven cones coincide with the z-axis so that $\theta = \phi$ in (3-4), (3-6), and (3-7). Since the driver cone must roll freely within a driven cones, it is essential that its radius of curvature be less than that of the driven cones all along their lines of contact. This will be the case if the radius of curvature at point 2 in Figure 9 is equal to or less than that of the driven cone at that point.

To satisfy this condition it is necessary to evaluate equation (3-5) at $\theta = \phi = 0$, Figure 10, at the z-value for point 2. A convenient means of doing this and selecting both cones is to plot equation (3-5) as a function of z, as in Figure 11. Examination of Figure 11 shows that satisfactorily mating cones

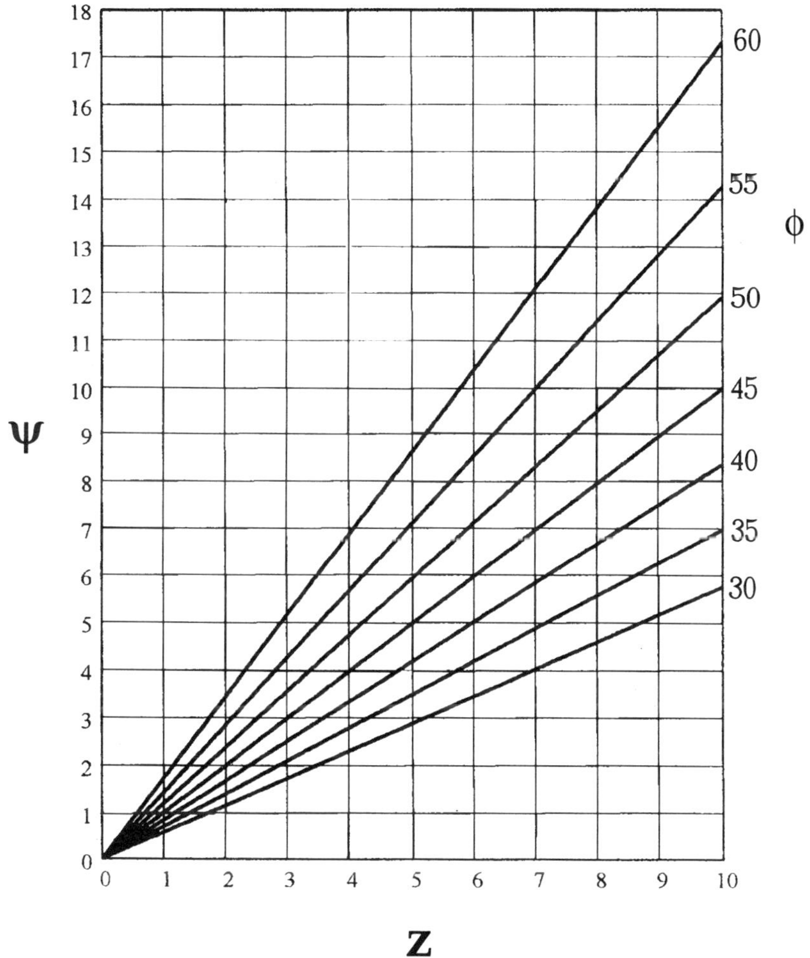

FIGURE 11 Plots of the radius of curvature ρ as a function of distance z from the apex of cones having the included half-angle θ shown. The units of ρ are the units of z.

whose axes of symmetry are mutually perpendicular may be selected in a five-step process. The first step is to choose two half-angles that add to 90°, such as $\phi = 30°$ and $\theta = 60°$. The second step is to select a generator length for the driver cone, z (the one with the smaller included angle), and to read up from that particular z-value to the line corresponding to the half-angle, θ, for that cone. That gives the radius of curvature at that value of z for a cone whose half-angle is θ. The next step is to decide whether the contacting driven cone should have the same radius of curvature at the inner contact point (point 2 in Figure 9) or a larger radius of curvature.

Both choices have consequences. The same radius of curvature may give a slightly broader contact strip about the contact line due to compression of the lining, at least near point *B*. A larger radius of curvature may give greater assurance of no interference.

If the same radius of curvature is selected, step 4 is to move toward the left along the line $\rho = \rho_0$ to the line for the half-angle ϕ of the driven cone. Last, read down to the ordinate to find the corresponding value of z on the driven cone. This completes the fifth step if the same radius of curvature was selected. Otherwise, it is completed by choosing a z-value for a larger ρ-value on the line for the corresponding ϕ.

Selecting $\theta = \phi = 45°$ is a special case. After choosing a particular value of z for the driving cone, select a larger value of z and a correspondingly larger value of ρ for the driven cones in order to allow the driver cone to roll freely inside the driven cones.

In either case, choosing ρ at point 2 in Figure 9 ensures that the driver cone will roll freely inside the driven cone, because ρ of the driver cone decreases as z moves toward the driver cone's apex and ρ for the driven cone increases as z moves outward, away from its apex. This may be verified by plotting x as a function of y from equation (3-4) when written in the form

$$x = \left[(y \sin\theta - z\cos\theta)^2 \tan^2\phi - (y\cos\theta - z\sin\theta)^2\right]^{1/2} \tag{3-8}$$

to get the curves shown in Figure 12 for the z-values selected. As pictured in Figure 12, when the smallest radius of curvature of the driven cone at every point along the contact line (centerline of a contact strip) is larger than the largest radius of curvature for the driver cone at that same point, there is no interference between them at locations away from the contact line because the cone surfaces move away from each other, as indicated by their values as the x-coordinate increases.

Selecting the length of the cylindrical section of the double cone may be accomplished using formula (3-9). It may be written from inspection of Figure 13, which shows the cross sections of half of the driver double cone and half of a driven cone on the left-hand side of the driver cone. From

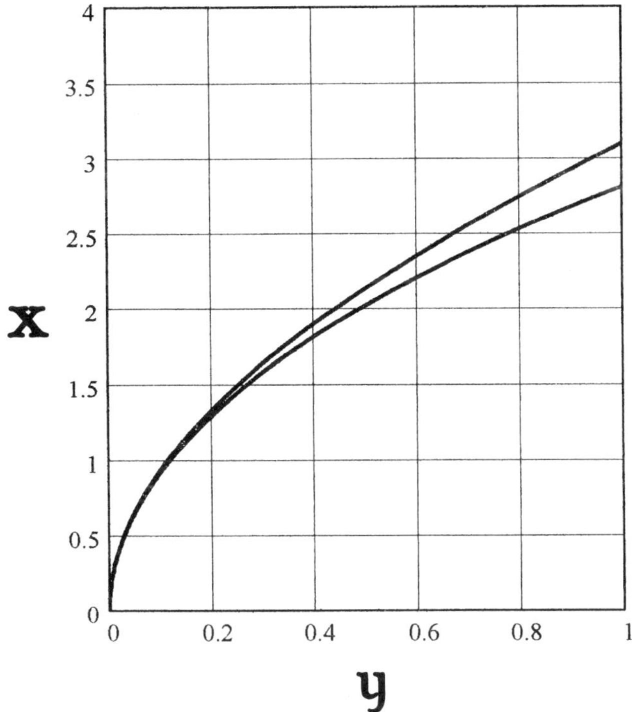

FIGURE 12 Intersection curves of contacting cones with a plane perpendicular to the contact line in terms of the coordinates in Figure 10. Upper curve, $z = 3$ in., $\phi = 55°$ (driver); lower curve, $z = 6$ in., $\phi = 35°$ (driven).

the dimensions shown in that figure, where b is the length of the contact strip, c is the length of the cylindrical section between cones, l_{n1} is the length of a generator on the driven cone, and D is the vertical distance that the driven cone must move vertically to go from contacting the upper driver cone to contacting the lower driver cone, it is evident that D may be written as $D = 2(l_{n1} - b) \sin\theta - c$. Thus,

$$c = 2(l_{n1} - b)\sin\theta - D \tag{3-9}$$

where $(l_{n1} - b)$ is the z-coordinate along the generator of the driven cone to that point where the contact line between the driver and driven cone begins.

Torque that could be transferred to a driven cone for a given dynamic friction coefficient may be estimated from the geometry shown in Figure 9. If the pressure is uniform along the contact line over a small width w, then the

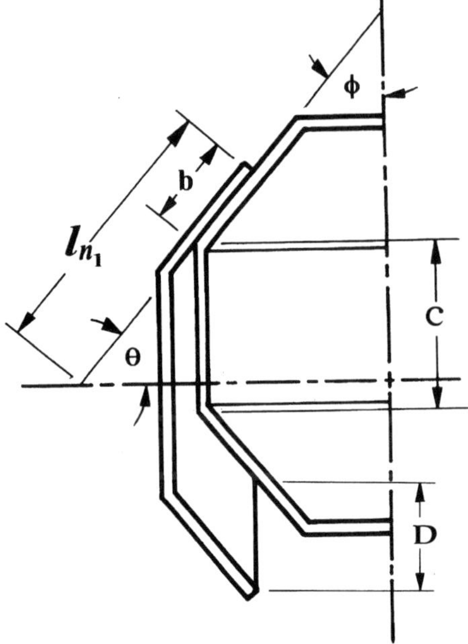

FIGURE 13 Cross sections of a driven cone and half of the driver cone in contact.

maximum torque T that can be contributed by the pressure over an element of length of the contact line may be written as $dT = \mu r\, dN$, where dN is given by $dN = pw\, dl$, r is the radius to the element of length dl, p denotes the pressure, and w represents the width of the lining that is compressed along the contact strip. From Figure 9 it follows that normal force N is related to vertical force V according to $N = V \cos\theta = V \sin\theta$. Thus,

$$dT = \mu r\ dN = \mu pwr\ dl \tag{3-10}$$

where from Figure 9 $dl \sin\theta = dr$, so integration from l_1 to l_2 is equivalent to integration from r_1 to r_2. Hence, from equation (3-10),

$$T = \frac{\mu pw}{\sin\phi}\int_{r_1}^{r_2} r\, dr \tag{3-11}$$

which integrates to

$$T = \frac{\mu pw}{2\sin\phi}\left(r_2^2 - r_1^2\right) = \frac{\mu pw}{\sin\phi}\,\frac{(r_2 + r_1)}{2}(r_2 - r_1) \tag{3-12}$$

Substitution from the relations for r_1 and r_2 yields

$$r_2 - r_1 = (l_2 - l_1)\sin\phi \qquad \text{and} \qquad r_2 + r_1 = (l_2 + l_1)\sin\phi$$

So with $pw(l_2 - l_1) = N = V \sin\phi$ we have

$$T = \mu V \frac{l_2 + l_1}{2} \sin^2\phi \tag{3-13}$$

IV. EXAMPLE I: BELT DRIVE, HINGED MOTOR MOUNT

Would you approve a motor mount as illustrated in Figure 1(a) for clutching and declutching? The mass of the motor is 18.6 kg and the center of the motor shaft is 12 cm above the bottom of the motor's base, which is 17.8 cm wide. Center-to-center distance between the shaft of the motor and the shaft of the driven machine is to be 50 cm. Belt tension is to be 298.5 N, the belt should be replaced after the center distance increases 2.4 cm, the angle of the line between centers may be from 15° to 20° with the horizontal, and the shaft of the driven machine is above and to the right of the motor shaft.

The gravitational force on the motor is given by $W = mg$ in terms of the mass of the motor and the acceleration of gravity. Thus

$$W = 18.6(9.8067) = 182.4 \text{ N}$$

to give $T/W = 1.637$. From the motor specifications, $b = 12$ cm and a must be equal to or larger than $17.8/2 = 8.9$ cm. As an aid to selecting a value for a, plot T/W as a function of ξ for $\phi = 20°$ and $30°$ and for $\theta = 20°$ and $30°$, as shown in Figure 14. Select $\phi = 20°$ and compare designs using $\theta = 20°$ and $\theta = 30°$. Use of $\phi = 15°$ was rejected in order to avoid excessive belt tension as the belt stretches. Consider $\xi = 0.2$ to have a larger T/W ratio and consider $\xi = 0.685$, corresponding to $a = 13$ cm, to get a more compact mounting. Thus $a = 44.5$ mm for $\xi = 0.2$ and 13 mm for $\xi = 0.685$.

Since belt elongation during use can alter the geometry shown in Figure 15 by changing angle θ and thereby changing the belt tension, calculate the change in θ due to belt elongation as part of the evaluation of the motor mounting system. Denote the axis of rotation of the motor mount hinge by A, and let c be the center distance between the centerline of the motor shaft and sheave, or pulley, shown on the left-hand side of Figure 15, and the centerline of the input shaft of the driven machine.

From the geometry shown in Figure 15 it is evident that

$$\lambda = \theta + a\tan\xi \tag{3-14}$$

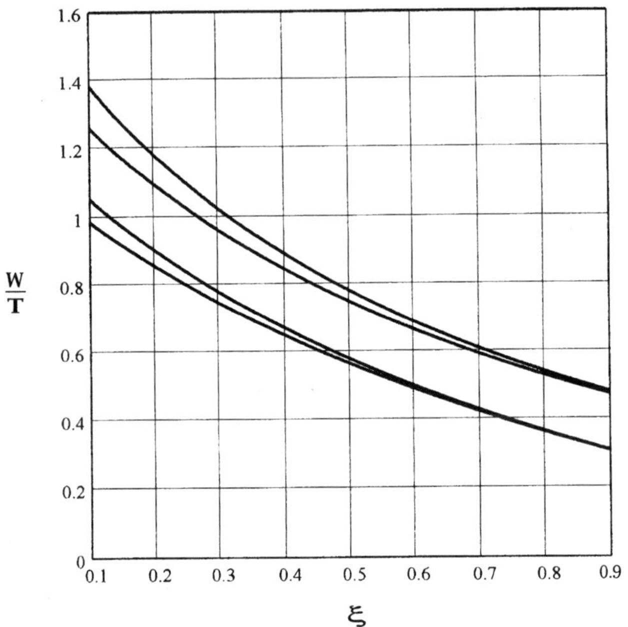

FIGURE 14 Variation of the ratio of the weight to belt tension, *W/T*, with the ratio $\xi = b/a$. Upper pair: top curve, $\theta = 20°$, $\phi = 15°$; bottom curve, $\theta = 30°$, $\phi = 15°$. Lower pair: top curve, $\theta = 20°$, $\phi = 20°$; bottom curve, $\theta = 30°$, $\phi = 20°$.

where $\alpha = \tan^{-1}\xi = a\tan\xi$. With this angle known, the law of cosines may be used to find length d from

$$d = \sqrt{h^2 + c^2 - 2hc\cos(\lambda + \phi)} \tag{3-15}$$

and from the law of sines,

$$\zeta = a\sin\left(\frac{c}{d}\sin(\lambda + \phi)\right) \tag{3-16}$$

Return to equation (3-15), with center distance c now replaced by c_s, the center distance when the belt is stretched, to calculate the increase in angle ζ which is equal to the decrease in angle θ, since $\theta + \alpha + \zeta$ is a constant. Thus

$$\Delta\theta = \cos^{-1}\left(\frac{h^2 + d^2 - c^2}{2hd}\right) - \zeta \tag{1-17}$$

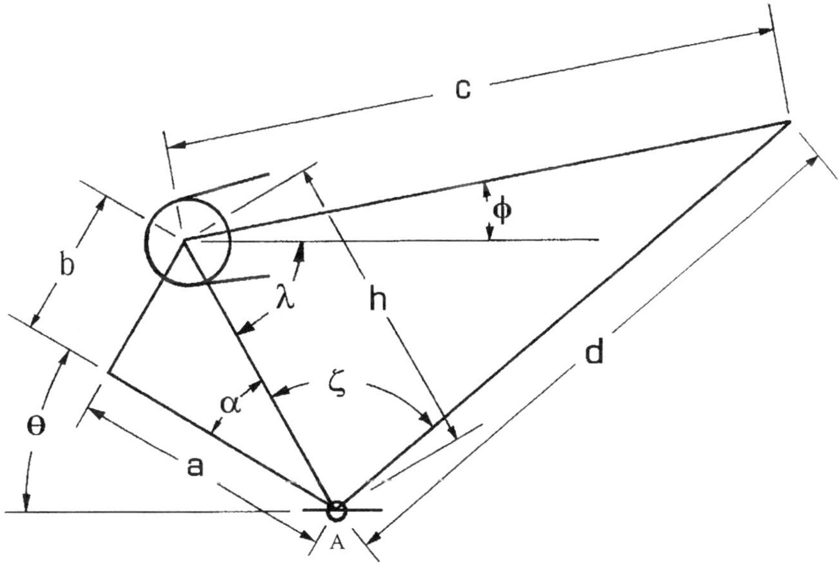

FIGURE 15 Hinged base and motor geometry.

in which ζ represents its value when the belt is new, i.e., as given by equation (3-16).

Call upon equation (1-1) to calculate T/W for the case where $\phi = \theta = 20°$ and $\xi = 0.2$, that is, for $a = 44.5$ cm, to find $T/W = 1.095$. Thus, the tension provided by the weight of the motor is only 199.73 N. Consequently, an additional mass of 9.21 kg must be added to the support to achieve the tension necessary to drive the load. Likewise for $\phi = 20°$ and $\theta = 30°$, the ratio $T/W = 0.434$, which means that weight of the motor alone can induce a tension of only 79.20 N. Thus an additional mass of 51.50 kg must be added. For simplicity of the following calculations, it will be assumed that the weight may be added such that the center of gravity remains along the centerline of the motor shaft.

Substitution into equations (1-14) through (1-17) for $\theta = \phi = 20°$ and $\xi = 0.2$ yields a reduction in θ of 3.930°, which reduces θ to 16.070°, so equation (1-1) gives $T/W = 1.095$. Using the augmented weight added to the motor weight in calculating belt tension when the belt center distance has increased to 52.4 cm gives a belt tension of 329.08 N. For the other design, in which $\xi = 0.685$, $\theta = 30°$, and $\phi = 20°$, equations (1-14) through (1-17) yield an angular reduction of 8.866°, which reduces θ to 21.134°. When the motor and its additional weight are in this position, the belt tension increases to 401.8 N.

If space is available, the choice of $\xi = 0.2$, in which $a = 44.5$ cm, would be preferred because the tension is less sensitive to belt elongation. Belt life may be enhanced in either choice by attaching the hinge to a movable base that can be periodically adjusted to hold θ near 20° as the belt stretches.

V. EXAMPLE 2: BELT DRIVE, SLIDING MOTOR MOUNT

Design a linkage similar to that in Figure 4 for a belt drive for a food grinder in which the operating tension is 173 lb. A line between the centers of the motor and generator shafts lies at an angle of 14° relative to the plane of the slide. The operator's lever arm, link r in Figure 4, should have 3- to 5-inch clearance between the free end of link l and the pin joint connecting it to link a. The belt that will be used stretches 1/32 of an inch when the tension is 173 lb on a freely turning sheave and a detent force of between 3 and 5 pounds.

Plotting F_o as a function of θ reveals that a detent effect is obtained, as is evident from Figure 16, for the parameters listed. Because the desired detent force is much less than 173 lb, initially select $\gamma = 180$; guided by Figure 5, select $\rho = 0.3$. Motivated by Figure 6, choose $\beta = 20°$; and from the

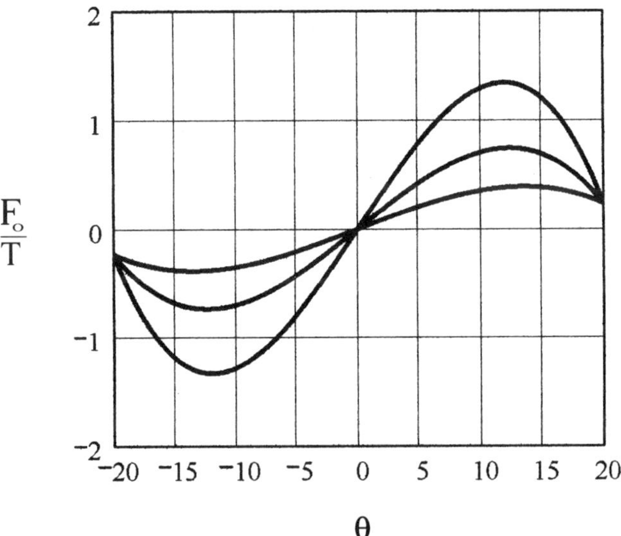

FIGURE 16 Variation of the ratio of operator force to belt tension, F_o/T, with angle θ. $\kappa = 1$, $\gamma = 1000$, $\rho = 0.3$, $\alpha = 14°$, and $\beta = 20°$ for all curves. $\gamma = 1000$ on the largest-amplitude curve, $\gamma = 500$ on the intermediate curve, and $\gamma = 200$ on the smallest-amplitude curve.

maximum in Figure 16, set $\gamma = 13.52°$, as read using the Mathcad Trace feature. Evaluation of equation (1-13) for $\kappa = 0.2$, $\gamma = 180$, $\rho = 0.3$, $\alpha = 14°$, and $\beta = 20$ for $\theta = 13.52°$ and for $\theta = 20°$ gives a detent force of 4.544 lb, which is within the acceptable range. Substitution of $\gamma = 180$ and $\varepsilon = 1/32$ in. into $a = \gamma\varepsilon = 180/32 = 5.625$ in. enables determination of r from $r = \rho a = 0.(5.625) = 1.688$ in. With length l given by $l = r/\kappa = 1.688/0.2 = 8.44$ in., it follows that the clearance given by $l - r = 8.440 - 1.688 = 6.752$ in. exceeds that specified.

This clearance requirement may be satisfied by increasing the magnitude of κ and reducing the magnitude of γ. Thus, if $\kappa = 0.28$ and $\gamma = 150$, the detent force becomes 4.702 lb, a is reduced to 4.688 in., and r becomes 1.406 in. These values give $l = 5.021$ in., so the clearance is $5.021 - 1.406 = 3.615$ in., which is within the desired range.

VI. EXAMPLE 3: CONE DRIVE

Select a cone drive for a combination golf card and a proposed congested area commuter cart for use in communities that accept them. Analysis of torque transmission on the basis of the dynamic coefficients of friction for acceptable linings indicates that a 2.00-in. overlap would be sufficient.

For comparison, consider one design with the driver cone having an apex half-angle of 40° and driven cones having apex half-angles of 50° and a second design in which both the driver and driven cones have apex half-angles of 45°. In both cases initially select a cone generator length of 6.00 in. to allow the overlap to be greater than 2.00 in. in the event that the prototype should require modification.

Begin with the 40°, 50° combination and turn to Figure 11 to select the dimensions of the cones by entering the curve at $z = 6$ and reading up to the 40° line. The principal radius of curvature at that point is 5.0346. Since z is measured in inches, the principal radius of curvature is 5.0346 in. Reading to the left at this value of ρ yields that at $z = 4.25$ in., the principal radius of curvature of the 50° half-angle cone is 5.0650 in. A plot of the x-dimension for each cone, shown in Figure 17(a), confirms that the two cones should roll without interference.

When both the driver and the driven cones have an apex half-angle of 45°, the driver cone may have a generator length of 6.00 in., but the driven cone generator length must be greater in order to have a larger principal radius of curvature. Since the selection of the radii of curvature in the 40°, 50° case differed by 0.0304 in., select the same difference in radii of curvature for the 45°, 45° degree choice, for comparison. From Figure 11 we find that along the 45° line, $z = \rho$; at $z = 6.000$ in., the principal radius of curvature is 6.000 in.; at $z = 6.0304$ in., $\rho = 6.0304$ in. Plotting of the x-dimensions of the two

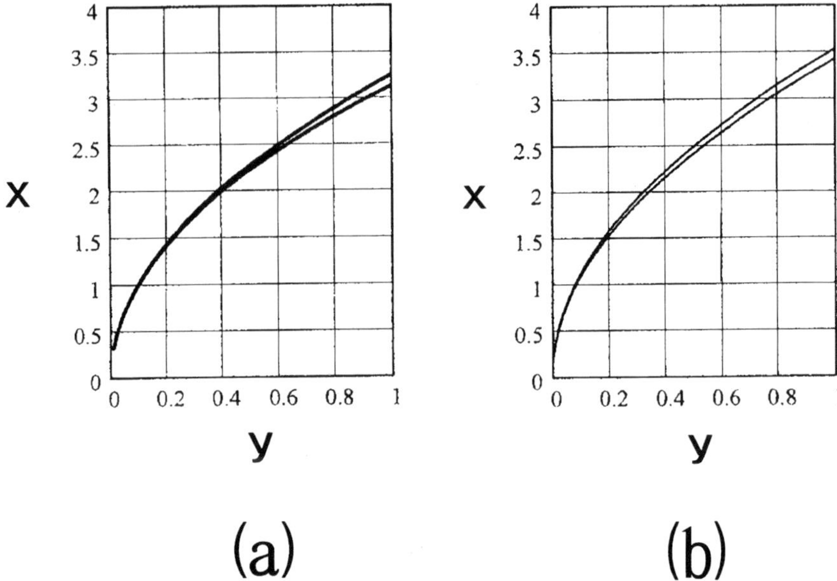

FIGURE 17 Intersection curves of a driver and a driven cone with a plane at the z-values listed. Graph (a): upper curve, $z = 4.25$ in., $\phi = 50°$; lower curve, $z = 6.00$ in., $\phi = 40°$. Graph (b): upper curve, $z = 6.20$ in., $\phi = 45°$; lower curve, $z = 5.8304$ in., $\phi = 45°$.

cones to the scale of Figure 17(a) shows, as might be expected, that the two curves are indistinguishable from one another, because at $y = 1$, their x-values differ by only 0.0088 in. Consequently, compression of the lining between the cones will produce a wider contact strip at this value of z. Calculation of the clearance at a point two in. into the overlap region (i.e., to a point $z = 6.2000$) produces a clearance of 0.107 in. at $y = 1$, which is evident on the scale of Figure 17(a) , as shown in Figure 17(b).

Length of the cylindrical section of the driven cone necessary to have a vertical motion of 0.5 in. for either of these designs may be calculated from equation (3-9). The results are that $c = 4.964$ in. for the 40°, 50° combination and $c = 8.028$ for the 45° pair. Thus, the 40°, 50° pair occupies a smaller volume, reduces the speed of the driven cone, may increase the torque to the driven cone, subject to the restrictions of the lining friction coefficient and the vertical force, and may have a smaller contact region between the cones. Conversely, the 45° pair occupies a larger volume, may give nearly a one-to-one speed ratio, and provides a larger contact area on a compressible lining between the cones.

VII. NOTATION

a,b	length (l)
c	center distance between shafts or cylindrical section, cone drive (l)
D	vertical displacement (l)
F	force (mlt^{-2})
H	horizontal force (mlt^{-2})
k	spring constant (mt^{-2})
l	length (l)
p	pressure ($ml^{-1}t^{-2}$)
R, r	radius (l)
T	torque (ml^2t^{-2}) or tension (mlt^{-2})
W	weight (mlt^{-2})
w	width (l)
α	angle sliding base
β	angle, sliding base
ε	belt elongation (l)
ζ	angle
θ	angle, hinged and sliding bases, driven cone
ϕ	angle, cone half-angle
λ	link angle, sliding base
μ	coefficient of friction
ω	angular velocity (t^{-1})
ψ	radius of curvature (l)

A. Dimensionless Ratios

Λ	W/T
κ	r/l
γ	a/ε
η	a/l
ξ	b/a
ρ	r/a

VIII. FORMULA COLLECTION

Belt drive, hinged base:

$$\frac{T}{W} = \frac{\cos\theta - \xi\sin\theta}{\sin(\theta + \phi) + \xi\cos(\theta + \phi)}$$

Belt drive, supported hinged base:

$$\frac{F}{T} = \eta[\sin(\theta + \phi) - \xi \cos(\theta + \phi) + \lambda(\cos \theta + \xi \sin \theta)]$$

where

$$\eta = \frac{a}{l} \qquad \Lambda = \frac{W}{T} \qquad \xi = \frac{b}{a}$$

Belt drive, sliding base:

$$\frac{F_o}{T} = \kappa\Big\{\cos \alpha + \gamma\left[1 + \rho^2 - 2\,\rho \cos\left(\beta - \sin^{-1}((\rho)\sin \beta)\right)\right]^{1/2}$$
$$- \gamma\left[1 + \rho^2 - 2\rho \cos\left(\theta - \sin^{-1}((\rho)\sin \theta)\right)\right]^{1/2}\Big\}$$
$$\times \frac{\sin\left[\theta - \sin^{-1}((\rho)\sin \theta)\right]}{\cos\left[\sin^{-1}((\rho)\sin \theta)\right]}$$

where

$$\gamma = \frac{a}{\varepsilon} \qquad \kappa = \frac{r}{l} \qquad \rho = \frac{r}{a}$$

Speed ratio, friction drive discs:

$$\omega_1 = \omega_0 \frac{R}{r}$$

Torque variation:

$$\frac{T_{\max}}{T_{\min}} = \frac{R + w/2}{R - w/2}$$

Friction drive, cone and plane curve of intersection:

$$x = \left[(y \sin \theta + z \cos \theta)^2 \tan^2 \phi - (y \cos \theta - z \sin \theta)^2\right]^{1/2}$$

Cone torque:

$$T = \mu V \frac{l_2 + l_1}{2} \sin^2 \phi$$

Cylindrical section length, driver cone:

$$c = 2(l_{n1} - b)\cos \theta - D$$

11

Fluid Clutches and Brakes

Fluid clutches and brakes may be divided into two groups: those containing a fluid only and those containing a mixture of fluids and solids. Those containing only a fluid rely primarily upon the mass of the fluid and secondarily upon its viscosity to transmit torque. Units containing both a fluid and a solid in a particulate form rely upon the suspended solids to provide the major bond between the components that either transmit or resist torque when under the influence of an external electromagnetic field.

The advantage of fluid clutches and brakes is that there is no lining to wear and replace. This, however, is obtained at the expense of some power loss in the transmission of torque and the distinct need for some sort of fluid cooling for both fluid clutches and fluid brakes. Moreover, occasional fluid seal replacement may also be required.

I. FLUID COUPLINGS AS CLUTCHES

Fluid couplings may serve as soft start clutches and as torque limiting clutches. A typical fluid coupling consists of an input shaft attached to an impeller and an output shaft attached to a runner, with both encased within a closed housing and oriented as shown in Figure 1. An impeller may differ from a runner in the shape of the radial vanes of the sort shown in Figure 2 and may be attached to, and rotate with, the housing that contains both the impeller and the runner. As indicated in Figure 1, the shafts are supported by bearings at the housing and by bearings at the far ends of each shaft that in turn are supported by an enclosure, as shown in Figure 3. Each impeller and

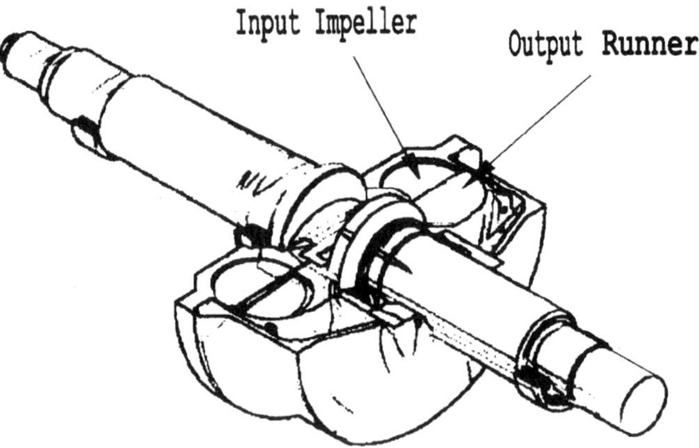

FIGURE 1 Cross section of a semitoroidal impeller and runner and their enclosure, or housing. (Courtesy TRI Transmission & Bearing Corp., Lionville, PA.)

runner consists of half of a torus, as shown in cross section in Figure 1, that is fitted with radial vanes that extend radially inward across the torus, as is evident in Figure 2. The location of the impeller and runner in a fluid coupling is also shown on the right-hand side of Figure 3 for a commercially available coupling that rests upon its oil reservoir, which is also known as a sump.

An internally driven pump located on the right-hand side of the outer housing is to pump fluid from the reservoir into the inner chamber that encloses the impeller and runner to provide a soft start over an interval of approximately five (5) seconds. Fluid from the reservoir must be circulated through a pumping and cooling system provided by the user. Standard cooling systems are generally not provided by the fluid coupling manufacturer because of the extensive variety of service conditions in which these coupling may be used.

Typically the heat to be dissipated is approximately three percent (3%) of the input power. Conversion between the power dissipated, in either watts or horsepower, and heat produced per unit time, as expressed in either large calories or Btu, is given by

$$1 \text{ Btu/sec} = 1.41391 \text{ hp}$$
$$1 \text{ kilocalorie/min} = 69.7333 \text{ W}$$

Transmitted power P is related to the input rpm (revolutions per minute) n according to the relation

$$P = P_0(n/n_0)^{\alpha} \tag{1-1}$$

FIGURE 2 Runner and shaft in a fixture used for dynamic balancing. Not all of the balancing equipment is shown. (Courtesy TRI Transmission & Bearing Corp., Lionville, PA.)

in which P_0 is a reference power and n_0 is a reference rpm. Both of them, along with exponent α, are dependent upon the fluid drive involved. Relation (1.1) may be displayed on log-log paper, as in Figure 4, for ease of selecting an appropriate fluid coupling without the use of pocket calculator or a computer to evaluate equation (1-1).

Use of Figure 4 is straightforward. For example, to select a coupling to be driven by an motor turning at 1160 rpm that is to transmit 150 hp, merely enter the graph at 1160 rpm and read up to 150 hp. As a guide to reading the

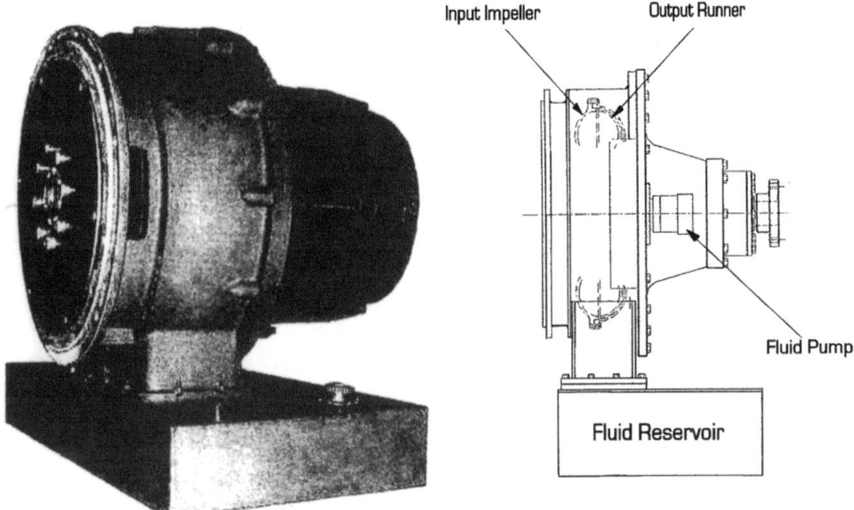

FIGURE 3 Fluid coupling designed for a sheave to be bolted to the face plate on the left. Dextron ATF, automatic transmission fluid, is the recommended fluid. (Courtesy TRI Transmission & Bearing Corp., Lionville, PA.)

logarithmic scale for power, notice that only the unlabeled 200-hp grid line lies between the labeled 100-hp and 250-hp grid lines. Hence, the point whose coordinates are 1160 rpm and 150 hp lies within the region of the model 230 coupling.

These and similar fluid couplings are suitable for use with crushers and chippers, with conveyors and similar materials handling equipment, as well as with portable equipment. They may also be used in series with marine drives to offer propeller protection.

Not all fluid couplings control their torque limits by adjusting the amount of fluid in the impeller chamber. One coupling manufacture produces a small coupling, shown in Figure 5, that is filled with fluid at all times; no pump or reservoir is needed. The housings rotate with the input shafts in both clutch and brake applications, so in both uses the attached cooling fins rotate to dissipate the heat generated by fluid losses.

Average heat loss drops from 240% for 0.125-hp continuous duty at 600 rpm to 30% for 5.0-hp continuous duty at 3600 rpm. Simplicity gained by pump and reservoir omission has been exchanged for these losses.

Typical applications include exercise machines, amusement rides, baking ovens, valve operations, crane trolleys, reversing carriages, and winding and unwinding equipment.

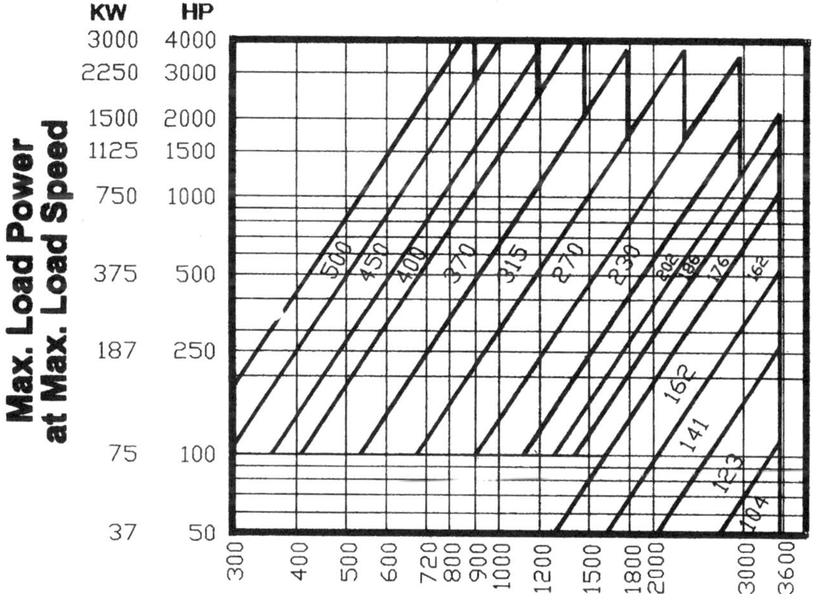

FIGURE 4 Output power as a function of input revolutions per minute. (Courtesy TRI Transmission & Bearing Corp., Lionville, PA.)

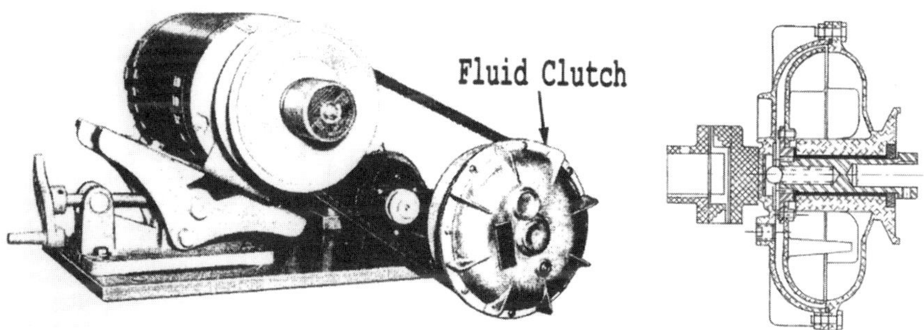

FIGURE 5 Photograph of a fluid clutch with input from an electric motor and a belt drive using the sheave that is a part of the right-hand side of the housing, shown in cross section. (Courtesy Fluid Drive Engineering Co., Inc., Burlingame, CA.)

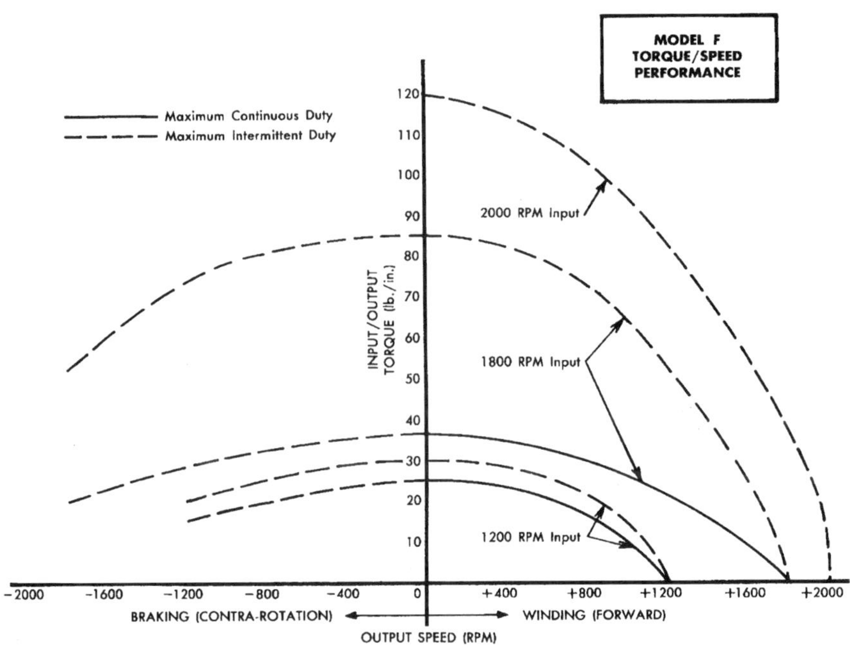

FIGURE 6 Clutch/brake torque/speed curve for the unit shown in Figure 5. (Courtesy Fluid Drive Engineering Co., Inc., Burlingame, CA.)

II. FLUID BRAKES: RETARDERS

Fluid retarders may be thought of as fluid couplings with the runner held stationary, which is, therefore, known as the stator. Figures 7(a) and (b) show opposites sides of a retarder that is equipped with a heat exchanger, an oil reservoir, or sump, and a remotely controlled valve that regulates the flow of oil from the sump into the chamber that encloses the impeller, or rotor, and the stator. The entire unit may be mounted in series with the primary shaft, as shown in Figure 7(d), for example, or it may be mounted on secondary shaft that maintains a given speed ratio relative to the primary shaft.

Removal of the bolts shown in Figure 7(b) and setting that section to the side reveals the internal construction, as shown in Figure 7(c). The rotor that rotates with the input shaft is shown on the right-hand side in Figure 7(c) and the stator is shown on the left-hand side of that figure. Both are mounted in the housing above its portion of the sump. The elbow on the lower left side of the housing section, Figure 7(c), that holds the stator carries external coolant from the heat exchanger that extends from the lower part of the housing, as shown in Figure 7(a). The flow control valve assembly also is shown at the top of the retarder in Figure 7(a).

FIGURE 7 (a) and (b): External views of a retarder. (c) Internal construction. (d) Retarder mounted in series with the shaft upon which it acts. (Courtesy Voith Transmissions, Inc., Sacramento, CA.)

No fluid is in the rotor/stator chamber when the retarder is not in use. Activating the retarder causes fluid to be forced from the sump into the rotor/stator chamber using air from the vehicle's air compressor as regulated by the valve assembly that in turn is controlled electrically by the driver in selecting the amount of braking desired. As in the case of a fluid coupling, the torque capacity of the retarder is determined by the amount of fluid in the chamber that encloses the rotor and the stator.

Retarder performance curves shown in Figure 8, display the retarding moment as a function of the rotor speed and the amount of fluid in the rotor/stator chamber. Curves 1 through 5 that arise from the origin in Figure 8 and ascend with increasing rotational speed ω are plots of the work done on the retarder as kinetic energy is imparted to the fluid by the rotor as given by

$$W = KE = \frac{I\omega^2}{2} \tag{2-1}$$

in which I denotes the moment of inertia of the fluid that is set into motion by the rotor and ω denotes its rotational speed in radians/second. Curves 6, 7,

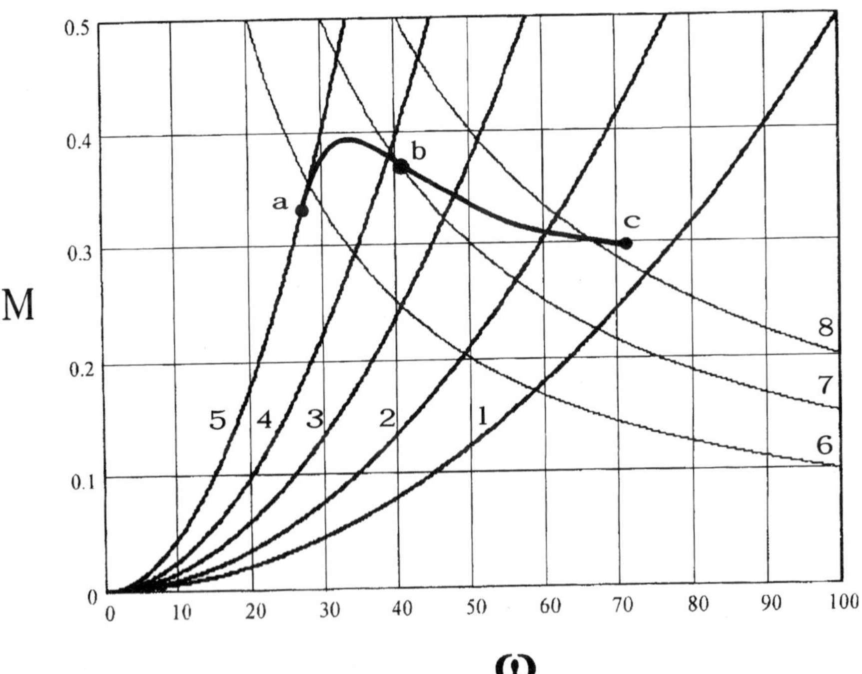

FIGURE 8 Retarding moment M as a function of rotor angular velocity ω.

and 8 that descend from the top of the figure toward the right hand side with increasing ω represent the moment M that is associated with each of the curves of constant power P according to the relation

$$M = \frac{P}{\omega}. \tag{2-2}$$

Both torque, or moment, and kinetic energy may be plotted on the same graph, of course, because they have the same units; namely, ml^2t^{-2}, in terms of the mechanical units mass m, length l, time t.

When the rotor/stator chamber is partially filled the retarding moment increases with rotor speed along a curve similar to curve 1 in Figure 8. Increasing the amount of fluid in the rotor/stator chamber causes the retarding moment to grow more rapidly with rotor speed ω, as represented by curves 2, 3, and 4 for intermediate fluid volumes. Whenever the chamber is filled the torque-speed curve may be represented by curve 5 in Figure 8.

Point a is reached on curve 1 when the rotor, which also acts a pump, forces more oil out through the stator than the air pressure on the sump can force into the rotor/stator chamber; i.e., the rotor induced pressure exceeds the air pressure in the sump that forces fluid into the chamber.

That portion of the curve that includes the maximum between a and b is determined by the design, position, and dimensions of the inlet and outlet throttles of the system.

The latter portion of the performance curve between points b and c is determined by the number and diameters of the outlet ports in the stator in combination with the flow resistance in the piping circuit to, from, and within the heat exchanger that transfers heat to the coolant that circulates through vehicle's radiator*.

Moment M is related to the resisting torque, T_r, that the retarder applies to the primary shaft according to

$$T_r = (\omega/\omega_r)M = (n/n_r)M, \tag{2-3}$$

where n represents the rotational speed of the retarder's rotor in revolution/minute and where ω_r and n_r represent the rotational speed of the primary shaft in radians/second and in revolutions/minute respectively. Clearly $n/n_s = 1$ when the retarder acts on the primary shaft directly, as in Figure 7(d).

Depending upon the model, retarders as described here may provide either a torque up to 4000 Nm (2950.4 ft-lb) at rotor speeds up to 2800 rpm or a torque up to 3200 Nm (2360.2 ft-lb) at rotor speeds up to 5000 rpm. Other

*This explanation of retarder operation was provided by Rainer Kläring of Voith Turbo GmbH & Co. KG. Any errors in the explanation are due entirely to the author.

combinations of torque and speed characteristics are also available, as well as a retarder that uses water as its working fluid.

Energy, E, to be dissipated by the retarder in slowing a vehicle may be estimated from the work done on the vehicle and the change in kinetic and potential energy; namely,

$$E = \frac{1}{2}m(v_1^2 - v_2^2) + mg(h_1 - h_2) + W_o \tag{2-4}$$

in which m represents the mass of the vehicle plus its load, v_1 and v_2 represent the initial and final velocities during the time that the retarder is engaged, g denotes the acceleration of gravity, h_1 and h_1 represent the initial and final elevation changes during the time that the retarder was engaged, and W_o denotes the work done on the vehicle while the retarder was active.

III. MAGNETORHEOLOGICAL SUSPENSION CLUTCH AND BRAKE

Magnetorheological suspensions have been referred to as magnetorheological fluids even though the fluid itself is not magnetorheological. It is the suspension of magnetically susceptible particles, such as carbonyl iron, in the fluid that causes the mixture to become a magnetorheological suspension, or a magnetorheological fluid. The first magnetorheological suspension was demonstrated by Rabinow and Winslow in 1948 and termed a *magnetic fluid clutch*, made from a suspension of carbonyl iron* in silicone oil and kerosene [1]. Application of a magnetic field causes the iron particles to converge along the lines of flux, which in turn increases the flux density. In the case of a brake, the braking action is due to increased magnetic attraction between stator and rotor. The same principle applies to a clutch, except that the attraction is between the input rotor and the output rotor. The concentration of particles along the flux lines also may retard fluid motion to some extent, and thereby aid somewhat in both the braking and clutching actions.

Settling of the suspended material is apparently not a problem because the suspended material is remixed by the motion of the clutch or brake. However, having a fluid that displays a low viscosity when the clutch or brake is disengaged is important in order to reduce operating losses when they are inactive.

Subsequent development of the magnetorheological fluids seems to have been concentrated in the area of finding or developing fluids whose

---

**The Handbook of Chemistry and Physics* (CDC Press) lists three forms of carbonyl iron. $FE(CO)_4$, $FE(CO)_5$, and $FE(CO)_9$.

viscosity does not change due to high shear stress, and perhaps compressive stress, over time. (Some earlier fluids were reported to have reached the vis cosity of shoe polish due to stress over time.) This thickening was thought to be due to spailing of a thin, brittle surface layer on the carbonyl iron. Presently available magnetorheological fluids that have been developed to ameliorate this problem are said to be able to sustain 10^7 J/cm^3 before becoming unusable [2].

A small, commercially available, brake that employs a magnetorheological suspension is shown in Figure 9. Its maximum torque is approximately 5.6 N-m (about 50 in.-lb), and, because it contains a fluid, it provides a small torsional load that is less than approximately 0.3 N-m (2.7 in.-lb) when the brake is not engaged.

The requisite magnetic field is supplied by an electric current of 1.0 A or less in a circular coil that induces the magnetic field shown in the schematic cross section of the brake and coil in Figure 10. This excitation produces a linear relation between the braking torque and the electric exciting current within the range from 0 to 1.0 A, as shown in Figure 11.

The operating temperature range of the brake is from about −30°C to 70°C, corresponding to −20°F to 160°F. Notice that a residual torque capability of 0.3 N-m is available at zero curent, probably due to fluid viscosity as augmented by either the suspended or precipitated particles.

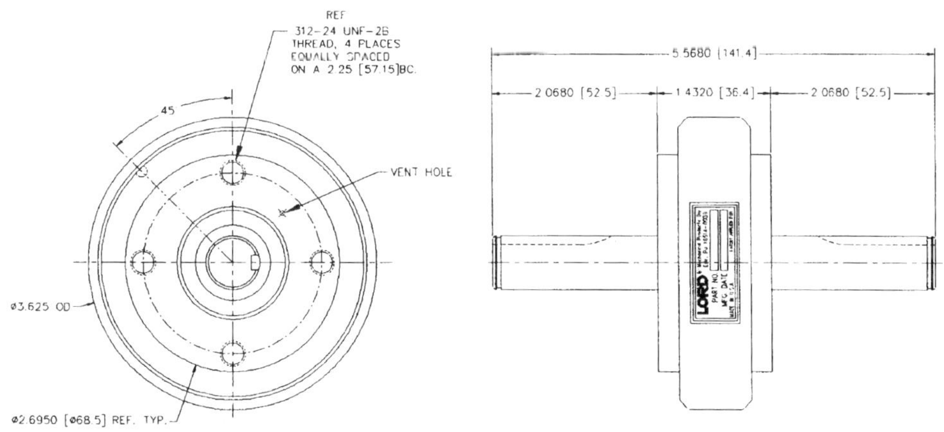

FIGURE 9 Magnetorheological brake. Omitted: power cord attached to housing. (© 2002 Lord Corporation. All rights reserved. Lord Corp., Materials Division, Cary, NC.)

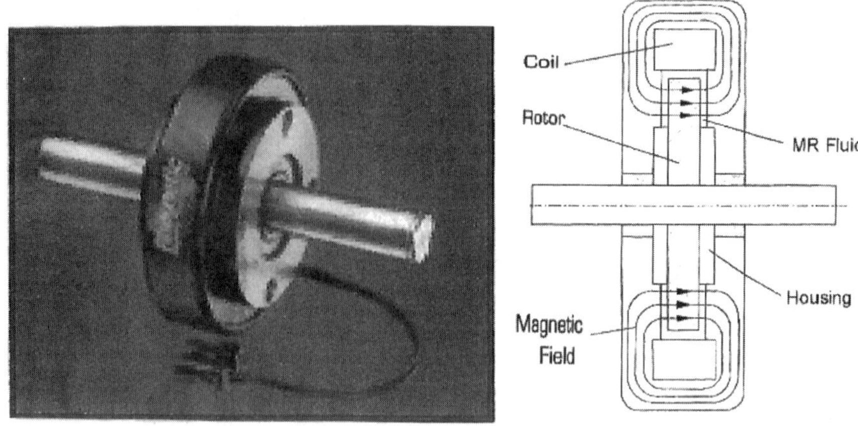

Figure 10 Photograph and schematic cross section of a magnetorheological brake. (© 2003 Lord Corporation. All rights reserved.)

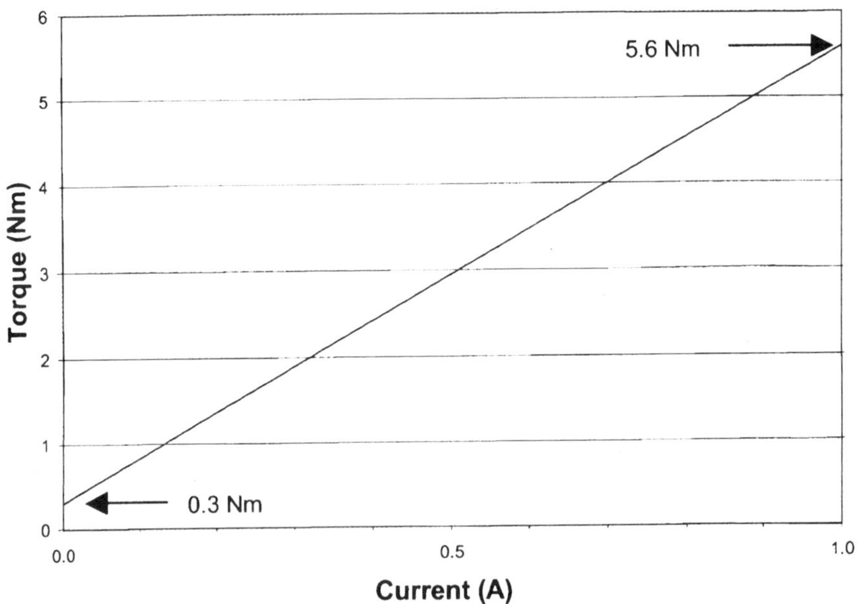

Figure 11 Typical torque in newton-meters vs. electric current in amps. It should not be used for specifications. (© 2002 Lord Corporation. All rights reserved. Lord Corp., Materials Division, Cary, NC.)

IV. NOTATION

g	acceleration of gravity (lt^{-2})
h	height (l)
KE	kinetic energy (ml^2t^{-2})
m	mass (m)
n, n_0	rpm (t^{-1})
P, P_0	power (ml^2t^{-3})
PE	potential energy (ml^2t^{-2})
t	time (t)
v_1, v_2	velocity (lt^{-1})

V. FORMULA COLLECTION

Power transmitted:

$$P = P_0\left(\frac{n}{n_0}\right)^{\alpha}$$

Energy dissipated:

$$E = \text{KE} + W_0 + \text{PE} = \frac{1}{2}m\left(v_t^2 - v_2^2\right) + mg\,\Delta h + W_0$$

Power dissipated:

$$P = \frac{\text{KE}}{t}$$

REFERENCES

1. Magnetic Fluid Clutch (1948). Technical News Bulletin, National Bureau of Standards, 32/4, pp. 54–60.
2. Carlson, J. D. (July 9–13, 2001). What Makes a Good MR Fluid, presentation at 8th International Conference on Electrorheological (ER) Fluids and Magnetorheological (MR) Suspensions, Nice, France.

12

Antilock Braking Systems

Antilock braking systems (also known as antiskid braking systems) for vehicles are discussed here because they represent perhaps the most involved commonly used systems for automatic brake control. The data collection, analysis, and system design involved may suggest initial procedures to be followed for clutch and brake automation in other applications.

Design of an antilock system (ABS) for highway vehicles requires decisions to what is to be measured, how it is to be measured, and how to use the data to prevent skidding. These systems are different from the early antilock systems in that they are computer based, so they collect and process more data.

The first patent for antilock brakes was granted in Germany in 1905 [1], and the first antilock brakes for railroad cars were available in 1943 [2]. Electronic control of antilock brakes was widely incorporated into aircraft by 1960 [3] in order both to control aircraft skidding and to prevent excessive wear to the tires on the landing gear of large aircraft. Although it may be difficult to specify when the first extension to highway vehicles began, Ford and Kelsey Hayes produced an ABS system for the rear wheels only of the 1969 Thunderbird [4]. Introduction of what was said to be modern electronically controlled ABS for passenger cars was by Daimler-Benz [5] and BOSCH [6] in 1978.

Because of the proprietary nature of the available antiskid and traction control systems, the latter portion of this chapter, dealing with antiskid braking and traction control systems, will be a combination of information from the literature and of conjecture regarding the possible techniques available for achieving brake control.

I. TIRE/ROAD FRICTION COEFFICIENT

Antilock brake control for stopping a vehicle in what is intended to be a straight-line path clearly requires some method for detecting the skid, or slip, of each wheel, for assimilating the data from all wheels, for analyzing this data to estimate the vehicle's motion, and for selecting the appropriate commands to be sent to each wheel or set of wheels both to stop the vehicle and to maintain stability.

Figure 1(a) portrays the condition in which there is no slip between the wheel and the road. Under these conditions, a wheel of radius r rotating with angular velocity ω_0 about its axis of rotation (the centerline of the axle to which it is attached) at any instant also rotates about its instantaneous center (the idealized point where it contacts the road as though there were no tire

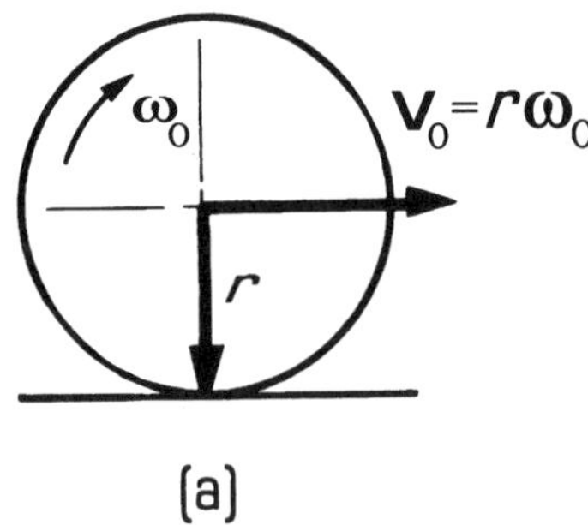

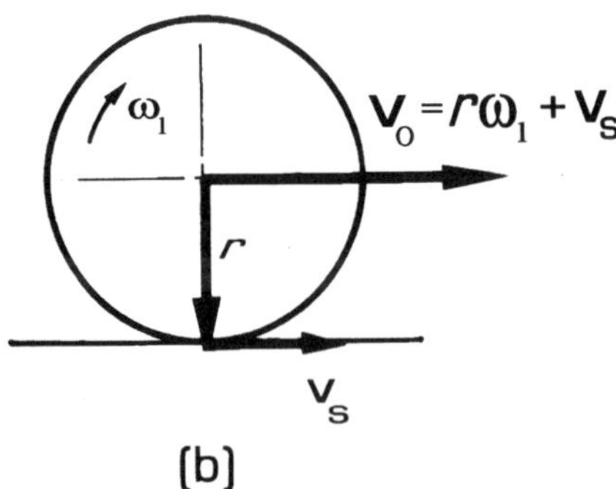

Figure 1 Velocity v_0 is the vehicle velocity as calculated (a) for a wheel rolling with angular velocity ω_0 without slip and (b) for rolling with angular velocity ω_1 and with slip velocity v_S.

deformation) with angular velocity ω_0. Hence, calculation of the rotation about the instantaneous center reveals that the axle moves horizontally with velocity v_0, as given by

$$v_0 = r\omega_0 \tag{1-1}$$

If there is slip between the wheel and the road, as in Figure 1(b), and if v_1 denotes the velocity of the axle with respect to the point where the wheel contacts the road, then the velocity of the axle relative to that point is given by

$$v_1 = r\omega_1 \tag{1-2}$$

where ω_1 is the angular velocity of the wheel about its axis of symmetry, which is perpendicular to the plane of the wheel. Thus, if the wheel slips with velocity

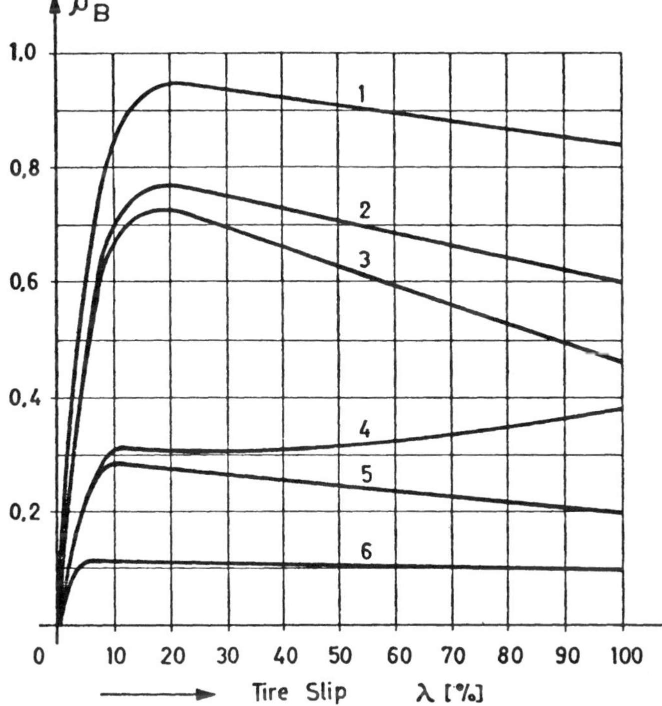

FIGURE 2 μ_B as a function of λ for (1) dry asphalt, (2) wet asphalt, thin water film, (3) wet asphalt, thick water film, (4) fresh snow, (5) packed snow, (6) glare ice. The positive slope of curve 4 with increasing λ is due to snow build-up in front of the tire as its rotation slows to zero.

v_s (i.e., the point where the wheel contacts the road moves with velocity v_s), then the velocity v_0 of the vehicle relative to the road is given by

$$v_0 = v_1 + v_s \tag{1-3}$$

Wheel slip during braking is commonly described by the slip ratio λ, as defined by

$$\lambda = \frac{v_s}{v_0} = \frac{v_0 - v_1}{v_0}. \tag{1-4}$$

The slip ratio is frequently presented as a percentage, $\lambda(\%) = 100\lambda$, as in Figure 2.

For reasons that may include tire flexibility, tension and torsion of the tread within the contact patch, and the continual replacement of material within the tire's contact patch, the complex nature of the tire's contact with the road within the contact patch means that the coefficient of friction, here represented by μ_B, does not immediately jump from its static to its dynamic value, as illustrated in Figure 2 [7]. That portion of each curve between $\lambda = 0$ and the maximum, except for curve 4, may be considered a stable region, in that initial braking causes the friction coefficient to increase so that increased brake pressure within this region is effective in reducing vehicle velocity. The region beyond the maximum in μ_B may be considered a region of instability, because, except for curve 4, increased brake pressure to further slow wheel rotation becomes increasingly ineffective in slowing the vehicle itself due to a decreasing friction coefficient. Returning to curve 4, its local maximum is also followed by a region of instability, but that region is followed by a stable region caused by the build up of snow in front of the wheel as its rotation slows.

II. MECHANICAL SKID DETECTION

Early antilock braking systems used annular disks that were friction driven to rotate with each wheel during normal acceleration and deceleration but that would slip as frictional resistance was overcome during abnormal or panic breaking, as a means of detecting wheel deceleration. Whenever the wheel would decelerate beyond a certain threshold, the disk that was concentric with it would continue rotating and thereby trip some mechanism that would reduce brake pressure. This technique, or a modification of it, was the only practical means of detecting wheel deceleration prior to the introduction of microprocessors. It was also relatively inexpensive and therefore its use continued through 1968, and perhaps beyone, for some inexpensive European automobiles. An example was the Lucas Girling Stop Control System (SCS), which is explained in the paragraphs below Figures 3–5, taken from Ref. 8, which describe the modulator. It was designed for front wheel drive (FWD) vehicles and employed only two modulators, one on each front wheel. Each modulator controlled its front wheel and the diagonally opposite rear wheel through a proportioning valve, as required by European regulations. Displayed components in these figures are

1. Drive shaft
2. Flywheel
3. Flywheel bearing
4. Ball and ramp drive
5. Clutch
6. Flywheel spring
7. Dump valve
8. Dump valve spring

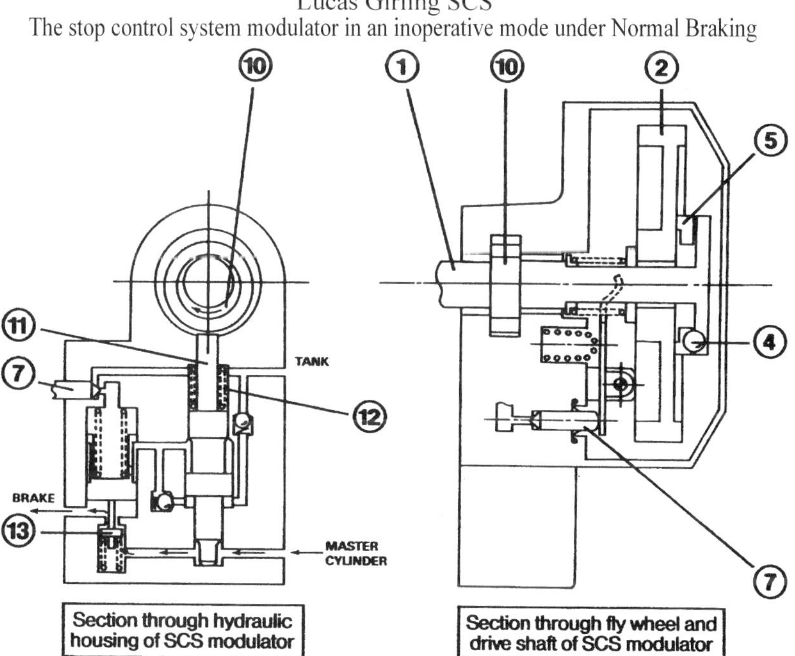

FIGURE 3 Flywheel and valve positions for the Lucas Girling SCS during normal braking.

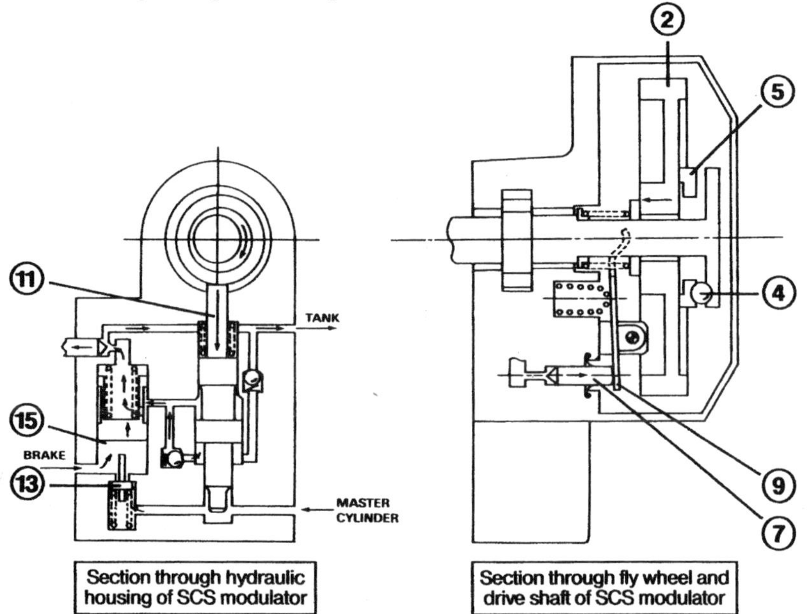

Figure 4 Flywheel and valve positions for the Lucas Girling SCS during panic braking.

9. Dump valve lever
10. Eccentric cam
11. Pump piston
12. Piston spring
13. Cutoff valve
14. Deboost piston spring
15. Deboost piston
16. Cutoff valve spring
17. Pump inlet valve
18. Pump outlet valve

Since the text below each figure was reproduced directly from Ref. 8. Figures 9 and 10 mentioned in Figure 4 correspond to Figures 3 and 5 as reproduced here.

All systems using rotating disks that must move axially to engage the brake control mechanism are handicapped by the time required to accelerate

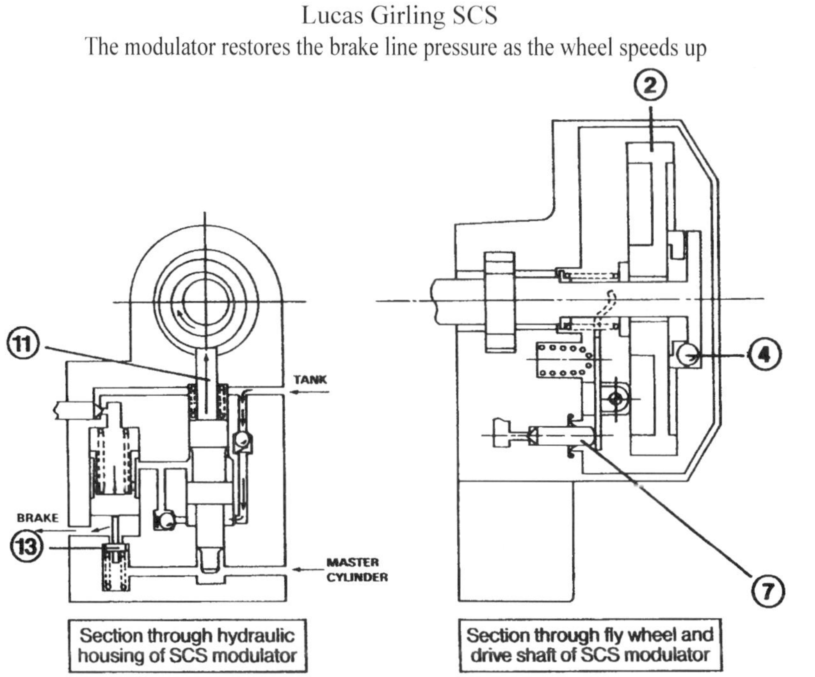

Lucas Girling SCS
The modulator restores the brake line pressure as the wheel speeds up

2
11
TANK
4
BRAKE
13
MASTER CYLINDER
7

Section through hydraulic housing of SCS modulator

Section through fly wheel and drive shaft of SCS modulator

FIGURE 5 Flywheel and valve positions for the Lucas Girling SCS during return to normal braking.

the mass of the disk laterally over the required distance s. This relationship is qualitatively similar to that for the distance traveled by a mass m that is accelerated from rest by a force F over time t:

$$\frac{x}{s} = \frac{F}{2m}(x, y)^2 \tag{2-1}$$

where x $(0 \leq x \leq s)$ is that portion of distance s traveled during time t $(0 \leq t \leq \tau)$, where t is the corresponding portion of the activation time t (see Figure 6). Thus, in the first half of the required time, the mass has moved only one-fourth of the required distance.

Faster response may be had by using electrical wheel-speed sensors that measure wheel speed and send that data to a small, dedicated computer known as an *electronic control unit*, or an ECU.

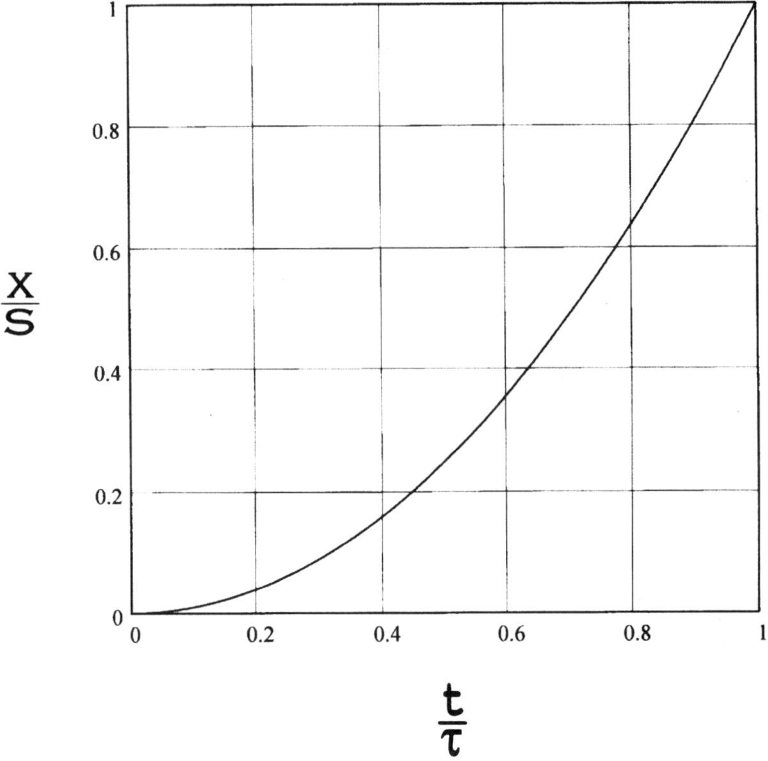

FIGURE 6 Graph of x/s as a function of t/τ from equation 12-1.

III. ELECTRICAL SKID DETECTION: SENSORS

Development of relatively inexpensive microprocessors, accelerometers, and electromagnetic wheel-speed sensors that could be incorporated into automotive controls permitted more precise measurement of wheel speed and, hence, vehicle speed, acceleration, and deceleration along with rapid detection of and improved response to individual wheel deceleration associated with wheel skid.

Addition of a small dedicated computer known as an electronic control unti, or an ECU, to an antilock system allows the correlation of data from wheel-speed sensors on each of all four wheels into a preprogrammed decision and control process. Presently each wheel-speed sensor consists of two components: a permanent bar magnet with a coil of wire wrapped around it and a sensor ring, as shown in Figure 7. The sensor ring rotates with the

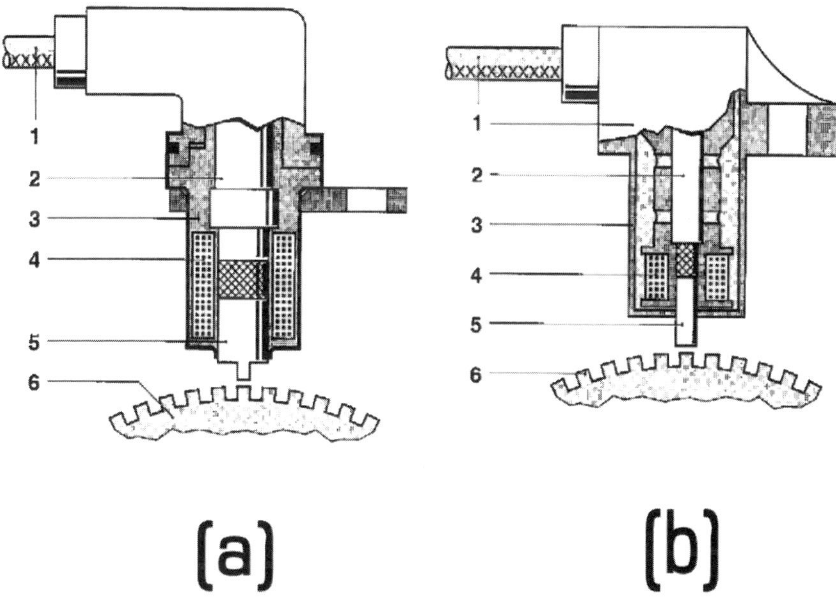

FIGURE 7 Sensor (a) has a chisel pole pin and sensor (b) has a cylindrical pole pin. The components in both: (1) electric cable, (2) permanent magnet, (3) housing, (4) winding, (5) pole pin, and (6) sensor ring. (Courtesy Robert Bosch GmbH, Stuttgart, Germany.)

vehicle wheel while the permanent magnet and its housing remain fixed relative to the vehicle's frame. As the wheel and the attached sensor ring rotate together, the magnetic field associated with the permanent magnet changes as a pole piece approaches and leaves each tooth on the toothed sensor ring. A fluctuating current is generated in the coil as the magnetic field fluctuates, with each fluctuation corresponding to the passage of a tooth.

These sensors also may be in the wheel bearings, in the differential, or on any other component whose rotation maintains a constant relationship to the wheel's rotation.

IV. ELECTRICAL SKID DETECTION: CONTROL

The ECU calculates wheel speed by counting the fluctuations per unit of time and differentiates the speed to calculate wheel acceleration or deceleration, wherein deceleration is handled as negative acceleration. In the absence of independent data on the motion of the vehicle itself, data from the wheel speed sensors must be used to estimate vehicle speed. When all wheels give the same vehicle speed, to within a specified error limit, that common speed is taken to be the vehicle speed. When all wheels do not give the same speed, wheel slip is assumed. The problem, of course, is to decide which wheel is slipping.

Typically the ECU in a front wheel drive vehicle with an antilock brake system will evaluate two data sets, one for the right front wheel and the left rear wheel and the other for the left front wheel and the right rear wheel. A typical rear wheel drive vehicle will also evaluate two data sets but one set will be for the front wheels and the other will be for the rear wheels. In either case, most systems test for wheel slip by compare diagonally opposed wheels in one of two ways: one is for the ECU control algorithm to use the signal from the faster of the two wheels as a reference speed for brake pressure modulation, known as the *select-high* method, the other is for the ECU to use the signal from the slower of the two wheels as the reference speed, known as the *select-low* method.

The proprietary control program, or algorithm, reacts once slip is detected. If the only input data is wheel speeds and their calculated acceleration/deceleration, the program may recall from permanent memory the greatest wheel acceleration/deceleration that is possible under zero-slip conditions. Hence, greater acceleration or greater deceleration (more negative acceleration) at a particular wheel indicates slip at that wheel.

Part of the ECU calculations is that of associating a wheel's rotational speed with the optimum wheel slip from equation (1-4) for λ between values λ_1 and λ_2, in which λ_1 may be 10% and λ_2 may be 20%, for example. This may be achieved by returning to equation (1-4) and solving for v_1 and then replacing v_1 and v_0 with the associated values of $r\omega_1$ and $r\omega_0$, respectively, where r is the wheel radius, to get

$$\omega_1 = \omega_0(1 - \lambda_1)$$

Likewise,

$$\omega_2 = \omega_0(1 - \lambda_2) \tag{4-1}$$

Since $\lambda_1 < \lambda_2$, it follows that $\omega_1 > \omega_2$ during braking. Thus, whenever the angular velocity ω of the wheel is such that it lies between ω_1 and ω_2, that is, whenever

$$\omega_1 \geq \omega \geq \omega_2$$

the slip velocity of the wheel is optimum, so the braking pressure will be held constant.

If $\omega \geq \omega_1$ (i.e., if the angular velocity of the wheel is large enough relative to ω_0 for the slip velocity to be small enough to lie between 0 and λ_1), the brake pressure may be increased because doing so will move the slip velocity into the optimum regions. If $\omega \leq \omega_2$ (i.e., if the angular velocity of the wheel is so small relative to ω_0 that the slip velocity is large), the brake pressure will be reduced in an attempt to move the slip velocity back to the optimum region.

Figure 8 represents most, if not all, ECUs that calculate angular acceleration from the measured wheel angular velocity in order to anticipate velocity changes in the next few milliseconds. This ability to anticipate velocity changes accounts for the superior performance of an electronically

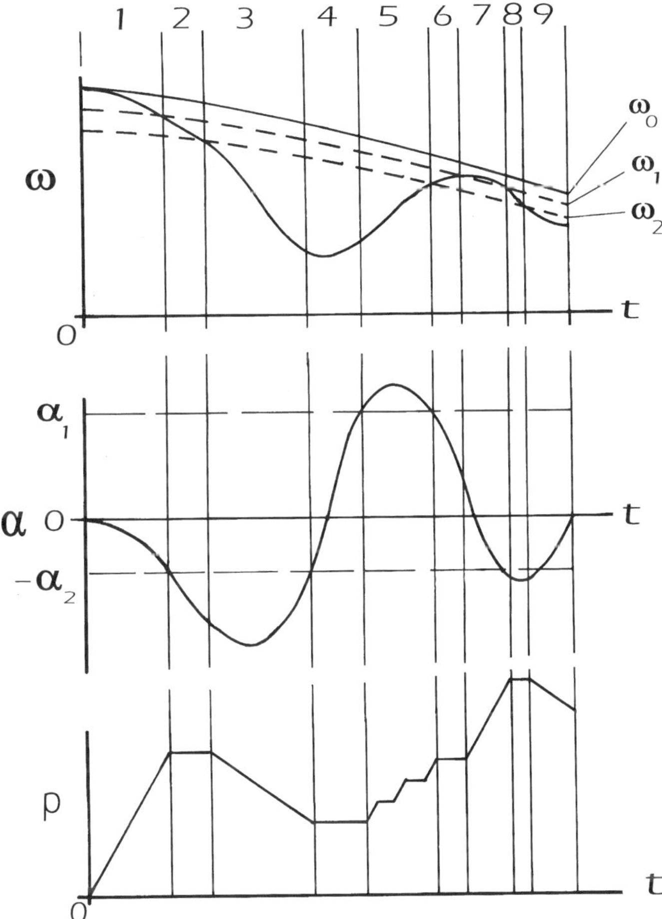

FIGURE 8 Estimated wheel reference angular velocity ω, optimum slip limits ω_1 and ω_2, angular acceleration α with decision limits α_1 and α_2, and brake pressure *p*, all as a function of time *t*.

controlled ABS over a less expensive one that relies upon a rotating annular ring to activate braking after velocity changes have begun.

An already noted, in an ABS that has no independent means of finding the vehicle velocity, the ECU memory in many such systems may contain typical data for the decrease in velocity as the brakes are applied for a selected road condition, as represented by the upper curve in Figure 2. The bottom graph in Figure 8 shows the pressure changes as commanded for a wheel by an ECU that does not alter the reference angular velocity ω_0 for zero slip while the ABS is in control of braking. The dashed lines labeled ω_1 and ω_2 bound the range of ω within which μ_B is at or near its maximum value.

ABS control is triggered by the wheel deceleration in region 1, which exceeds the reference deceleration α_2 (i.e., negative acceleration is less than $-\alpha_2$) as it crosses into the optimum slip region, region 2, for braking where angular velocity ω is larger than ω_1 [9]. At this point the ECU calls for constant brake pressure until either the acceleration reverses or the angular velocity falls below ω_2, which is the case in this instance. Once ω is below ω_2 in region 3, the brake pressure is reduced until the acceleration increases enough to again be greater than $-\alpha_2$. In region 4, the brake pressure stays constant until the acceleration is larger than α_1, which indicates that the wheel is speeding up and wheel slip is being reduced to the point that it may again enter the optimum region. Thus the pressure is increased in region 5 in small steps, and the acceleration is checked after each step before commanding the next step. Wheel slip enters the optimum slip range in region 6, and brake pressure is again held constant. Once the wheel's angular velocity in region 7 rises above ω_1, it is in the stable region of Figure 6, and the brake pressure may be increased until the slip velocity enters the optimum range between ω_1 and ω_2 in region 8, where the ECU again holds the pressure constant. In region 9, ω is below ω_2, so brake pressure is reduced.

Similar logic holds in systems that employ accelerometer information to indicate actual vehicle response to the braking action of all wheels [10]. Since vehicle velocity and acceleration are determined independently from each wheel's angular velocity and angular acceleration, the road conditions at each wheel associated with the curves in Figure 2 may be estimated from calculations of μ_B and its gradients as a function of λ.

With this information, the reference curve for ω may be continuously updated to give better data on wheel slip, as displayed in Figure 9, which in turn should usually yield shorter stopping distances when used with equally well-programmed ECUs. As in Figure 8, the ω_0 curve represents the angular velocity of the particular wheel, which is directly related to the velocity of the vehicle when there is zero slip between the wheel and the road.

Again the ABS is activated at the beginning of region 1 when the acceleration falls below $-\alpha_2$, at which point the angular velocity ω exceeds ω_1

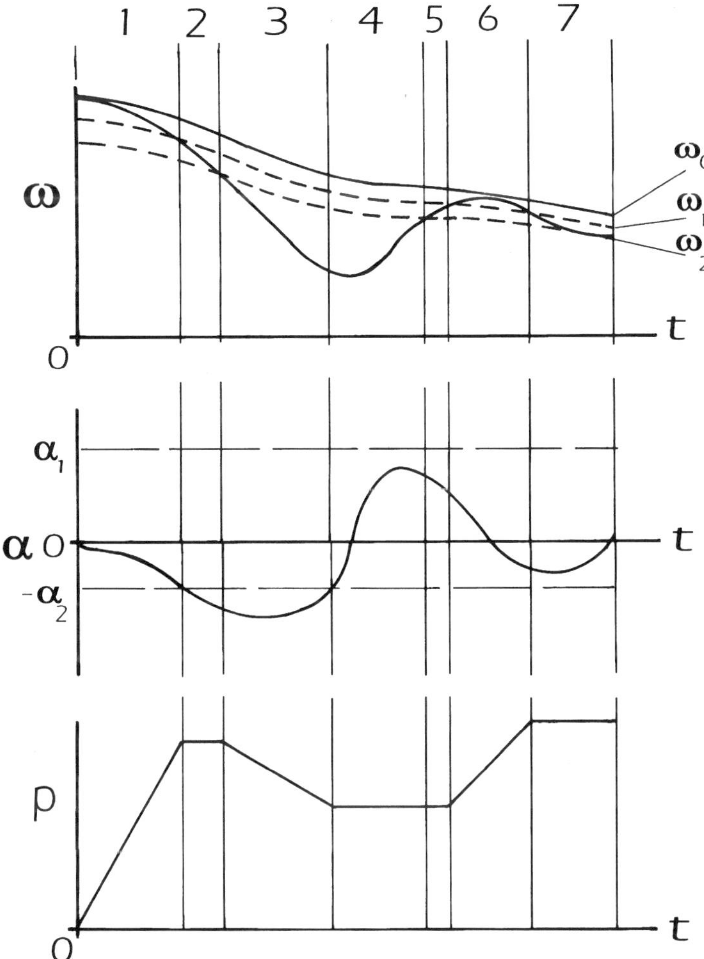

FIGURE 9 Wheel reference angular velocity ω based upon accelerator data, optimum slip limits ω_1 and ω_2, angular acceleration α with limits α_1 and α_2, and brake pressure p, all as a function of time t.

and the ECU holds the brake pressure constant throughout region 2. Because braking of all four wheels has caused the vehicle to slow, as detected by one or more accelerometers, the reference angular velocity has decreased in region 2 and continues to decrease in regions 3 and 4 due to the action of the remaining wheels, even though this wheel continues to slip. Brake pressure is reduced in region 3 because $\omega < \omega_2$ and $\alpha < -\alpha_2$. Brake pressure is held constant in region 4 because the acceleration is larger than $-\alpha_2$. It remains constant in region 5 because the velocity is between ω_1 and ω_2, and it increases in region 6 because the velocity has increased to a value greater than ω_1, which places it in the stable region where μ_B increases as the brake pressure increases. Finally, brake pressure remains constant through region 7 because ω has again moved into the optimum range.

Control systems described in Figures 8 and 9 are but two of many possible control methods. The system implied in Figure 8 may be improved by an ECU that remembers all of the road conditions shown in Figure 2 and compares all wheel velocities and accelerations to select the appropriate curve. For example, if all wheels decelerate quickly so that the negative angular acceleration falls below $-\alpha_2$ for all wheels, the ECU may assume the vehicle is on glare (black) ice, curve 6.

The decision process itself, as illustrated in Figures 8 and 9, is independent of whether or not that particular wheel is driven. Driven wheels have a larger effective moment of inertia than do undriven wheels, however, which will slow their response time to braking relative to an undriven wheel. Clearly their effective moment of inertia will depend upon what components are automatically disengaged during brake application. Some ECUs may be designed to account for this by receiving data that indicate when the wheels are engaged to a drive train.

Additional control may be had if an ECU has input from accelerometers, such as shown in Figure 10, that are positioned to measure both longitudinal and transverse acceleration. With both longitudinal and transverse accelerometer data, the ECU can compare accelerations to detect both spinning and transverse sliding during panic stops and other driving maneuvers and thereby can better maintain stability to the extent possible with brake control alone.

An example of the analysis associated with a particular accelerometer arrangement may be had by supposing that four accelerometers are arranged with their sensitive axes lying in a common horizontal plane that is parallel to the plane of motion of the vehicle. Also suppose that they are placed in the vehicle with two at the right front and two at the left front, each at distance R from an arbitrary reference point P in the plane, as shown in Figure 11. Accelerometers in each pair are positioned such that their sensitive axes are mutually perpendicular: One is parallel to the longitudinal axis of the vehicle, and the other is perpendicular to the longitudinal axis of the vehicle.

Let A and α denote the linear acceleration at point P and the angular acceleration about P, respectively, let a_{1L} and a_{1T} represent the accelerations measured in the longitudinal and transverse directions, respectively, at location 1, and let a_{2L} and a_{2T} represent the accelerations measured in the

FIGURE 10 Single-axis accelerometer capable of measuring acceleration from $\pm 20g$ to $\pm 500g$, depending upon the model, which may be attached to the vehicle. (Courtesy GS Sensors, Ephrata, PA.)

longitudinal and transverse directions, in that order, at location 2. Let Ω denote the angular velocity of the spin of the vehicle about point P.

In terms of the dimensions shown in Figure 11, the accelerations measured by accelerometers a_{1L}, a_{1T}, a_{2L}, and a_{2T} are given by

$$a_{1L} = A \cos\theta + R\Omega^2 \cos\phi - R\alpha \sin\phi \tag{4-1}$$

$$a_{1T} = A \sin\theta + R\Omega^2 \sin\phi + R\alpha \cos\phi \tag{4-2}$$

$$a_{2L} = A \cos\theta + R\Omega^2 \cos\phi + R\alpha \sin\phi \tag{4-3}$$

$$a_{2T} = -A \sin\theta + R\Omega^2 \sin\theta - R\alpha \cos\phi \tag{4-4}$$

From the measured accelerations, the ECU can provide

$$\Omega = \sqrt{\frac{a_{1T} + a_{2T}}{2R \sin\phi}} \tag{4-5}$$

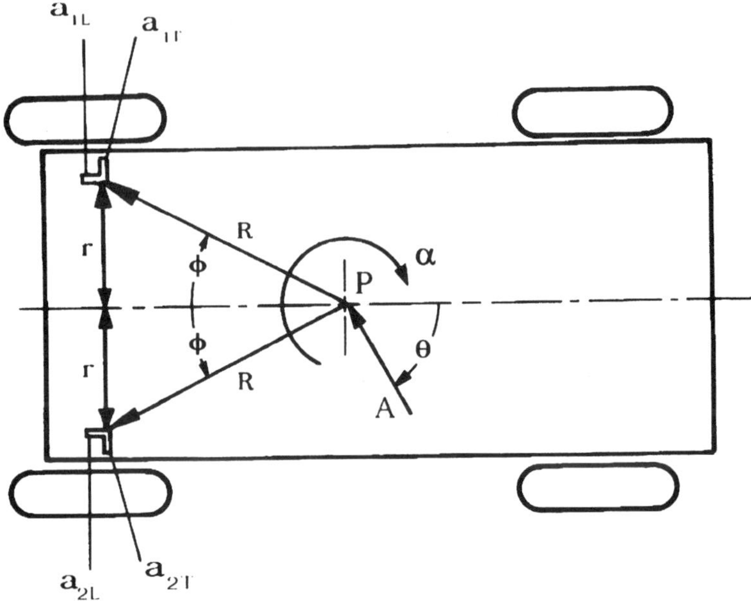

FIGURE 11 Plan view of accelerometer locations in a vehicle. Accelerometers a_{1L} and a_{2L} record positive in the forward direction (motion right to left); accelerometers a_{1T} and a_{2T} record positive outward (away from the centerline).

and

$$\alpha = \frac{a_{2L} - a_{1L}}{2\ R \cos \phi} \tag{4-6}$$

Note that angular velocity Ω was found without integration of the angular acceleration α with respect to time. However, because it appears in equations (4-1) through (4-4) as Ω^2, only its magnitude may be calculated. This is consistent with the centrifugal force being independent of the direction of rotation.

By defining B_1 and B_2 as

$$B_1 \frac{a_{1L} + a_{2L}}{2} - R\Omega^2 \cos \phi \tag{4-7}$$

$$B_2 = \frac{a_{1T} - a_{2T}}{2} - Ra \cos \phi \tag{4-8}$$

the acceleration A may be found from

$$A = \sqrt{B_1^2 + B_2^2} \tag{4-9}$$

and angle θ may be calculated from

$$\theta = a\tan 2(x, y)\frac{180}{\pi} \tag{4-10}$$

in which $a$tan 2 (x,y) calls a program that calculates θ from sin θ and cos θ and places the result in the proper quadrant in a coordinate system in which $\theta = 0$ on the positive x-axis and θ is positive in the counterclockwise direction. In equation (4-10),

$$x = a_{1L} + a_{2L} - 2R\Omega^2 \cos \phi \tag{4-11}$$

and

$$y = a_{1T} - a_{2T} - 2R\alpha \cos \phi \tag{4-12}$$

Finding the linear velocity of the vehicle does require integration. A simple means of numerical integration is for the ECU to remember the last calculated velocity v_{i-1} at time t_{i-1} and thereby calculate the latest velocity v_i at time t_i from

$$v_i = \frac{A_i + A_{i-1}}{2}\Delta_t + v_{i-1} \tag{4-11}$$

in terms of the time interval $\Delta t = t_i - t_{i-1}$ between accelerometer readings. An even simpler formula may be to replace the average acceleration over the sampling period $[(A_i + A_{i-1})/2]$ with the last measured acceleration A_i. This second choice emphasizes the last acceleration measurement and in some cases may be equally satisfactory as formula (4-11). Whether to use either of these formulas or another numerical formula depends upon the vehicle's inertia, braking response times, and perhaps other factors, such as the sampling time and the expected acceleration range, along with the expected accumulated error over the time that the ABS operates.

Depending upon the ECU program, control may be improved by merging this information with present data on the steering angles and angular rates of the front wheels. Similar data from all wheels may be used for vehicles in which all wheels steer.

Commands from the ECU are transmitted to the brakes through either hydraulic lines or electrical lines, which may differ in number from the data lines that run from the wheel sensors to the ECU. For example, in front wheel drive automobiles, all wheels are fitted with wheel sensors, but the ECU typically may have only two command lines, or circuits: One controls the right front and left rear wheels and the second controls the left front and right rear wheels. Rear wheel drive automobiles typically may have one command line to the front wheels and one to the rear wheels. Four-wheel-drive vehicles may have four control lines: one to each wheel. Data lines and control lines generally are more in number and the ECU programs are usually more elaborate for trucks and buses.

A system that includes the control both of braking and of acceleration during driving is commonly referred to as *fraction control system*, or TCS.

Although Figure 2 also may apply to vehicle acceleration, equation (1-4) must be modified to read

$$\lambda = \frac{v_1 - v_0}{v_0} = \frac{\omega_1 - \omega_0}{\omega_0} \tag{4-11}$$

if λ is to remain positive. Here v_1 represents the velocity the vehicle would have achieved during acceleration if there were no slip and v_0 represents the actual vehicle velocity. Associated wheel angular velocities are represented by ω_1 and ω_0.

For additional information on published details of ABS and TCS designs, see publications from the Society of Automotive Enginers (SAE), such as Ref. 11, from BOSCH publications, such as Ref. 9, and from other automotive engineering journals.

Example 1

To demonstrate that equation (4-5), (4-6), (4-9), and (4-10) yield the angular and linear accelerations and the magnitude of vehicle rotation, suppose that an accelerating vehicle's motion at a particular instant is equivalent to the center of gravity's accelerating at 14.9 ft/sec^2 at an angle of $-8°$ relative to the vehicle centerline and to the vehicle rotating with an angular velocity of 1.9 rad/sec and an angular acceleration of -1.7 rad/sec^2 about its center of gravity. Let the center of gravity be such that $R = 3$ ft and $\varphi = 18°$.

Substitution of $A = 14.9$, $R = 3$, $\Omega = 1.9$, $\varphi = 18°$, and $\theta = -8°$ into equations (4-1) through (4-4) yields

$$a_{1L} = 18.076 \text{ ft/sec}^2 \qquad a_{1T} = -13.574 \text{ ft/sec}^2$$

$$a_{2L} = 25.736 \text{ ft/sec}^2 \qquad a_{2T} = -2.692 \text{ ft/sec}^2$$

Substitution of these values into (4.5), (4-6), (4-9), and (4-10), yields

$$A = 14.9 \text{ ft/sec}^2 \quad \Omega = 1.9 \text{ rad/sec} \qquad \alpha = -1.7 \text{ rad/sec}^2$$

$$\theta = -8°$$

Example 2

During braking, a vehicle decelerates with a motion that is equivalent to a linear deceleration of 20 m/sec^2 along a line oriented at 30° relative to the vehicular centerline, an angular acceleration of 11.7 rad/sec^2, and an angular velocity of −4.2 rad/sec. Accordingly, enter $A = -20$ m/sec², $R = 1$ m, $\Omega = -4.2$ m/sec, $\varphi = 18°$, and $\theta = 30°$ into equations (4-1) through (4-4). The result is

$$a_{1L} = -4.159 \text{ m/sec}^2 \qquad a_{2L} = 3.072 \text{ m/sec}^2$$
$$a_{1T} = 6.578 \text{ m/sec}^2 \qquad a_{2T} = 4.329 \text{ m/sec}^2$$

Substitution of these accelerations into equations (4.5), (4-6), (4-9), and (4-10), yields

$$A = 20 \text{ m/sec}^2 \qquad \Omega = 4.2 \text{ m/sec} \qquad \alpha = 11.7 \text{ rad/sec}^2$$
$$\theta = -150°$$

which indicates that the foregoing relations correctly evaluate the input data to describe the vehicle's motion except for the algebraic sign of Ω.

Accelerations and velocities used in our two examples are for demonstration only. Most ABS and TCS programs analyze and process input data rapidly enough that the controlled vehicle follows the driver's input commands (stopping, turning, accelerating, and backing) to the extent that deviations between the commands and the response of the vehicle are unnoticed under most conditions.

V. NOTATION

A, a	linear acceleration (l/t^2)
F	force (ml/t^2)
m	mass (m)
R, r	radius (l)
s	stopping distance (l)
t	time (t)
v	linear velocity (l/t)
x	distance (l)
α	angular acceleration ($1/t^2$)
λ	tire slip (1)
φ	angle (1)
θ	angle (1)
τ	stopping time (t)
Ω	angular velocity (t^{-1})
ω	angular velocity (t^{-1})

VI. FORMULA COLLECTION

Slip velocity, linear:

$$v_s = v_0 - v_1$$

Slip velocity, angular:

$$\omega_s = \omega_0 - \omega_s$$

Acceleration:

$$a_{1L} = A \cos\theta + R\Omega^2 \cos\varphi - R\alpha \sin\varphi$$
$$a_{1T} = A \sin\theta + R\Omega^2 \sin\varphi + R\alpha \cos\varphi$$
$$a_{2L} = A \cos\theta + R\Omega^2 \cos\varphi + R\alpha \sin\varphi$$
$$a_{2T} = -A \sin\theta + R\Omega^2 \sin\varphi - R\alpha \cos\varphi$$

Angular velocity:

$$\Omega = \sqrt{\frac{a_{1T} + a_{2T}}{2R \sin\phi}}$$

Angular acceleration:

$$\alpha = \frac{a_{2L} - a_{1L}}{2R \sin\phi}$$

Linear acceleration:

$$A = \sqrt{B_1^2 + B_2^2}$$

Linear velocity:

$$v_i = \frac{A_i + A_{i-1}}{2}\Delta t + v_{i-1}$$

REFERENCES

1. Lieber, H., Czinczel, A. (1987). Antiskid System for Passenger Cars with a Digital Electronic Control Unit. In: Martin, J. M., Gritt, P.S., eds. *Anti-Lock Braking Systems for Passenger Cars and Light Trucks-A Review*. Warrendale, PA: Society of Automotive Engineers, pp. 233–239.
2. Buckman, L. C. (1998). *Commercial Vehicle Braking Systems: Air Brakes, ABS and Beyond*. Warrendale, PA: Society of Automotive Engineers.
3. (1987). Martin, J. M., Gritt, P. S., eds. Preface, *Anti-Lock Braking Systems for Passenger Cars and Light Trucks—A Review*. Warrendale, PA: Society of Automotive Engineers.
4. Douglas, J. W., Schafer, T. C. (1987). The Chrysler "Sure-Brake"—The first Production Four-Wheel Anti-Skid System. In: Martin, J. M., Gritt, P. S., eds. *Anti-Lock Braking Systems for Passenger Cars and Light Trucks—A Review*. Warrendale, PA: Society of Automotive Engineers.
5. Burckhardt, M. (1979). Erfahrungen bei der Konzeption und Entwicklung des Mercedes-Benz/Bosch-Anti-Blockier-Systems (ABS) ATZ, 81:5.

6. Lieber, H. (1979). Antiblockiersystems für Personenwagen mit digitaler Elek tronik-Aufbau und Funktion, ATZ, 81:11.
7. Müller, P., Czinczel, A. (1987). Electronic Anti-Skid System—Performance and Application. In: Martin, J. M., Gritt, P. S., eds. *Anti-Lock Braking Systems for Passenger Cars and Light Trucks—A Review*. Warrendale, PA: Society of Auto motive Engineers, pp. 25–33.
8. Newton, W.R., Riddy, F.T. (1987). Evaluation Criteria for Low-Cost Anti-Lock Brake Systems for FWD Passenger Cars. In: Martin, J. M., Gritt, P. S., eds. *Anti-Lock Braking Systems for Passenger Cars and Light Trucks-A Review*. Warrendale, PA: Society of Automotive Engineers, pp. 277–287.
9. (1994). Bauer, H., ed. *Brake System for Passenger Cars*. Trans. P. Girling, Robert Bosch GmbH.
10. Yoneda, S., Naitoh, Y., Kigoshi, H. (1987). Rear Bruke Lock-Up Control System of Mitsubishi Starlou. In: Martin, J. M., Gritt, P. S., eds. *Anti-Lock Braking Systems for Passenger Cars and Light Trucks—A Review*. Warrendale, PA: Society of Automotive Engineers, pp. 249–255.
11. Schwarz, R., Willimowski, M., Isermann, R., Willimowski, P. (1997). In: *Overview of ABS/TCS and Brake Technology*. Warrendale, PA: Society of Automotive Engineers, pp. 123–133.

13

Brake Vibration

Formal studies of brake vibration appear to have been first reported in 1935 by Lamarque and Williams, who were concerned with brake squeak [1]. Subsequent theoretical studies provided a mathematical formulation of the problem and further experimental data. Holography and finite element analyses have provided a unified description of the behavior of the brake assembly (drum, backplate, shoes, and lining in the case of drum brakes and disk and caliper in the case of disk brakes) as it vibrates and have shown that brake vibration is the result of an interplay between the variation of the coefficient of friction as a function of the relative velocity between the brake pad the friction surface (disk or drum) and the masses, equivalent springs, and dampers that comprise the associated mechanical system.

I. BRIEF HISTORICAL OUTLINE

Lamarque and Williams [1] appear to have been the first to suggest that brake vibration was due to a stick-slip frictional phenomenon dependent on the coefficient of friction decreasing as the relative velocity between the friction surfaces increased. Although not explained in detail, the implication was that the brake would engage and the associated mechanical system would deform slightly under the applied load to the point where the shoe and drum configuration would change enough for the elastic forces to cause the shoe and brake to disengage momentarily. Once disengaged, the elasticity of the mechanical system would cause the shoe and drum to snap back to their undistorted configuration fast enough to lower the friction between contacting surfaces sufficiently for the components to nearly reassume their original configuration and the process to repeat. In the years that followed, many investigations were conducted to better understand the role of the factors that affected brake and clutch vibration. Only a few of the many contributors to the literature of brake vibration will be discussed explicitly in this section. This partial listing is sufficient, however, to portray the general trends in the years after 1935.

An experimental investigation by Hollmann [2], reported in 1954, indicated that the tendency for frictional vibration increased with the contact pressure and that it also increased with the temperature of the friction materials up to 100°C but then decreased rapidly as the temperature rose above 100°C. After studies of brake squeal on railway vehicles, Broadbent [3] also concluded that brake vibration was dependent on the mechanical linkage used and was associated with a friction coefficient that decreased with increasing velocity between shoe and brake. He also observed that no chatter was found when wooden brakes whose friction coefficient increased with increased slip speed were used. Similar results were found by Sinclair [4], who also reported that the frequency of oscillation was strongly dependent on the equivalent mass and spring constant of the mechanical system that held the brake lining in place in the slider block configuration used in the laboratory model.

Spurr [5] rejected the notion that it was necessary for the friction coefficient to decrease with increasing slip speed in order to have frictional vibration. His experiments with railway block brakes on a railway wheel indicated that vibration was independent of the change in the friction coefficient with slip speed but that it was more likely if the friction coefficient was large. The stick-slip theory of friction appeared to hold as the driving means, however, and Spurr also observed that the frequency of oscillation depended on the associated mechanical system used to force the brake block against the wheel.

In the discussion of Spurr's paper, F. R. Murray introduced what has been called the sprag theory in British literature, involving a sprag (a cantilevered beam pressed against a moving surface), as shown in Figure 1. A somewhat similar configuration was used by Jarvis and Mills [6] in their theoretical and experimental simulation of a caliper disk brake. Their analysis was based on the classical analysis of the normal deflection of a circular disk in terms of Bessel functions and an assumed deflection of the sprag with coefficients chosen to match experimentally observed values. Substitution of these deflections and their time derivatives into the Lagrangian equations of motion resulted in a set of nonlinear partial differential equations whose solution was approximated by what they termed the "slowly varying amplitude and phase" method described in Ref. 7. The solution found in this

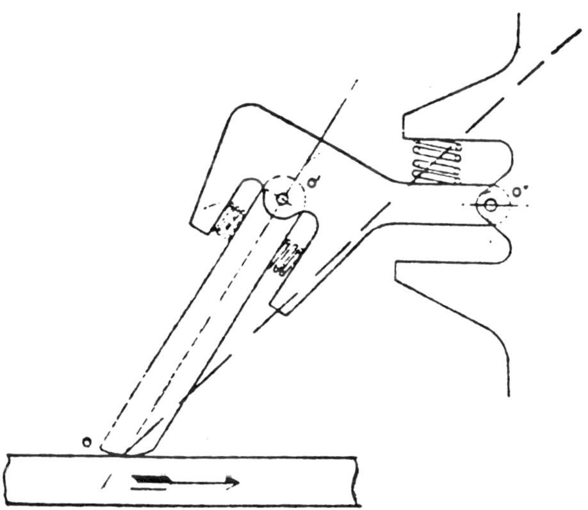

Figure 1 The sprag mechanism suggested by Murray in Ref. 5.

manner contained an exponential term which became infinite for certain lateral displacement distributions along the sprag and certain forces between the disk and the sprag. These combinations of sprag deflections and disk forces defined a region of instability which was interpreted to signify large sprag and disk vibration. As formulated, the solution depended only on the mass and elastic properties of the disc and sprag and was not influenced by variation of the coefficient of friction with slip velocity.

Although the solution so obtained agreed with the measured regions of instability in terms of the slope of the sprag relative to the plane of the disk, the shape of the calculated boundary curve for the region of instability differed markedly from the experimentally measured curve. The measured curve was concave upward but the theoretical curve was concave downward.

Based on these results, Jarvis and Mills concluded that brake vibration could be avoided by careful design of the disk and caliper without specifying a particular variation for the coefficient of friction. In the published discussion Spurr agreed that brake vibration could be controlled by careful design of the associated mechanical system.

In spite of the comments by Broadbent, Sinclair, Spurr, and Jarvis and Mills on the effect of the associated mechanical system in determining the vibration resulting from the friction excitation, various authors, such as North [8], have considered the papers by Spurr and by Jarvis and Mills as expounding a theory different from those of Broadbent, Sinclair, and others, who discussed the effect of a negative slope for the curve of the friction coefficient as a function of the relative speed between friction surfaces, the slip speed. A more careful reading of their papers, however, shows that while they may have devoted more space to a discussion of friction characteristics, they were definitely aware of the importance of the response characteristics of the brake's activation system in determining the nature of the resulting oscillations. There has, therefore, been general agreement as to the nature of the problem even though the aspects emphasized have changed with time.

Analysis of the mechanical system studied by Spurr, by Jarvis and Mills, and by North [9] was extended by Millner [10] to consider the effect of a single pad at the contact region of the sprag and the disk. This analysis, which was based on a highly simplified lumped-parameter model, implied that nonlinear pad compression may cause the brake to vibrate within discrete bands of actuating pressure.

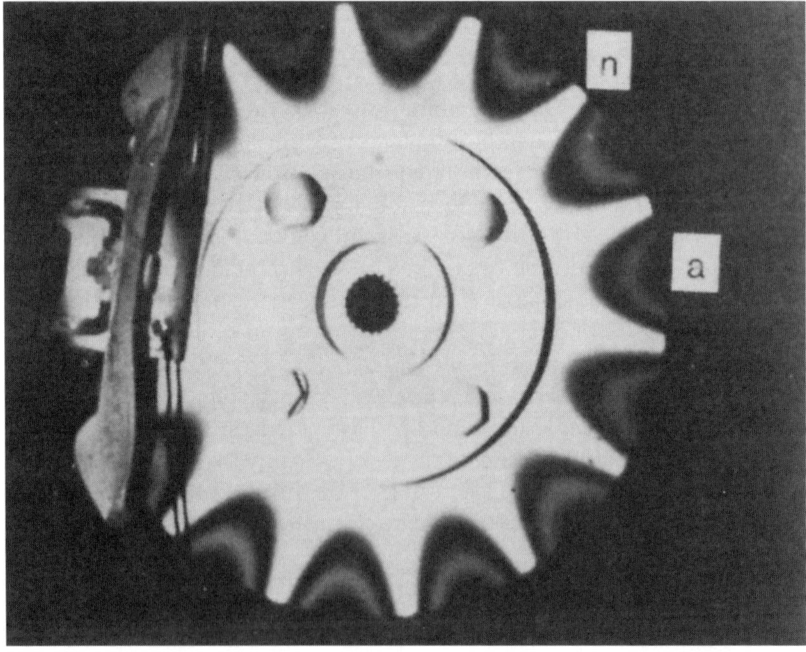

FIGURE 2 Vibration mode of a yoke-type disk brake at 10 kHz. (Reprinted with permission; © 1984 Society of Automotive Engineers, Inc.)

II. RECENT EXPERIMENTAL DATA

Using experimental techniques not available in the 1960s for the examination of both disc and drum brake vibration, Felske, Hoppe, and Matthäi employed holographic interferometry to demonstrate conclusively that it is the caliper vibration that is the major contributor to brake noise from disk brakes [11] and the backplate vibration that is the major contributor from drum brakes [12]. Typical standing wave shapes, or the nodal patterns, for the disk are shown in Figures 2 and 3. Vibration of the caliper is shown in Figures 4 and 5. The alternating black and white line boundaries represent contour lines, or elevation lines, on the caliper and disk and consequently measure the deflection of the disk and caliper in a direction perpendicular to the plane of the photograph, as indicated in Figure 6 for an antinode on the disk.

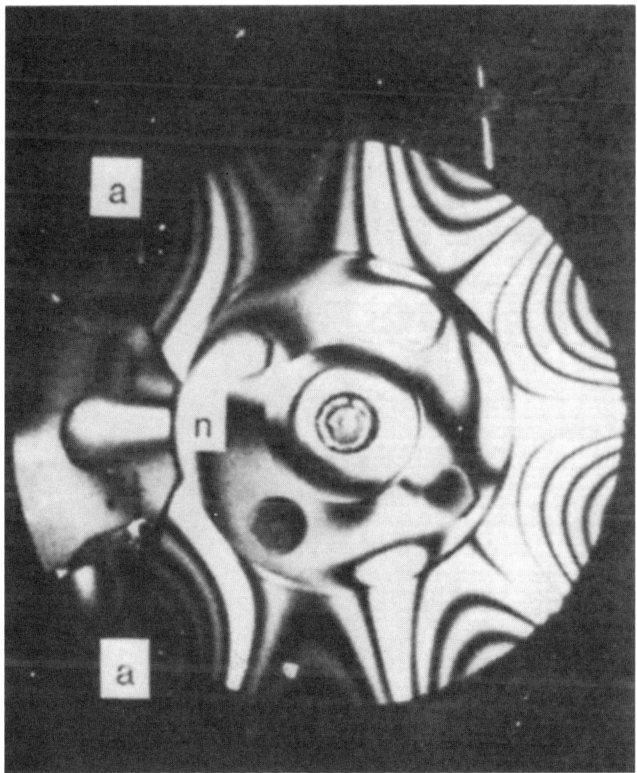

FIGURE 3 Vibration of a first-type disk brake at kHz. (Reprinted with permission; © 1984 Society of Automotive Engineers, Inc.)

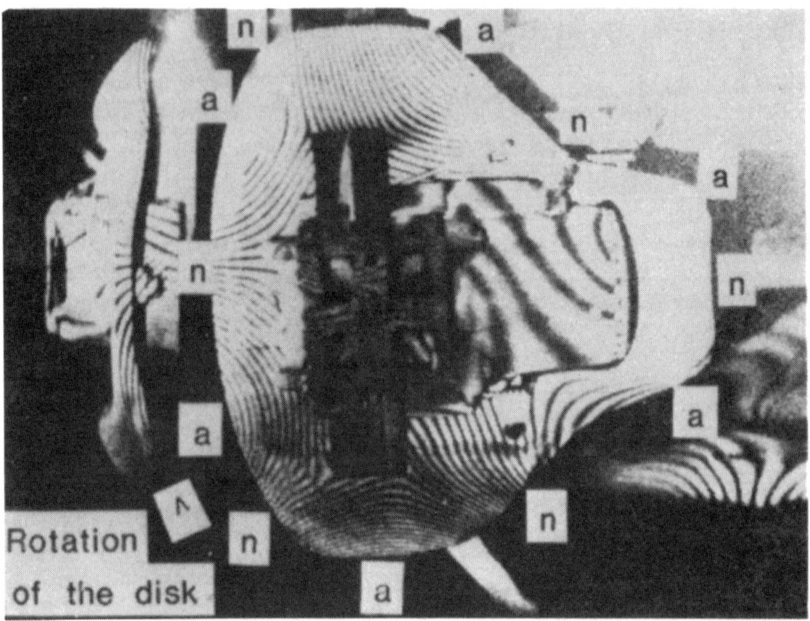

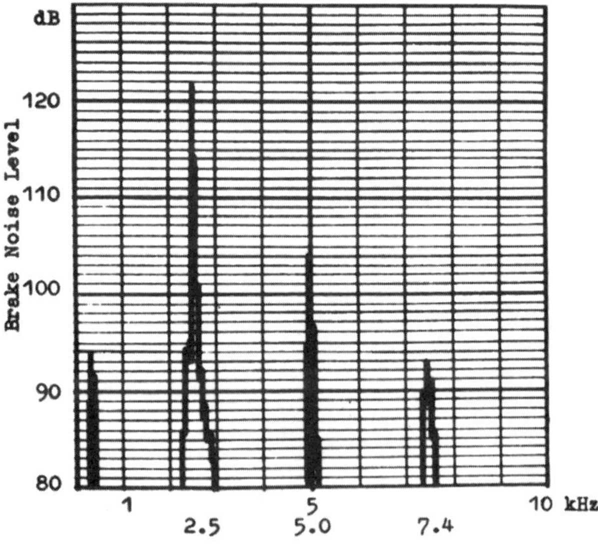

FIGURE 4 Reconstruction of a double-pulsed hologram of a squealing yoke-type caliper exposed at a noise level of 120 dB at 2.5 kHz along with its frequency spectrum. (Reprinted with permission; © 1984 Society of Automotive Engineers, Inc.)

Typical modes of backplate vibration in the first mode are shown in Figure 7 and its nodal lines, indicated by dashed curves in Figure 7, are shown alone in Figure 8. Figures 9 and 10 show a backplate before and after a raised portion (Figure 11) was added to reduce the frequency of the noise generated.

III. FINITE ELEMENT ANALYSIS

Murakami, Tsudada, and Kitamura [13] reported on a finite element analysis of automotive disk brakes to compare the calculated resonance frequencies with previous measurements of brake squeal on a chassis dynamometer and to associate them with calculated deformation of the brake components. Secondary low-frequency squeal from 2 to 3 kHz and primary high-frequency squeal from about 5.5 to 10.5 kHz, as shown in the histogram in Figure 12, correlated well with the clustering of frequencies found for the brake disk, cylinder, pad, and torque member, shown in Figure 13. Calculated disk and caliper frequency modes where verified by holographic interferometry and by accelerometer measurements in the case of the caliper, or cylinder.

Although the cluster of frequencies correlated with the primary and secondary squeal regions, several component resonant frequencies between 3 and 5.5 kHz did not result in brake squeal in this frequency range. This implied that the driving force, the stick-slip phenomenon, did not excite these frequencies and that they were not excited by structural coupling. Linearized, seven-degree-of-freedom analysis of the system using the lumped-parameter model, as shown in Figure 14, gave solutions of the form

$$\theta = Ae^{\alpha t}\sin(\omega t + \phi) \tag{3-1}$$

for each of the seven variables in which exponent α, which determined the regions of instability in the Jarvis and Mills analysis, was termed the *squeal index*. This analysis indicated that the squeal index was related to the negative gradient of the friction coefficient, as illustrated in Figure 15, which also shows test results for two test brake lining pads. These results also correlate with Spurr's finding that the squeal probability increased as the friction coefficient increased. The low squeal index between 3.0 and 5.5 kHz and its slope also correlate with the histogram shown in Figure 12, which implies that negative slope of the friction coefficient versus velocity curve is one of several significant parameters in the generation of brake squeal. Analysis of the influence of the torque member, shown in Figure 16, indicated that its influence was also minimum in the vicinity of 3 kHz, as shown in Figure 17, which also accounts for the low squeal amplitude between 3.0 and 5.5. kHz.

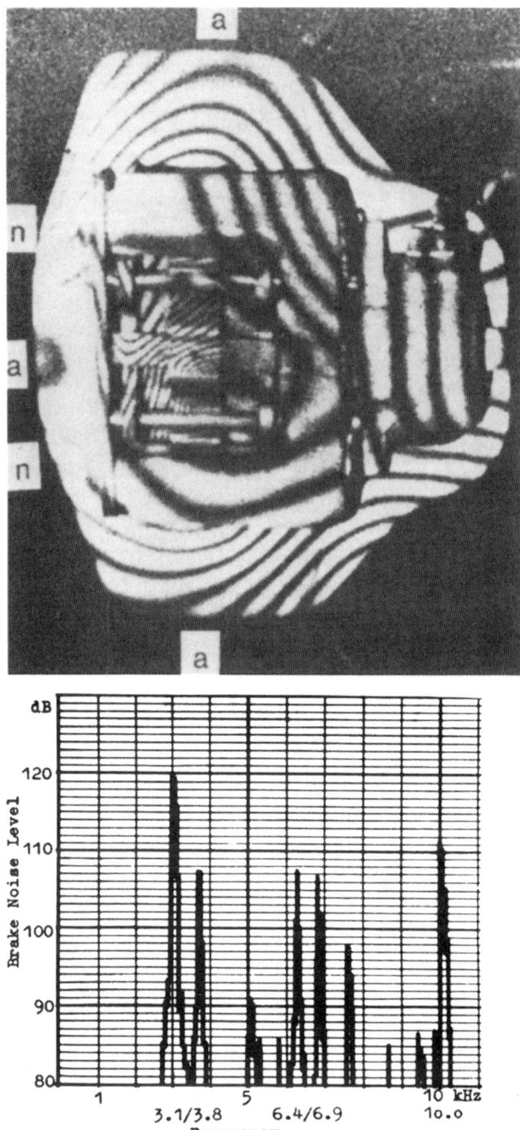

FIGURE 5 Vibration mode of a squealing yoke-type caliper exposed at 120 dB at 3.1 kHz along with its frequency spectrum. Reconstruction of a double-pulsed hologram. (Reprinted with permission; © 1984 Society of Automotive Engineers, Inc.)

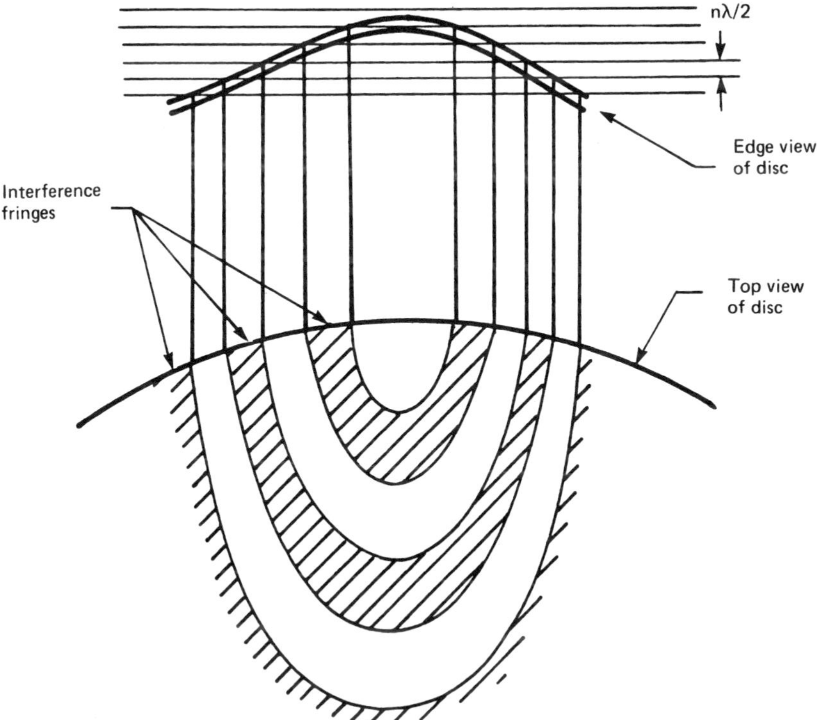

FIGURE 6 Relation between contour lines in an interference hologram and the displacement perpendicular to the plane of the interference pattern.

Calculated mode shapes for the disk, pad, cylinder, and torque member are displayed in Figures 18 through 21.

IV. CALIPER BRAKE NOISE REDUCTION

An example of caliper brake redesign in an attempt to reduce brake noise is that shown in Figure 22. According to the manufacturer, the one-piece sound insulator and backplate shown in that figure are central to its noise reduction. Details are not given because the design is patented and proprietary. Examination of the cross section shown in Figure 22, however, suggests that the physical dimensions and the material properties of the different regions in

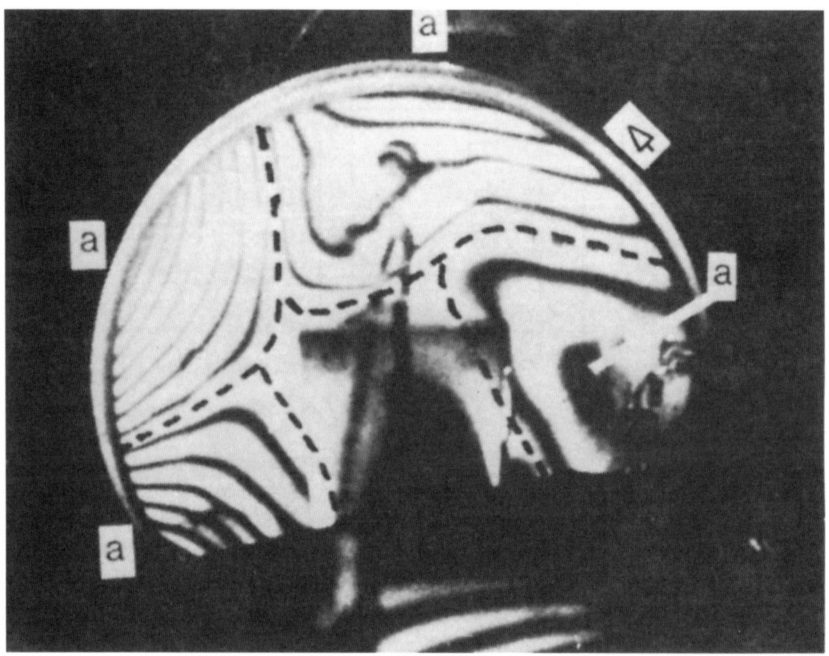

FIGURE 7 First vibration mode of a backplate on 200 × 400 mm drum brake at 1.1 kHz. Broken nodal lines superimposed on reconstructed double-pulsed hologram. (Reprinted with permission; © 1984 Society of Automotive Engineers, Inc.)

the insulator strip above the friction material may serve to dampen and suppress high-frequency vibration, and the pad material, central plate, and grooves may aid to suppress low-frequency vibration.

Grooves in brake pads of caliper brakes also aid in removing water when the brake operates in wet conditions.

Longer pad life is said to be another advantage of this design, because the insulator and backplate tend to absorb and dissipate the heat generated during stopping over a greater surface than in caliper brake pads that are not of this design. This improved heat dissipation is said to be an advantage because the heat generated causes the pad material to deteriorate.

According to U.S. patent 5,433,194, the proprietary lining material, which may also contribute to the noise reduction, is composed of organic, carbonaceous, metal, and mineral particles, rubber/resin curatives, and a corrosion inhibitor, in several different proportions.

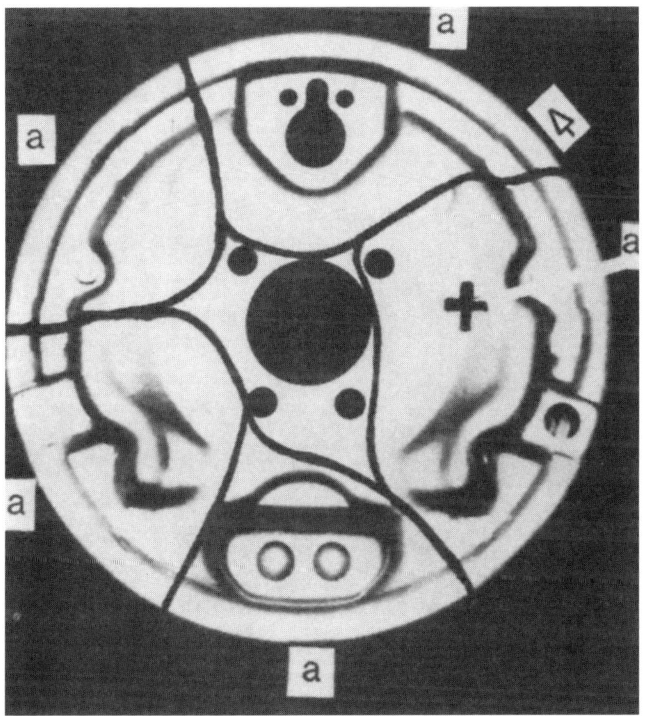

Figure 8 Nodal lines for the first vibration mode superimposed on a photograph of the entire back plate. (Reprinted with permission; © 1984 Society of Automotive Engineers, Inc.)

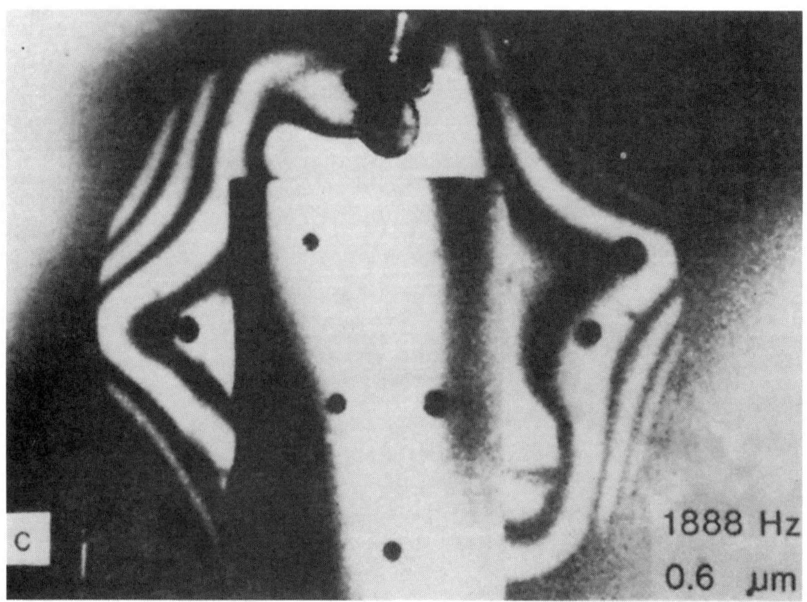

FIGURE 9 Backplate modes at 1888 Hz for a 180 × 30 mm drum brake. Noisy configuration. (Reprinted with permission, © 1984 Society of Automotive Engineers, Inc.)

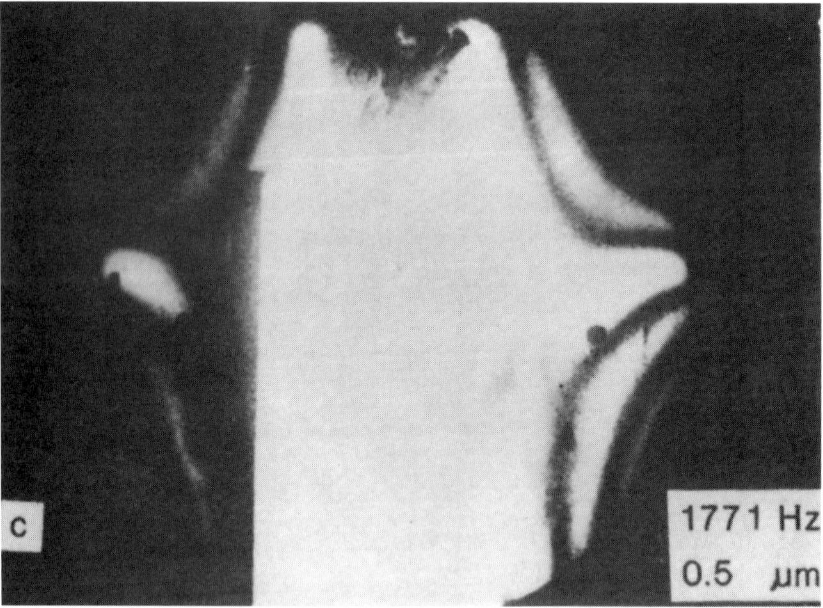

FIGURE 10 Backplate modes at 1771 Hz after backplate modification for a 180 × 30 mm drum brake. Quiet configuration. (Reprinted with permission; © 1984 Society of Automotive Engineers, Inc.)

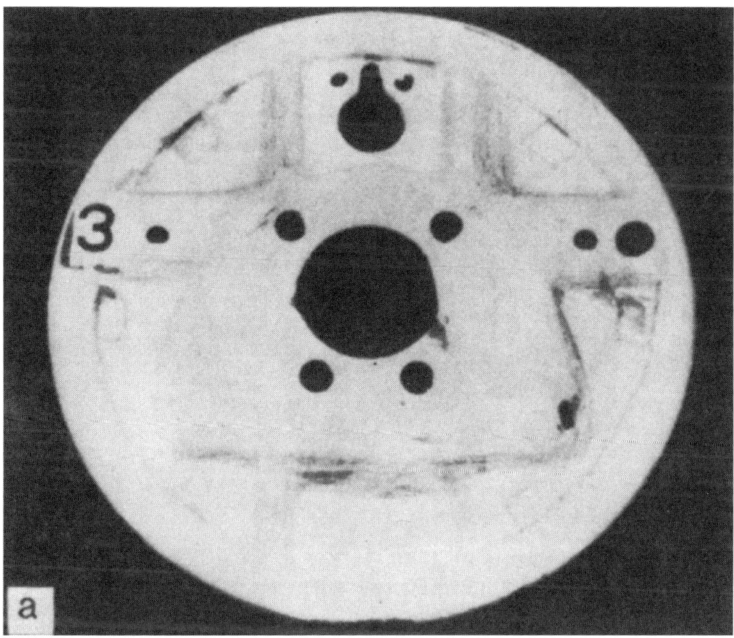

FIGURE 11 Backplate modified by adding stiffening ridges and a central raised portion, which reduced squeal by 6.5 dB at low frequency (820–990 Hz) and 4.8 dB at high frequency (1681–1888 Hz).

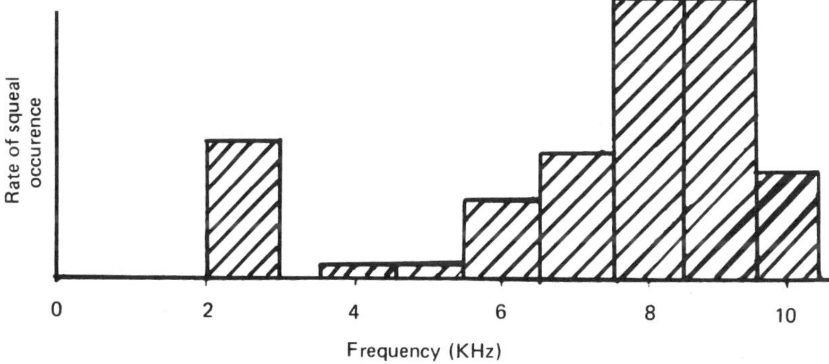

FIGURE 12 Squeal histogram for a vehicle test on a chassis dynomometer.

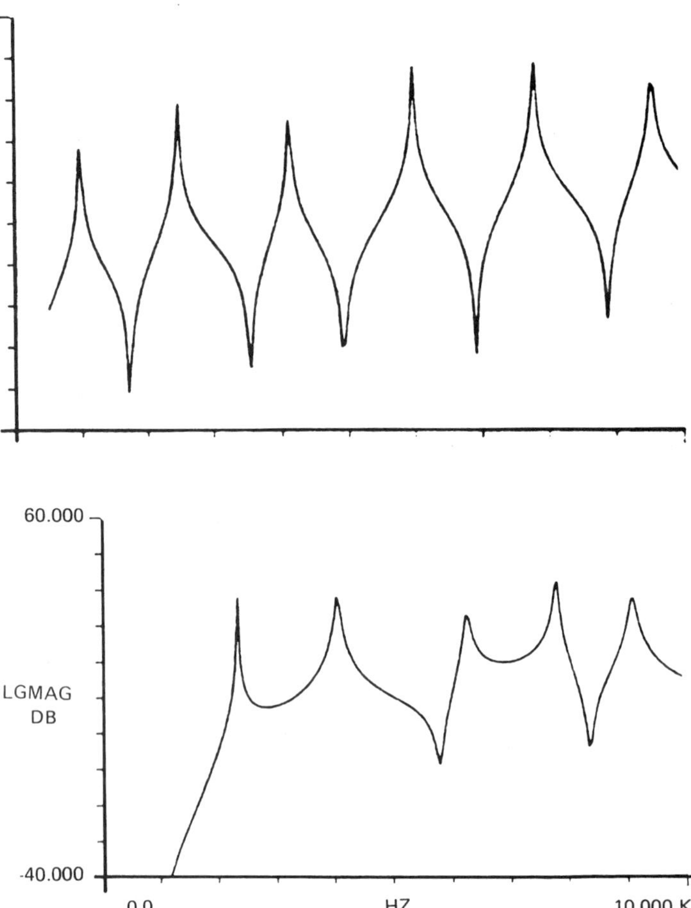

FIGURE 13 The natural, or resonant, frequencies of each component of the automotive disk brake tested.

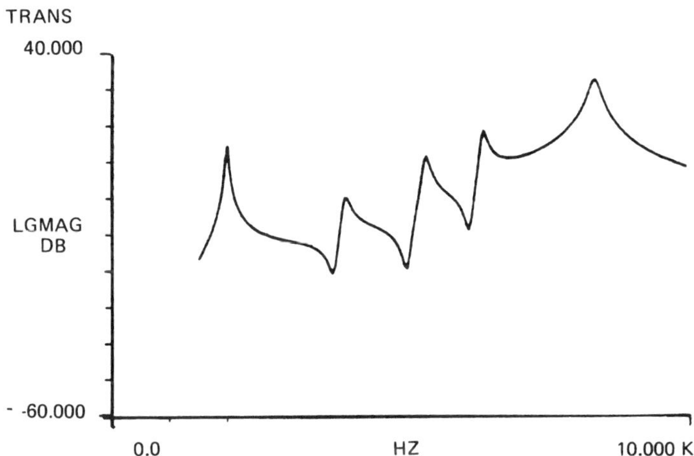

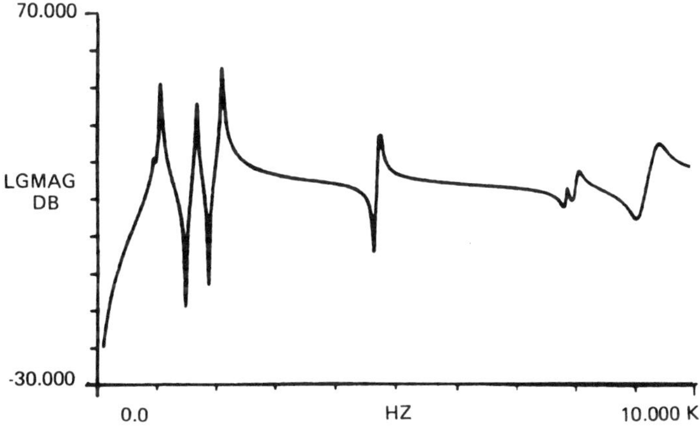

Figure 13 Continued.

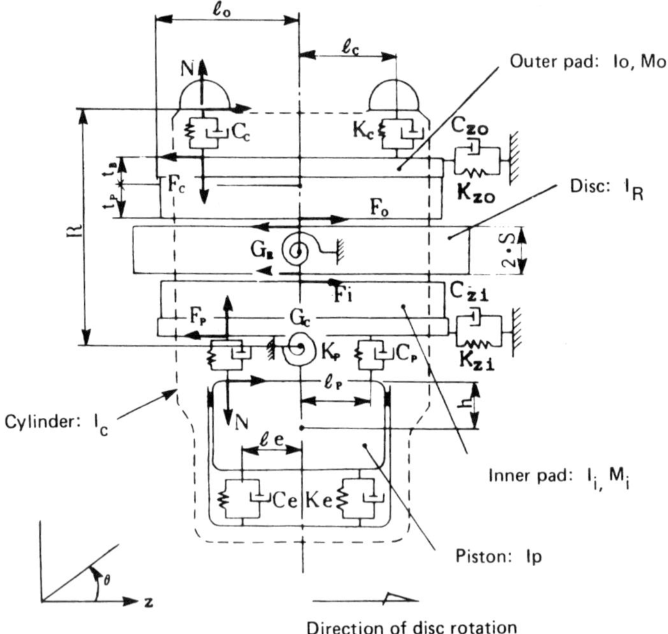

Figure 14 Sketch of a caliper and disk showing the spring and dashpot associated with each of the masses involved.

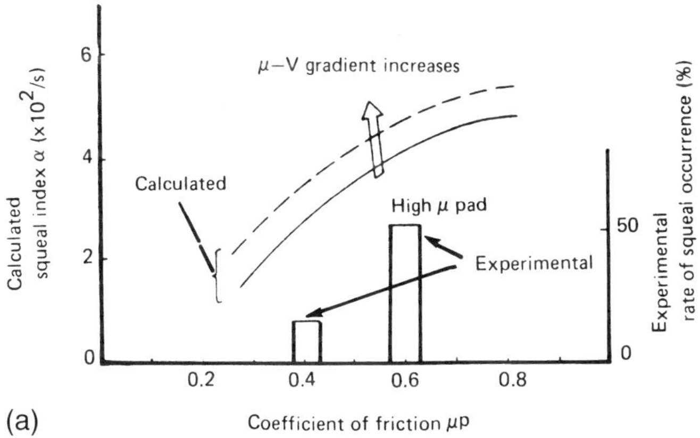

(a)

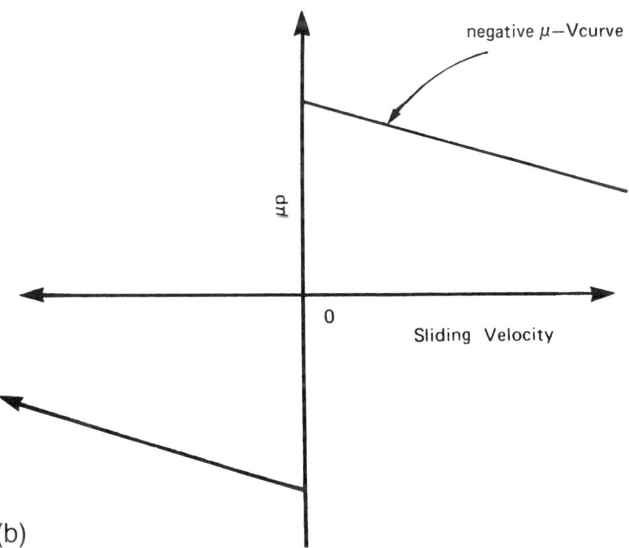

(b)

FIGURE 15 (a) Variation of squeal index α with variation in the friction coefficient; (b) assumed dependence of the friction coefficient μ on the relative, or slip, velocity.

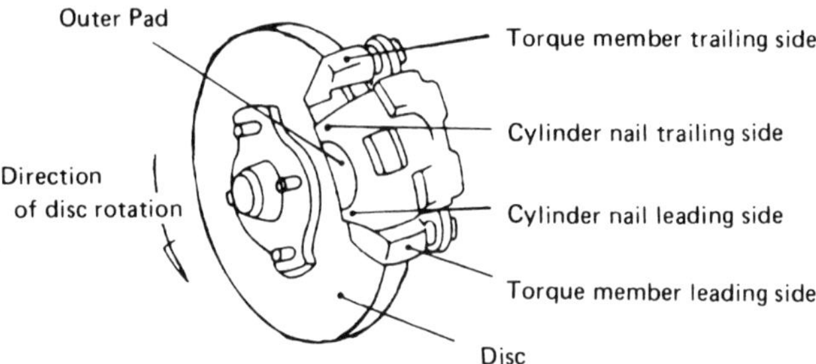

FIGURE 16 Sketch of the caliper and disk brake assembly.

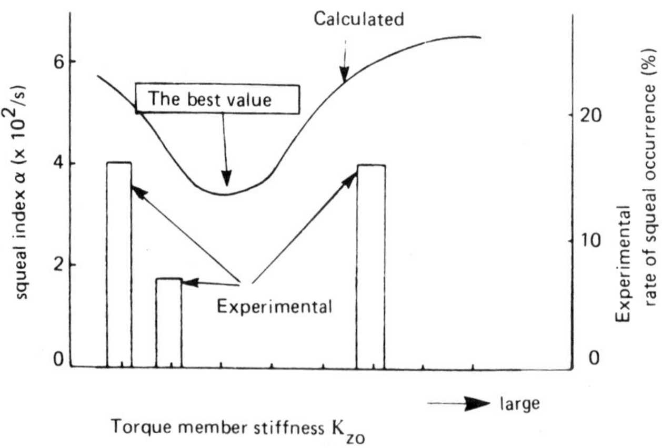

FIGURE 17 Dependence of the squeal index on the torque member stiffness.

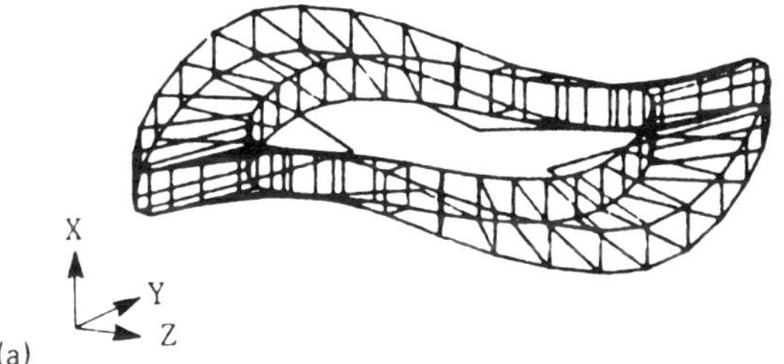

(a)

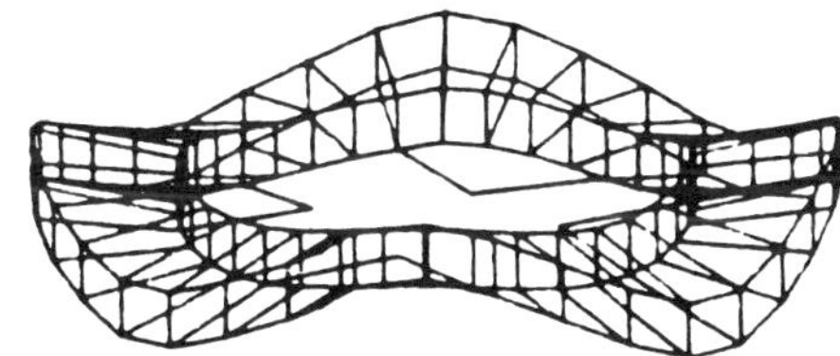

(b)

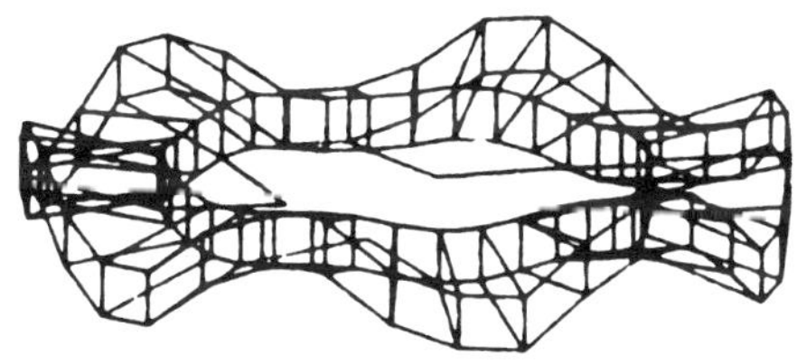

(c)

	Modal analysis	FEM	Error
(a)	2.47 kHz	2.47	0%
(b)	4.14	4.25	+2.7%
(c)	7.82	8.31	+6.3%

FIGURE 18 Examples of calculated mode shapes of the disk.

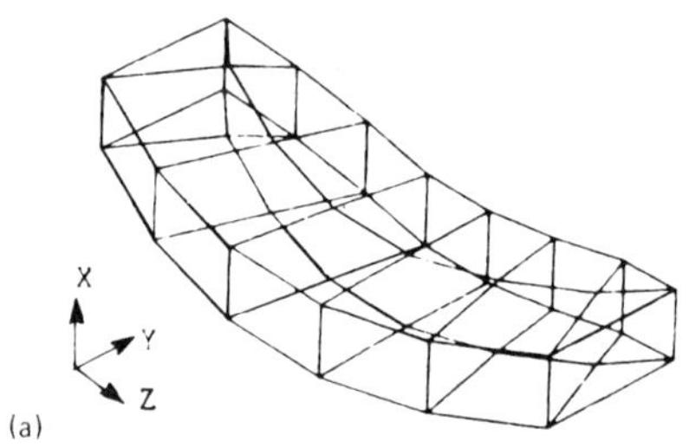

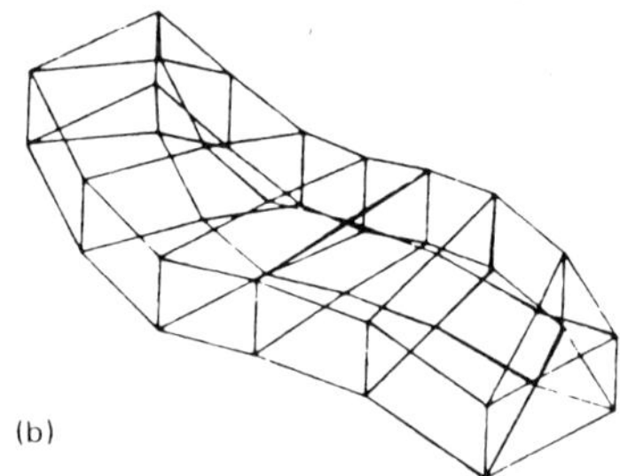

	Modal analysis	FEM	Error
(a)	4.06 kHz	3.82	-6.1%
(b)	8.36	9.06	+8.4%

FIGURE 19 Examples of mode shapes for the brake pad.

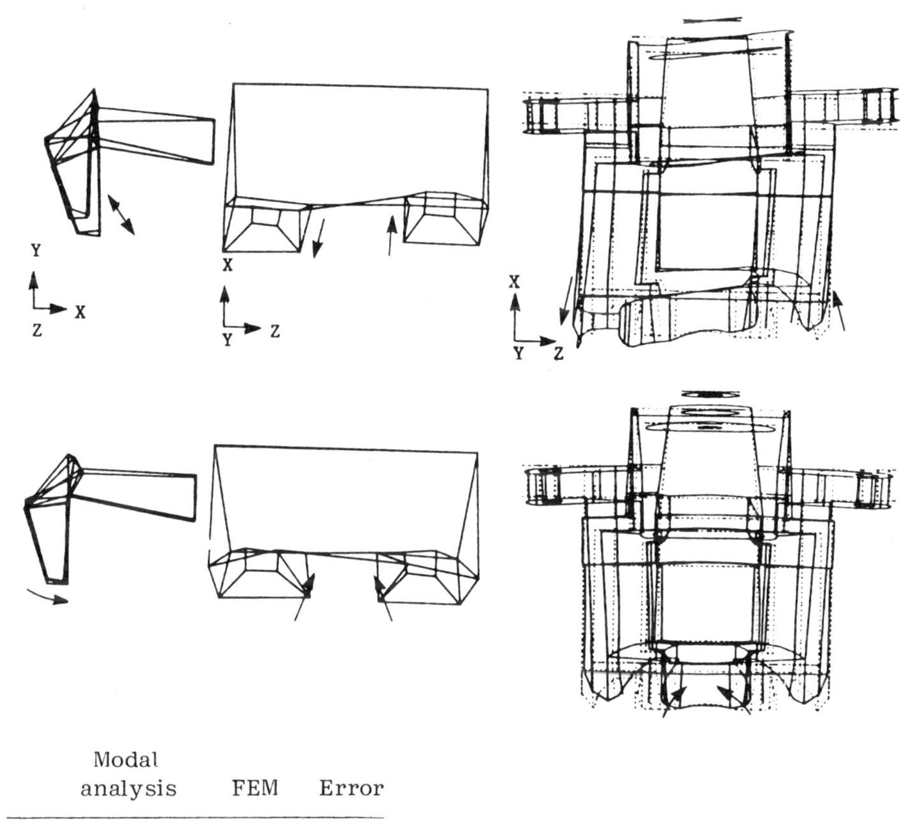

	Modal analysis	FEM	Error
(a)	2.32 kHz	2.45	+5.6%
(b)	3.97	4.00	+0.8%

FIGURE 20 Examples of the mode shapes for the cylinder portion of the caliper. (a) and (b) Modal analysis (left); FEM (right).

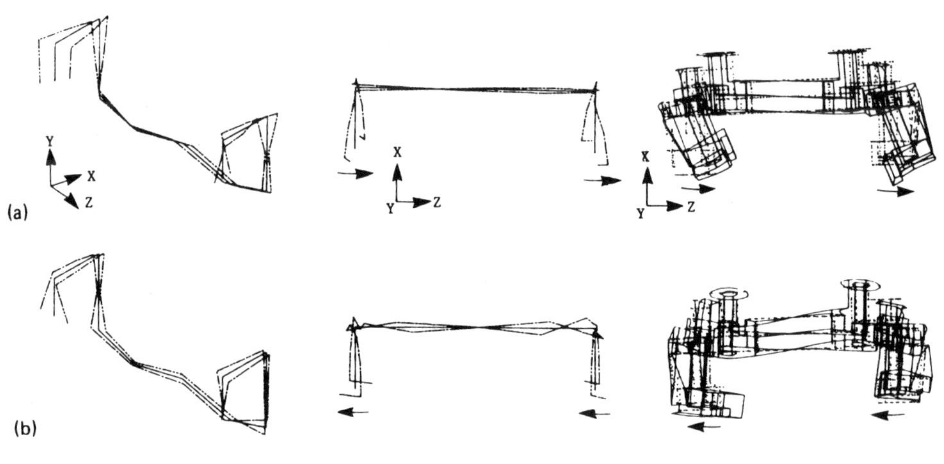

	Modal analysis	FEM	Error
(a)	2.14 kHz	2.14	0%
(b)	8.06	8.17	+1.3%

FIGURE 21 Examples of mode shapes for the torque member of the caliper. (a) and (b) Modal analysis (left); FEM (right).

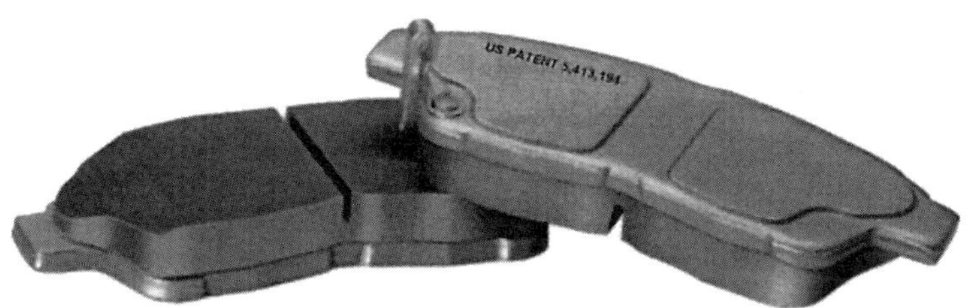

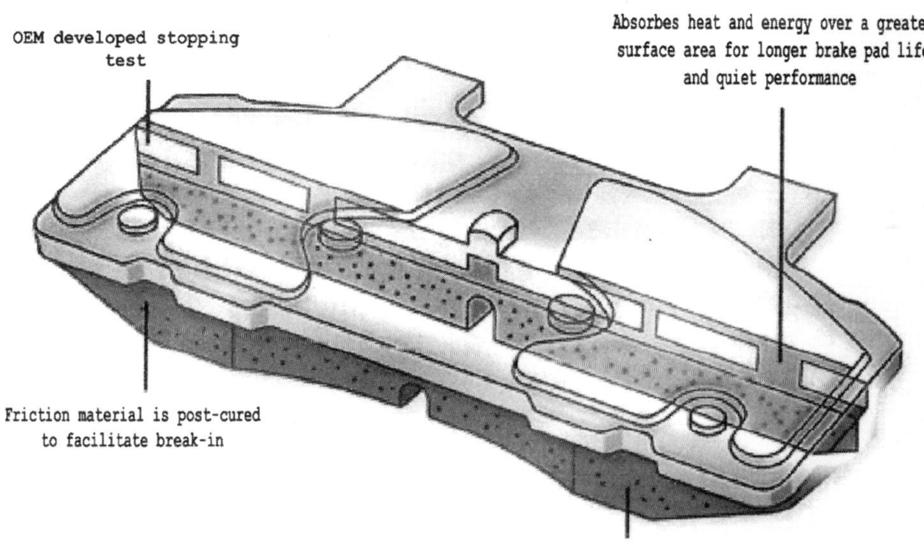

FIGURE 22 Photograph and cross section of a ThermoQuiet brake pad. (Courtesy Wagner Brake Products, Federal-Mogul Corp., Southfield, MI.)

REFERENCES

1. Lamarque, Williams. (1935). *Of Manufacturers and Operators and Some Preliminary Experiments, 1938 Research Report 8500B*. The Institution of Automobile Engineers, Research and Standardization Committee [British].
2. Hollman. (March 1954). *Dipl. Ing., Bremsgeräusche, ihre Ursachen und einige Weg zu ihrer Verhütung, A.T.Z.* 56(3):65–67.
3. Broadbent, H. R. (1956). Forces on a brake block and brake chatter. In Proceedings of the Institution of Mechanical Engineers. Vol. 170, pp. 993–1008.
4. Sinclair, D. (June 1955). Frictional vibrations. In Journal of Applied Mechanics 22:207–212.
5. Spurr, R. T. (1961–1962). A theory of brake squeal. In Proceedings of the Automobile Division of the Institution of Mechanical Engineers, pp. 33–52.
6. Jarvis, R. P., Mills, B. (1963–64). Vibrations induced by dry friction. *Institution of Mechanical Engineers.* Vol. 178, 32(pt. 1):847–866.
7. Kryloff, N., Bogoliuboff, N. (1947). *Introduction to Nonlinear Mechanics (translated from Russian by S. Lefschertz)*. Princeton, NJ: Princeton University Press.
8. North, M. R. (1968). *A survey of published work on vibrations in braking systems, Bulletin No. 6*. Motor Industry Research Association, pp. 8–12.
9. North, M. R. (1972). Ph.D. thesis. Loughborough University.
10. Millner, N. (1978). An analysis of disc brake squeal, Paper 780332. *Tansactions of the Society of Automotive Engineers* 87:1565–1575.
11. Felske, A., Hoppe, G., Matthäi, H. (1978). Oscillations in squaling disk brakes–Analysis of vibration modes by holographic interferometry, SAE Paper 780333. *Transactions of the Society of Automotive Engineers,* 87:1575–1576.
12. Felske, A., Hoppe, G., Matthäi, H. (1980). A study on drum brake noise by holographic vibration analysis, SAE Paper 800221. *Transactions of the Society of Automotive Engineers,* 89:1092–1110.
13. Murakami, H., Tsunada, N., Kitamura, T. (October 1–4, 1984). A study concerned with a mechanism of disc-brake squeal, SAE Technical Paper 841233, Passenger Car Meeting, Dearborn, MI.

14

Engineering Standards for Clutches and Brakes

I. SAE STANDARDS

Fifteen of these standards for the year 2003 that pertain to brake or clutch components are briefly described in the remainder of this chapter. Other standards apply to brake line hoses, fluids for hydraulically activated brakes, brake cables, clutch flywheels, and so on.

Descriptions of these standards are arranged in the numerical order of the standards. This order differs from the coded order found in the *SAE Handbook*, wherein the first digits are listed in numerical order, followed by the second digits, and so on. In that system the order would be SAE J1079, SAE J1087, SAE J160, SAE 286, SAE J866.

Society of Automotive Engineers (SAE) standards may be purchased from the Society of Automotive Engineers, 400 Commonwealth Drive, Warrendale, PA, 15096-0001. All of them are collected in the *SAE Handbook*, which is also available for sale from the society. They are frequently updated to reflect current engineering practice and some, such as SAE J659, have been cancelled.

A. SAE J160

This seven-page standard, entitled *Dimensional Stability of Friction Materials When Exposed to Elevated Temperatures*, is to "establish a common laboratory method for determining the dimensional stability of friction materials when exposed to elevated thermal conditions." Dimensional changes in a lining material may cause unintended contact in either clutches or brakes between mating surfaces, with the obvious possibility of serious consequences. This standard is to provide a specific measure of dimensional changes under well-defined conditions.

B. SAE J286

This five-page standard, entitled *SAE No. 2 Clutch Friction Test Machine Guidelines*, is to specify the requirements for a friction test machine to evaluate friction characteristics of wet friction clutch systems using automatic transmission fluids. It also outlines testing procedures to be employed.

C. SAE J373

This two-page standard, entitled *Housing Internal Dimensions for Single- and Two-Plate Spring-Loaded Clutches*, gives the minimum internal dimensions to provide recommended clearance for these clutches relative to SAE clutch numbers and to SAE flywheel housing numbers.

D. SAE J379

This four-page standard, entitled *Gogan Hardness of Brake Lining*, requires use of a commercially available Gogan Model 911 direct-reading hardness-testing machine or an equivalent to perform testing in which a shallow surface deformation induced by a penetrator of specified geometry serves as a quality check of a brake lining. It is to be a quality and processing check by a lining manufacturer for each thickness, configuration, and formulation. Loads for the scale symbols shown are given, along with the associated recommended ranges of Gogan numbers.

E. SAE J380

This three-page standard, entitled *Specific Gravity of Friction Material*, is also to serve as a quality check by providing another indicator of the consistency of the formulation and processing of friction materials. It describes the equipment to be used and the procedures to be followed. Acceptable range for each product is to be established by the manufacturer. See ASTM B 376 for a corresponding test for sintered metal powder friction materials.

F. SAE J621

This four-page standard, entitled *Industrial Power Take-Offs with Driving Ring-Type Overcenter Clutches*, includes a drawing of a driving ring-type overcenter clutch, along with a table of recommended dimensions for industrial power takeoffs. See SAE 1079 for related speed testing of these clutches.

G. SAE J661

This seven-page standard, entitled *Brake Lining Quality Test Procedure*, defines a uniform laboratory procedure for obtaining and recording friction and wear properties of brake lining material using a typical, commercially available, friction materials test machine as illustrated in the first two figures in the standard. Its intended uses are for quality control by a manufacturer and for lining evaluation by buyers of brake lining material. Grids for plotting the friction coefficient as a function of temperature for first fade, first recovery, second fade, second recovery, and wear are shown in the master form plot sheet on page 7.

H. SAE J662

This two-page standard, entitled *Brake Block Chamfer*, shows the chamfering in two figures, with appropriate instructions.

I. SAE J663

This two-page standard, entitled *Rivets for Brake Linings and Bolts for Brake Blocks*, is presented in three tables and three figures for rivet and bolt configurations and dimensions.

J. SAE J840

This ten-page standard, entitled *Test Procedures for Brake Shoe and Lining Bonds*, "covers equipment and procedures for qualification of bonded or integrally molded drum and disc shoe and lining assemblies." It includes bonded and molded shear test procedures and related equipment, complete with drawings.

K. SAE J866

This two-page standard, entitled *Friction Coefficient Identification System for Brake Linings*, gives the code letters for friction coefficient ranges from below 0.15 to over 0.55. It scheduled to be withdrawn immediately upon adoption of standards SAE 1802 and SAE 2430.

L. SAE J1087

This seven-page standard, entitled *One-Way Clutches—Nomenclature and Terminology*, with eight figures, is "intended to establish common nomenclature and terminology for automotive transmission one-way clutches." This terminology also applies to one-way clutches for other applications, such as jet engine and turbine starters.

M. SAE J1079

This two-page standard, entitled *Overcenter Clutch Spin Test Procedure*, sets forth a test procedure either to find the rotational velocity at which a driving ring type of overcenter clutch will burst or to verify that it will not burst at a specified speed. The required test equipment is described in SAE J1240. See SAE J621 for a description and figure of a driving ring-type overcenter clutch.

N. SAE J1499

This four-page standard, entitled *Band Friction Test Machine (SAE) Test Procedure*, refers to a SAE band friction test machine and gives the test procedure to be followed. The band friction test machine involved is neither shown nor described in this standard.

O. SAE J1646

This 18-page standard, entitled *Glossary of Terms—Lubricated Friction Systems*, contains graphs, definitions, and equations related to these terms as used in the testing automatic transmission clutch plates, band brakes, and other friction systems that operate in a fluid environment. Terms and graphs used to present test results are also described and defined.

II. AMERICAN NATIONAL STANDARDS INSTITUTE (ANSI)

The American National Standards Institute, whose acronym is ANSI, collects and compiles standards from between 60 and 70 professional societies (The number depends upon whether different committees within an organization are counted separately) in the United States and also cooperates with international standards organizations. Many of these standards include the acronym of the authoring organization, and many, but not all, of these standards are sold to users through its subsidiary organization, the National Standards System Network, whose Web address is www.nssn.org.

ANSI itself appears to have authored some of these standards, such as ANSI B11.3—2002, Machine Tools—Safety Requirements for Power Press Brakes.

III. OTHER STANDARDS ORGANIZATIONS

Notice that some of the acronyms for the following organizations may not represent the initials of that organization. For example, the ISO standards from the International Organization for Standardization are published in three languages: French, English, and Russian. The order of the majuscules ISO in its acronym does not correspond to the order of the leading characters in any of these three languages.

Australia: Standards Australia (SAA).
Canada: Standards Council of Canada (SCC).
Europe: European Committee for Standardization (CEN)
Finland: Finnish Standards Association (SFS).
France: Association Francaise de Normalisation (AFNOR).
Germany: Deutsches Institut für Normung (DIN).
Italy: Eute Nazionale Italiano di Unificazione (UNI).
Japan: Japanese Industrial Standards (JIS).
Korea: Korean Standards Association (KSA).
Malaysia: Standards and Industrial Research of Malaysia (SIRIM).
Netherlands: Nederlands Normalisatie-Instituut (NEN).
Norway: Norwegian Standards Association (—)
Slovenia: Standards and Metrology Institute (SMIS).
Taiwan: Bureau of Standards and Metrology Institute (BSMI).
United States: Department of Defense (DoD—STD—), (MIL—).

A. Selected Standards from these Other Organizations

The titles of the following standards were copied from the NSSN Web site as a partial indication of the topics and applications considered. At this time of this writing, the standards listed by NSSN may be found by going to www.ossn.org, where a left-click of a computer mouse on the words *Search for Standards*, which are directly under the words www.NSSN.ORG at the top left of the screen, will call a search form to the screen. Entering the word *clutch* in the *Find By Title Word* block and leaving all other blocks blank will produce the name of all of the standards available to NSSN that have the word *clutch* in the title. A left click on the document number of a particular standard will then yield a screen with instructions for obtaining a copy of that standard.

JIS D 4421:1996: Method of Hardness Test for Brake Linings, Pads and Clutch Facings for Automobiles.
JIS D 0152:1997: Clutches and Brakes—Vocabulary.
ASME B5.55M: Specification and Performance Standard, Power Press Brakes R(2002).
BSMI B2004400: Hand Brakes for Bicycles.
BSMI B4003700: Axles with Brakes of Trailer for Power Tiller.

BSMI B7005800: Method of Test for the Rear Brakes of Bicycles.
BSMI D3010900: Method of Test for Air Brakes of Automobiles.
CEN EN12622: Safety of Machine Tools—Hydraulic Press Brakes.
CEN PREN 12622: Hydraulic Press Brakes—Safety.
CEN PREN 1292-2: Safety Requirements for Passenger Transportation by Rope—General Provisions—Part 2: Additional Requirements for Reversible Bicable Aerial Ropeways Without Carrier Truck Brakes.
DIN 11742: Agricultural Machinery: Internal Expanding Brakes for Trailers, Dimensions.
DIN 154332-2: Power Transmission Engineering: Disc Brakes, Brake Linings.
DIN 15435-2: Power Transmission Engineering: Drum Brakes, Brake Shoes.
DIN 15435-2: Power Transmission Engineering: Drum Brakes, Brake Linings.
DIN 22261-4: Excavators, Spreaders and Auxiliary Equipment in Opencast Lignite Mines—Part 4: Hoisting Winch Brakes.
DIN 22261-5: Excavators, Spreaders and Auxiliary Equipment in Opencast Lignite Mines—Part 5: Slewing Brakes and Overload Protecting Devices.
DIN 25108: Suburban Railways: Interface and Outline Dimensions for Magnetic Track Brakes.
DIN 27205-4: State of Railway Vehicles—Brakes—Part 4: Non-ventilated Axle-Mounted Brake Discs.
DIN 27205-5: State of Railway Vehicles—Brakes—Part 5: Wheel-Mounted Brake Discs.
ISO 8710:1995: Motorcycles—Brakes and Braking Devices—Test and Measurement Methods.
ISO 8709:1995: Mopeds—Brakes and Braking Devices—Test and Measurement Methods.
ISO 7176-3:2003: Wheelchairs—Part 3: Determination of Effectiveness of Brakes.
KSA B4045: Mechanical Multiple-Disc Clutches (Wet Type).
KSA B4080: Mechanical Multiple-Disc Clutches (Wet Type).
KSA B4081: Oil Hydraulic Multiple-Disc Clutches (Wet Type).
KSA B4082: Electromagnetic Multiple-Disc Clutches (Wet Type).
DoD-STD-2144 NOT 2: Induction Clutches, Low-Magnetic-Field Design of.
MIL-C-18087A: Clutches for Propulsion Units and Auxiliary Machinery, Naval Shipboard.

Bibliography

1. Limpert, R. (1999). *Brake Design and Safety*. 2nd ed. Warrendale, PA: Society of Automotive Engineers.
2. Peeken, H., Troeder, Christoph. (1986). *Elastische Kupplungen: Ausführungen, Eigenschaften, Berechnungen*. New York: Springer-Verlag.
3. Shaver, R., Shaver, F. R. (1997). *Manual Transmission Clutch Systems*. Warrendale, PA: Society of Automotive Engineers.
4. Winkelmann, S., Harmuth, H. (1985). *Schaltbare Reibkupplungen: Grundlagen, Eigenschaften, Konstruktionen*. New York: Springer-Verlag.

Index